Energy Efficiency, Conservation and Management

Energy Efficiency, Conservation and Management

Edited by **Lucas Collins**

SYRAWOOD
PUBLISHING HOUSE
New York

Published by Syrawood Publishing House,
750 Third Avenue, 9th Floor,
New York, NY 10017, USA
www.syrawoodpublishinghouse.com

Energy Efficiency, Conservation and Management
Edited by Lucas Collins

International Standard Book Number: 978-1-68286-240-7 (Hardback)

Printed in China.

Contents

Preface

Scientists and energy experts across the globe are devising new measures to address the apparent energy crisis through measures for energy efficiency and conservation. This book discusses the modern concepts and practices in the field of energy efficiency and conservation. It contains some of the most significant topics related to the field of energy management such as energy consumption trends in different regions of the world, enhancing energy extraction processes, etc. Students and researchers will find the case-studies included in this book very helpful.

All of the data presented henceforth, was collaborated in the wake of recent advancements in the field. The aim of this book is to present the diversified developments from across the globe in a comprehensible manner. The opinions expressed in each chapter belong solely to the contributing authors. Their interpretations of the topics are the integral part of this book, which I have carefully compiled for a better understanding of the readers.

At the end, I would like to thank all those who dedicated their time and efforts for the successful completion of this book. I also wish to convey my gratitude towards my friends and family who supported me at every step.

Editor

1

A green supply chain network design model for enhancing competitiveness and sustainability of companies in high north arctic regions

Hao Yu[1], Wei Deng Solvang[1], Chen Chen[2]

[1] Department of Industrial Engineering, Narvik University College, Postboks 385 Lodve gate 2, NO-8505 Narvik, Norway.
[2] Northern Research Institute Narvik AS.Postboks 250, NO-8504 Narvik, Norway.

Abstract
To survive in today's competitive and ever-changing marketplace, companies need not only to engage in their products and/or services, but also to focus on the management of the whole supply chain. Effectively managing and balancing the profitability and interconnection of each player in the supply chain will improve the overall supply chain surplus as well as individual profit. However, it is extremely difficult to simultaneously optimize several objectives in design and planning of a supply chain, i.e., cost-minimization, risk-minimization, responsiveness-maximization, etc., which are somehow conflict with one another. Furthermore, the natural and infrastructural challenges in high north arctic regions make it become much more difficult and complicated to design and develop cost-efficient, highly responsive, environmentally friendly, and sustainable supply chain network. In order to provide companies in high north arctic regions with decision support tool for the design and planning of theirs supply chain networks, a green supply chain network design (GrSCND) model is formulated in this study based on multi-objective mixed integer programming (MIP). The optimal trade-off among several conflicting objectives is the focus of this GrSCND model aiming to enhance both competitive competence and sustainability of companies and supply chains operated in high north regions. In addition, a numerical experiment is also given to present a deep insight of the GrSCND model.

Keywords: Green supply chain; Network model; Competitiveness; Sustainability; High north Arctic regions.

1. Introduction

To survive in today's competitive and ever-changing marketplace, companies need not only to engage in their products and/or services, but also to focus on the management of the whole supply chain. A typical supply chain includes raw material/component supplier, manufacturer, distributor, retailer, and customer [1]. Effectively managing and balancing the profitability and interconnections of each player in the supply chain will improve the overall supply chain surplus as well as individual profit. Conventionally, the objective of supply chain network design is to maximize the overall profit generated through balancing the total costs and responsiveness to customer needs. A poor responsiveness to meet the customer needs will decrease customer satisfaction, and therefore increase the risk of losing sales. In order to achieve high responsiveness to the rapid-changing market, a more flexible manufacturing system should be applied, which sacrifices economies of scale and results in high production and transportation costs. The break-even point which optimizes the overall supply chain performance in terms of both cost and responsiveness has been extensively addressed in previous studies through bi-objective programming.

However, for the companies and supply chains operated in high north arctic regions, more challenges, i.e., inhospitable and extreme climate, absence or poor infrastructure [2], and complicated terrain and environment, make it very difficult to deliver high responsive products and/or services with low costs, and relatively high supply chain risks are imposed as well. Besides, environmental issues, i.e., vulnerable eco-environmental system and higher sensitivity to greenhouse gas emissions, must be taken into account in the decisional process of supply chain network design (SCND) considering that CO_2 emissions have increased rapidly over past decades. Furthermore, population density in high north arctic regions is extremely low (For instance, the population density in three counties located in northern Norway is $7/km^2$ in Nordland, $6/km^2$ in Tromsø, and $2/km^2$ in Finnmark [3]), hence, the transportation of small amount of raw materials and/or finished products over very long distance is quite common in this sparsely populated area, which dramatically increases the costs of transportation. Due to the aforementioned reasons, the supply chain network faces more challenges than those which are operated in densely populated areas [4].

In order to tackle those challenges and provide decision supporttool for the companies and supply chains operated in high north arctic regions, we aim in our study to develop the theoretical framework and computational model for green supply chain network design (GrSCND) in order to enhance both competitive competence and sustainability of companies of this area. The proposed theoretical framework and computational model aim to optimize the overall supply chain performance through balancing the trade-off among costs, risks, and greenhouse gas (GHG) emissions. In addition, the adopted methodology for model formulation is based on multi-objective mixed integer programming (MIP), and an numerical experiment is also given to present a deep insight and applicability of the GrSCND model developed in this research.

The rest of this article is organized as follows. Section 2 provides an extensive literature review of green supply chain management (GrSCM) and GrSCND models. Section 3 formulates the theoretical framework and computational model for GrSCND in high north arctic regions, and the method for model solution is also given in this section. Section 4 presents the numerical experiment, and section 5 concludes this article with a future outlook.

2. Literature review

The concept of green supply chain management (GrSCM) has been introduced and extensively studied for almost two decades. The first attempts to define GrSCM can be found in late 1990s (see ref. [5]), and the most cited definition of GSCM [6] is given by Srivastava [7] which defines GrSCM as "Integrating environmental thinking into supply-chain management, including product design, material sourcing and selection, manufacturing processes, delivery of the final product to the end customers as well as end-of-life management of the product after its useful life." GrSCM is also referred as environmental logistics [8], green logistics [9], sustainable supply chains [10], and sustainable supply network management [11, 12], and a number of review articles contributed to both theoretical and practical development of GrSCM are recently published by Seuring and Muller [13], Carter and Rogers [14], Sarkis et al. [15], Ali and Searcy [6], and Ashby et al. [16]. To achieve GrSCM, two types of "greenness" are divided by researchers [7]: green product design [17] and green operations, and the green operations, i.e., network design problem [18-20], sustainable waste management [20-22], and material flow [22] of a supply chain, are the focus of this research.

Network design is the logical place at which strategic decisions should be made for GrSCM [23]. Designing the physical network structure of a supply chain is called supply chain network design (SCND) [24]. Due to its significant influence on supply chain's performance, resilience, profits, and competitive competence [25], SCND is believed to be one of the most important strategic decisions in supply chain management, which affects the long-term profitability and sustainability of a supply chain. To take into account environmental or "green" thinking in SCND, a large number of articles have contributed to develop both theoretical and computational models for green supply chain network design (GrSCND). Wang et al. [26] develop a bi-objective optimization model for GrSCND, which aims to balance the trade-off between overall costs and environmental influence in terms of CO_2 emissions. The "Pareto optimal" solutions are employed for model computation, and a comprehensive numerical experiment is also conducted in this study. Elhedhli and Merrick [23] propose a mathematical model for reducing carbon emissions in GrSCND. The carbon emissions are monetized and converted into environmental pollution costs, and the model aims to minimize the overall system costs including fixed and variable facility costs, production costs, as well as environmental pollution costs. Govindan et al. [27] introduce a two-stage bi-objective location-routing model with time-windows for GrSCND, and the optimal balance of costs and greenhouse gas emissions is the goal of this model. The optimal supply chain network configuration is determined through selecting appropriate number and locations of facilities as well as the route within each stage. A large number of GrSCND models and practices incorporating cost objective with emission objective of greenhouse gas (GHG) can also be found in Yu and Solvang [20], Quariguasi-Frota-Neto et al. [28], Harris et al. [29], Ulbeda et al. [30], and Adballah et al. [31].

To consider different influencing factors in GrSCND other than GHG emissions, Jamshidi et al. [32] develop a bi-objective mathematical model for GrSCND, which simultaneously minimizes the overall system costs and environmental impacts. The environmental impacts in this study are measured by the amount of hazardous gases, i.e., CO, NO_2 and volatile organic particles, generated by facility operations and transportation of goods within the supply chain. Latha Shankar et al. [33] pose a bi-objective optimization model for strategic planning and material flow decisions of a three-echelon supply chain network. The focus of this model is the optimal balance between system operating costs and the fill rate of customer demands. Sheu and Lin [34] incorporate multi-objective mixed integer programming (MIP) and hierarchical cluster analysis method to configure and optimize global logistics network. The proposed model aims to minimize the network investments, while maximize the total profits generated by the supply chain and satisfaction rate of customer demands, and the weighted sum utility method is employed in this research for model computation.

To take into account of the changes in input parameters with time horizon, Yu et al. [22] formulate a multi-period dynamic model for managing and operating the reverse network of waste management system in an environmentally friendly manner. The proposed model aims to simultaneously minimize the system operating costs and environmental risks imposed by waste recycling and disposal through optimally managing the material flow between different facilities at each time period. A three-stage dynamic model for open-loop reverse supply chain and logistics network planning is developed by Ene and Ozturk [35], which aims to maximize the overall network costs of product recovery and disposal. Zeballos et al. [36] propose a multi-product and multi-period mathematical model for optimal planning of closed-loop supply chains through the minimization of net costs (expected costs minus expected revenue through recycling and remanufacturing), and both forward flows and reverse flows are formulated in this model. It is noted that the input parameters in this model are assumed to be stochastic in nature and therefore exist great uncertainties, and a reduced scenario tree is applied to achieve a reasonable representation of the original problem so that the model can be resolved.

Dealing with uncertainties in input parameters is another focus in GrSCND. Pishvaee and Razmi [37] formulate a fuzzy mathematical programming for GrSCND. This model aims to balance the trade-off between costs and environmental impact, and the environmental impact is measured by eco-indicator 99 which is a life cycle assessment-based (LCA-based) method. Further, an interactive fuzzy solution approach is also established for model computation. Ramezani et al. [38] develop a multi-stage, multi-period and multi-product optimization model for closed-loop SCND with fuzzy environment, and the goal of this model is to simultaneously minimize the costs, delivery time, and defects of raw materials acquired from suppliers. Amin and Zhang [39] propose a bi-objective model for closed-loop SCND with inexact input information on demands and return, and the balance between the minimization of costs and maximization of the use of environmentally friendly materials is the focus of this research.

Through the extensive literature review of GrSCND models and practices, two characteristics can be identified. One is most previous researches use bi-objective optimization approach in order to balance the trade-off between costs and environment impacts, and the other is the indicator of environmental impacts is most frequently measured by GHG emissions. Besides, other objectives i.e., amount of hazardous gases, customer satisfaction rate, etc., are formulated as well in some previous models, and the time-varying and uncertain parameters have also been extensively focused in GrSCND. There is no denying the fact that costs and GHG emissions are the most crucial influencing factors in GrSCND, but more focus and emphasis have to be attached to the risks and reliability of the supply chains operated in high north arctic regions where natural and infrastructural challenges, i.e., poor and limited transport access (e.g. railway transportation is unavailable in most arctic regions), significant influence of inhospitable and extreme climate (e.g. the road transportation may be closed for several days due to avalanche), etc., bring more complexities in GrSCND. A poorly planned supply chain network without considering supply chain risks in this area will result in extremely high costs, high risks, high GHG emissions and poor responsiveness, which will then lead to the failure of a company or a supply chain in pursuing long-term profitability and sustainability. Therefore, it is of significant importance to account supply chain risks in the decisional process of GrSCND in high north arctic regions, however, it is extremely difficult to find such an instance from previous researches. Therefore, in order to fill the literature gap, the theoretical framework and mathematical model for GrSCND of a three-stage supply chain operated in high north arctic regions are formulated in this paper so that the supply chain costs, GHG emissions and risks are simultaneously considered in GrSCND.

3. Model

3.1 Theoretical framework

In this section, the theoretical framework of a general three-stage forward supply chain is first formulated in Figure 1. As shown in the figure, the proposed theoretical supply chain network is comprised of four levels of entities: supplier, producer, warehouse and customer, and those entities are communicated and connected through three flows: material flow, information flow and capital flow. The material flow in this supply chain network starts from upstream raw material suppliers and moves via intermediate production plants and warehouses towards end customers, and the information and capital flow in opposite direction from end customers towards suppliers.

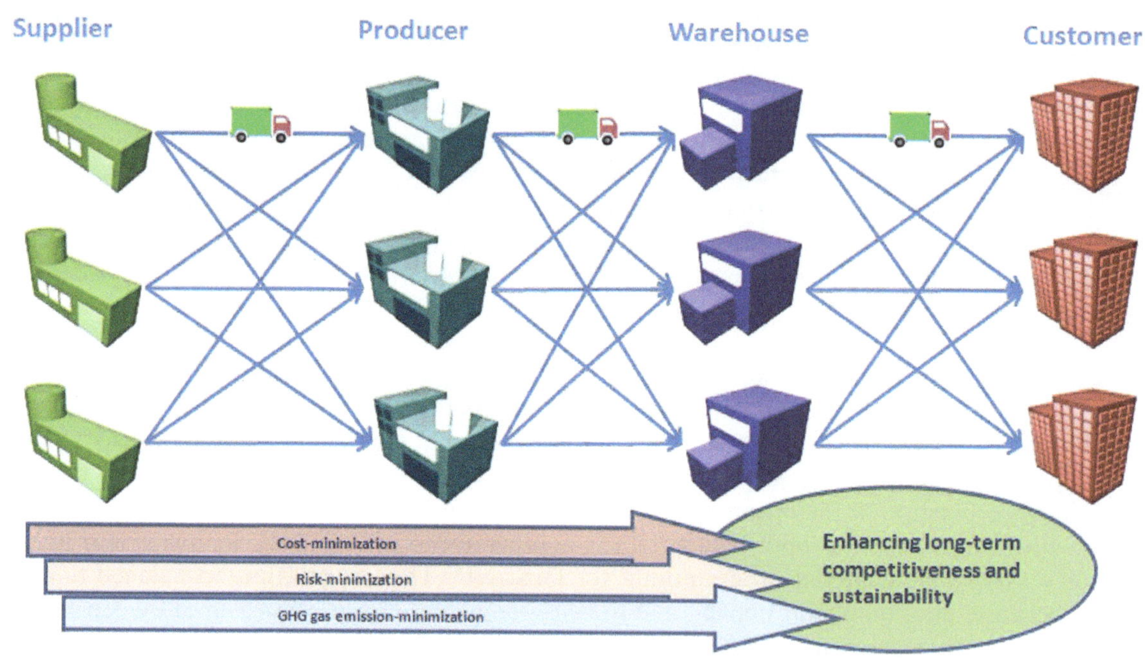

Figure 1. Theoretical framework of GrSCND of a three-stage supply chain operated in high north arctic regions

Conventionally, the focus of GrSCND is to simultaneously minimize the costs and GHG emissions of a supply chain, however, it is also of great significance to decrease the risks and increase reliability of a supply chain operated in high north arctic regions due to the complex natural and infrastructural challenges discussed in previous section. Therefore, in order to tackle this challenge, the optimal trade-off among cost-minimization, risk-minimization and GHG emission-minimization will be focused in this research so that long-term competitive competence, profitability and sustainability can be achieved.

3.2 Mathematical model

The proposed MIP model aims to determine, in an optimal manner, the number and locations of potential facilities, selection of suppliers, and the inter-facility material flow in each stage of a supply chain. The indices, input parameters and decision variables are first given as follows:

Indices	
s	The set of suppliers (s=1, 2, 3,…, S)
p	The set of candidate locations for production plants (p=1, 2, 3,…, P)
w	The set of candidate locations for warehouses (w=1, 2, 3,…, W)
c	The set of customers (c=1, 2, 3,…, C)

Input parameters	
PC_s	The unit purchasing costs for raw materials and components at supplier s
FC_p, FC_w	The fixed costs for production plant p and warehouse w
C_p, C_w	The unit operational costs (e.g. production costs, inventory costs, packaging costs, etc.) of production plant p and warehouse w
TC_{sp}, TC_{pw}, TC_{wc}	The unit transportation costs between supplier s and production plant p, production plant p and warehouse w, warehouse w and customer c
EMS_{sp}, EMS_{pw}, EMS_{wc}	The GHG emission factor between supplier s and production plant p, production plant p and warehouse w, warehouse w and customer c
DIS_{sp}, DIS_{pw}, DIS_{wc}	The transport distance between supplier s and production plant p, production plant p and warehouse w, warehouse w and customer c
LD_{sp}, LD_{pw},LD_{wc}	The average load of transport vehicles between supplier s and production plant p, production plant p and warehouse w, warehouse w and customer c
RK_s	The risk index of suppler s
RK_{sp}, RK_{pw}, RK_{wc}	The risk index of the transportation between supplier s and production plant p, production plant p and warehouse w, warehouse w and customer c
CD_c	The demands of customer c
IF	An infinite positive number
MPR_p	The material-to-product rate at production plant p which specifies how many materials are needed for producing one product
ITR_w	The inventory turnover rate at warehouse w which specifies the ratio of outgoing products and incoming products
CAP_s, CAP_p, CAP_w	The capacity of supplier s, production plant p, and warehouse w

Decision variables	
S_s	If S_s=1, supplier s is selected, and if S_s=0, otherwise
X_p	If X_p=1, candidate location p is selected for opening production plant, and if X_p=0, otherwise
X_w	If X_w=1, candidate location w is selected for opening warehouse, and if X_w=0, otherwise
AT_{sp}, AT_{pw}, AT_{wc}	The amount of raw materials or finished products transported between supplier s and production plant p, production plant p and warehouse w, warehouse w and customer c

$$\text{Min } OBJ1=\sum_{s=1}^{S} PC_s S_s \left(\sum_{p=1}^{P} AT_{sp}\right) + \sum_{p=1}^{P} X_p \left(FC_p + \sum_{s=1}^{S} C_p AT_{sp}\right) + \sum_{w=1}^{W} X_w \left(FC_w + \sum_{p=1}^{P} C_w AT_{pw}\right)$$
$$+ \sum_{s=1}^{S}\sum_{p=1}^{P} TC_{sp} AT_{sp} + \sum_{p=1}^{P}\sum_{w=1}^{W} TC_{pw} AT_{pw} + \sum_{w=1}^{W}\sum_{c=1}^{C} TC_{wc} AT_{wc} \tag{1}$$

$$\text{Min } OBJ2=\sum_{s=1}^{S}\sum_{p=1}^{P} EMS_{sp} \frac{AT_{sp} DIS_{sp}}{LD_{sp}} + \sum_{p=1}^{P}\sum_{w=1}^{W} EMS_{pw} \frac{AT_{pw} DIS_{pw}}{LD_{pw}}$$
$$+ \sum_{w=1}^{W}\sum_{c=1}^{C} EMS_{wc} \frac{AT_{wc} DIS_{wc}}{LD_{wc}} \tag{2}$$

$$\text{Min } OBJ3 = \sum_{s=1}^{S} RK_s S_s(\sum_{p=1}^{P} AT_{sp}) + \sum_{s=1}^{S}\sum_{p=1}^{P} RK_{sp} AT_{sp} + \sum_{p=1}^{P}\sum_{w=1}^{W} RK_{pw} AT_{pw} + \sum_{w=1}^{W}\sum_{c=1}^{C} RK_{wc} AT_{wc} \quad (3)$$

Eqs.(1), (2) and (3) are objective functions of this multi-objective MIP model for GrSCND in high north arctic regions. Eq. (1) is the cost-minimization objective function which takes into account the costs for supplier selection. The first part of this equation represents the purchasing costs of the raw materials from suppliers, and the second and third parts represent the fixed and operational costs of potential production plant and warehouse, and the last three parts represent the transportation costs in each stage. The purchasing costs and operational costs are directly proportional to the amount of raw materials and components purchased, and the transportation costs are directly proportional to the quantity transported in each stage. Eq. (2) is the GHG emission-minimization objective function. GHG emissions are very important environmental indicator especially for high north arctic regions where the GHG emissions have more negative influence on the ozone. In this model, the GHG emissions are directly proportional to the distance and amount transported, and it is inversely proportional to the load of transport vehicle. It is noted that the emission factor is applied for quantifying the equivalent GHG emissions, and it is determined by the type of transport vehicle, road condition as well as other influencing factors. Eq. (3) is the risk-minimization objective function in which the methodology developed by Yu and Goh [40] to quantify supply chain risks is employed and adapted accordingly. The first part of this equation represents the potential risks of supplier in fulfilling the demands of producer, and the other parts represent the potential transportation risks. The risk index of supplier is determined by inherent risks, supplier's capacity, supplier's reliability and reputation, and the risk index of transportation is influenced by transporter's reliability, probability of infrastructural risks, probability of natural disaster, etc. Besides, in order to fulfill the requirement for material flow, facility capacity as well as other restrictions, thirteen sets of model constraints are also formulated as follows.
Subject to:

$$CD_c = \sum_{w=1}^{W} AT_{wc}\ , \text{For} c = 1, \dots, C \quad (4)$$

$$ITR_w \sum_{p=1}^{P} AT_{pw} = \sum_{c=1}^{C} AT_{wc}\ , \text{For} w = 1, \dots, W \quad (5)$$

$$MPR_P \sum_{s=1}^{S} AT_{sp} = \sum_{w=1}^{W} AT_{pw}\ , \text{For } p = 1, \dots, P \quad (6)$$

$$\sum_{p=1}^{P} AT_{sp} \leq CAP_s, \text{For } s = 1, \dots, S \quad (7)$$

$$\sum_{s=1}^{S} AT_{sp} \leq CAP_p, \text{For } p = 1, \dots, P \quad (8)$$

$$\sum_{w=1}^{W} AT_{pw} \leq CAP_w, \text{For } w = 1, \dots, W \quad (9)$$

$$S_s \leq IF \sum_{p=1}^{P} AT_{sp}\text{ , For } s = 1, \dots, S \tag{10}$$

$$X_p \leq IF \sum_{p=1}^{P} AT_{sp}\text{ , For } p = 1, \dots, P \tag{11}$$

$$X_w \leq IF \sum_{p=1}^{P} AT_{pw}\text{ , For } w = 1, \dots, W \tag{12}$$

$$AT_{sp} \leq S_s X_p IF\text{, For } s = 1, \dots, S, p = 1, \dots . P \tag{13}$$

$$AT_{pw} \leq X_p X_w IF\text{, For } p = 1, \dots, P, w = 1, \dots . W \tag{14}$$

$$AT_{wc} \leq X_w IF\text{, For } w = 1, \dots, W, c = 1, \dots . C \tag{15}$$

$$S_s, X_p, X_w \in \{0, 1\}\text{, For } s = 1, \dots, S, p = 1, \dots, P, w = 1, \dots, W \tag{16}$$

Eq. (4) restricts the demands of each customer must be fulfilled. Eqs. (5) and (6) are the requirements of material flow balance, which specify the relationship between the amount of incoming raw materials and the quantity of outgoing finished products at production plant p and warehouse w. It is noted that the defect rate should be taken into consideration in determining the value of *MPR* at production plant p. Eqs. (7), (8) and (9) are capacity constraints, which restrict the maximum quantity served by supplier s, production plant p, and warehouse w cannot exceed their corresponding capacities. Eq. (10) restricts supplier s will not be selected if it doesn't supply raw materials or components to any producers. Eqs. (11) and (12) ensure the candidate locations for production plant p and warehouse w will not be selected if they do not perform any functions. Eq. (13) guarantees the producer p can be served by supplier s only when both supplier $s$and candidate location p for opening production plant are selected. Eq. (14) restricts the finished products from producer p can be stored at warehouse w only when both candidate location p for opening production plant and candidate location w for opening warehouse are chosen. Eq. (15) ensures the demands of customer c can be served by warehouse w only when candidate location w is selected for building new warehouse. Eq. (16) is the binary constraint of decision variables. In addition, all the indices and input parameters of this multi-objective MIP model for GrSCND belong to non-negative domain.

3.3 Model solution

In order to composite multiple objective functions with different measures of units, the weighted sum utility method developed by Sheu and Lin [34] is employed in this research to composite the three objective functions of this GrSCND model, and similar practices of this method can also be found in Yu et al. [22], Nema and Gupta [41], and Sheu [42]. Before the weighted sum utility method is formulated, the notations of some adjustable parameters, benchmark parameters and response variables are first given as follows.

Adjustable parameters	
WT_{OBJ1}, WT_{OBJ2}, WT_{OBJ3}	The weight of cost-utility, GHG emission-utility, and risk-utility

Benchmark parameters	
$OBJ1_{min}$, $OBJ2_{min}$, $OBJ3_{min}$	The individual minimum achievable value of cost-minimization objective, GHG emission-minimization objective, and risk-minimization objective
$OBJ1_{max}$, $OBJ2_{max}$, $OBJ3_{max}$	The individual maximum achievable value cost-minimization objective, GHG emission-minimization objective, and risk-minimization objective

Response variables	
$OBJ1$, $OBJ2$, $OBJ3$	The actual value of cost-minimization objective, GHG emission-minimization objective, and risk-minimization objective
UT_{OBJ1}, UT_{OBJ2}, UT_{OBJ3}	The individual cost-utility, GHG emission-utility, and risk-utility
UT	The composite utility

$$\text{Min}UT = WT_{OBJ\,1}UT_{OBJ\,1} + WT_{OBJ\,2}UT_{OBJ\,2} + WT_{OBJ\,3}UT_{OBJ\,3} \tag{17}$$

Eq. (17) is the objective function of the weighted sum utility method and aims to minimize the weighted sum utility of each objective function. The weight of each individual utility presents the relative importance of each objective function determined by decision-makers. Eqs. (18), (19) and (20) illustrate the method for calculating the individual utility of each objective function. In Eq. (18), $OBJ1_{max}$ minus $OBJ1_{min}$ denotes the theoretically maximum deviation between the maximum achieve costs and the minimum achievable costs, which can be used as the benchmark for calculating the individual utility, and $OBJ1$ minus $OBJ1_{min}$ represents the deviation between actual value and minimum achievable value. The numerator and denominator in this equation share the same unit, and the unit can then be eliminated, and this method also applies for Eqs. (19) and (20).Therefore, the individual utility of each objective function becomes unit less and can be directly summed by giving the corresponding weights. The summation of the weights of those three objectives in this model is regulated to 1, so the theoretically minimum achievable individual utility is 0 when the actual value OBJ equals to the minimum achievable value OBJ_{min}, and the theoretical achievable maximum individual utility is 1 when the actual value OBJ equals to the maximum achievable value OBJ_{max}.

$$UT_{OBJ\,1} = \frac{OBJ1 - OBJ1_{min}}{OBJ1_{max} - OBJ1_{min}} \tag{18}$$

$$UT_{OBJ\,2} = \frac{OBJ2 - OBJ2_{min}}{OBJ2_{max} - OBJ2_{min}} \tag{19}$$

$$UT_{OBJ\,3} = \frac{OBJ1 - OBJ1_{min}}{OBJ1_{max} - OBJ1_{min}} \tag{20}$$

4. Numerical experiment

In this section, a numerical experiment is given to present a deep insight of the proposed multi-objective MIP model for GrSCND in high north arctic regions. The numerical experiment is performed based upon a hypothetical case of a three-stage supply chain network, including supplier, producer, warehouse and customer, operated in high north arctic regions, and the producer sources from domestic and international suppliers, and it mainly serves local customers. In order to design and maintain an efficient and sustainable supply chain with relatively low risks, the supply chain manager has to make several crucial decisions, i.e., the number and locations of production plants and warehouses to be opened, selection of suppliers, the amount purchased from each selected supplier, the amount of finished products stored in which warehouse, and the customer demands are served from which warehouse. The proposed GrSCND model is applied for decision support in this case.

The hypothetical supply chain network is comprised of 7 raw material suppliers, 5 candidate locations for production plant, 5 candidate locations for warehouse, and 4 end customers. Table 1 gives the unit purchasing costs, capacity and risk index of each supplier s, and the fixed costs, unit operational costs and capacity of the candidate locations for production plant p and warehouse w are presented in this table as well. The material-to-production rates MPR_p of candidate location $p1$, $p2$, $p3$, $p4$, $p5$ are 0.8, 0.7, 0.8, 0.6 and 0.7, respectively. The inventory turnover rate ITR_w of each potential warehouse is assumed to be equal, and it is 0.8. It is noted that the units of input parameters are given as unit cost (uc), unit weight (uw) and unit distance (ud) to represent the genericity, and they can easily and accordingly specified into a certain measure of units in a real world case study.

Tables 2, 3 and 4 present the unit transportation costs, distance and risk index of the 1st stage inter-facility transportation between supplier s and producer p, the 2nd stage inter-facility transportation between producer p and warehouse w, and the 3rd stage inter-facility transportation between warehouse w and customer c, respectively. The transportation of raw materials from suppliers to producers is suppliers' responsibility in this supply chain. Because the same type of vehicles are used for transporting raw materials from one supplier to all the producers, the GHG emission factor EMS_{sp} and average load LD_{sp} are assumed to be equal in all the outbound transportation of supplier s, where EMS_{s1p}=0.7 1/ud, EMS_{s2p}=0.81/ud, EMS_{s3p}=0.8 1/ud, EMS_{s4p}=0.7 1/ud, EMS_{s5p}=0.9 1/ud, EMS_{s6p}=0.75 1/ud, EMS_{s7p}=0.6 1/ud, and LD_{s1p}=4 uw, LD_{s2p}=6 uw, LD_{s3p}=8 uw, LD_{s4p}=6 uw, LD_{s5p}=12 uw, LD_{s6p}=8 uw and LD_{s7p}=12 uw, respectively. The transportation of finished products in stages 2 and 3 is outsourced to a 3rd party logistics (3PL) company, and the same type of transport vehicles are used to perform the transportations, so all the GHG emission factors and average load in 2nd and 3rd stage inter-facility transportation are assumed to be equal, where EMS_{pw}=EMS_{wc}=0.8 1/ud and LD_{pw}=LD_{wc}=4 uw, respectively. It is noted that the GHG emission-minimization objective function *OBJ2* and risk-minimization objective function *OBJ3* are quantified through calculating the emission index and risk index, which are relative value. The optimal solution of objective function *OBJ2* and *OBJ3* are achieved through comparing different scenarios, and the absolute value of individual scenario is meaningless. Furthermore, objective function *OBJ2* and *OBJ3* are unitless, and the unit 1/ud of emission factors EMS_{sp}, EMS_{pw} and EMS_{wc}, and 1/uw of risk index RK_{sp}, RK_{pw} and RK_{wc} are applied in order to eliminate the units of Eqs. 2 and 3, respectively.

Table 1. Input parameters of supplier *s*, candidate locations for production plant *p*, and candidate locations for warehouse *w*

Supplier	Parameters			Producer	Parameters			Warehouse	Parameters		
	PC_s [uc][a]	CAP_s[uw][b]	RK_s		FC_p [uc]	C_p [uc]	CAP_p [uw]		FC_w[uc]	C_w [uc]	CAP_w [uw]
s_1	760	500	0.2	p_1	500000	750	300	w_1	220000	80	250
s_2	320	100	0.5	p_2	480000	870	250	w_2	290000	65	350
s_3	400	140	0.4	p_3	515000	745	400	w_3	175000	95	200
s_4	80	500	0.3	p_4	450000	960	350	w_4	240000	75	350
s_5	102	350	0.7	p_5	475000	905	325	w_5	310000	60	450
s_6	115	450	0.3								
s_7	110	400	0.5								

[a]uc=unit currency, the same abbreviation is also applied in subsequent parts of this section.
[b]uw=unit weight, the same abbreviation is also applied in subsequent parts of this section.

Table 2. The unit transportation costs, distance and risk index of the 1st stage inter-facility transportation between supplier *s* and producer *p*

Supplier	Parameter TC_{sp}[uc]					Parameter DIS_{sp} [ud][c]					Parameter RK_{sp} [1/uw][d]				
	p_1	p_2	p_3	p_4	p_5	p_1	p_2	p_3	p_4	p_5	p_1	p_2	p_3	p_4	p_5
s_1	80	75	95	45	60	8	7.5	9	3	5.5	0.8	0.6	0.85	0.5	0.55
s_2	102	90	75	40	65	9.2	7	6	3.5	5	0.95	0.8	0.7	0.5	0.55
s_3	55	60	70	65	65	4.5	5	6.5	5.5	5.7	0.45	0.75	0.6	0.65	0.55
s_4	80	90	95	40	75	8.2	8.7	9	3.2	9	0.65	0.85	0.9	0.5	0.65
s_5	55	105	95	45	75	4.5	9.7	8.8	3.2	6.5	0.45	0.95	0.8	0.55	0.8
s_6	58	90	75	102	70	5.9	8.8	7.2	9.8	6.4	0.6	0.7	0.8	0.95	0.65
s_7	75	45	60	95	55	7.1	3.8	6.2	10.1	6.7	0.7	0.5	0.75	0.85	0.6

[c]ud=unit distance, the same abbreviation is also applied in subsequent parts of this section.
[d]1/uw=1/unit weight, the same abbreviation is also applied in subsequent parts of this section.

Table 3. The unit transportation costs, distance and risk index of the 2nd stage inter-facility transportation between producer *p* and warehouse *w*

Producer	Parameter TC_{pw}[uc]					Parameter DIS_{pw} [ud]					Parameter RK_{pw} [1/uw]				
	w_1	w_2	w_3	w_4	w_5	w_1	w_2	w_3	w_4	w_5	w_1	w_2	w_3	w_4	w_5
p_1	65	55	40	50	55	7	6	5.5	6.5	6	0.8	0.6	0.5	0.5	0.55
p_2	45	75	55	50	45	5	5.5	4.5	5	4.5	0.4	0.9	0.6	0.6	0.5
p_3	75	45	50	55	65	7.5	5	5.3	6	6.5	0.8	0.5	0.5	0.5	0.7
p_4	70	55	45	65	75	7.2	6	4.8	6.2	7.3	0.8	0.7	0.5	0.6	0.7
p_5	45	75	55	60	50	5	5.7	5.5	5.5	5	0.5	0.75	0.6	0.6	0.5

Table 4. The unit transportation costs, distance and risk index of the 3rd stage inter-facility transportation between warehouse w and customer c

Warehouse	Parameter TC_{wc}[uc]				Parameter DIS_{wc} [ud]				Parameter RK_{wc} [1/uw]			
	c_1	c_2	c_3	c_4	c_1	c_2	c_3	c_4	c_1	c_2	c_3	c_4
w_1	80	82	75	65	9	9.5	8	7.5	0.9	0.8	0.7	0.6
w_2	55	90	95	45	6	9.5	10	4.5	0.6	0.85	0.9	0.55
w_3	75	60	55	65	7.5	5.5	5	6	0.7	0.6	0.6	0.6
w_4	65	70	75	95	7	6.5	7.2	9	0.65	0.65	0.7	0.85
w_5	45	55	85	80	5	6	9.5	9	0.5	0.55	0.9	0.85

In order to test the performance of the proposed multi-objective GrSCND model, the model is coded and resolved by using Lingo solver, and all the model computations are performed on a Inter(R) Core(TM)2 2.13 GHz computer with 2 GB RAM and 150 GB hard drive capacity under Windows 7 operating system. The tested weights of cost utility, GHG emission utility and risk utility are set to 0.4, 0.3 and 0.3, respectively. The time consumed and iterations performed to calculate individual maximum and minimum costs, individual maximum and minimum GHG emissions, individual maximum and minimum risks, and minimum overall utility are presented in Table 5, and the objective value of those scenarios are also given in this table. It is illustrated from the result, the calculation of individual cost-minimization objective and overall utility are much more complicated and time consuming than the calculation of GHG emission-minimization objective and risk-minimization objective due to the larger number of integer variables and nonlinear variables. Besides, it is also shown from the table that, in this GrSCND model, the calculation of maximum achievable value is much easier and less time consuming than the calculation of minimum achievable value.

Table 5. The objective value, time consumed and iterations performed of each scenario

Scenario	Objective value	Time (s)	Iterations
Maximum individual costs	5246727 uc	4	1421
Minimum individual costs	2859436 uc	8	58513
Maximum individual GHG emissions	3010.529	1	549
Minimum individual GHG emissions	1462.585	1	332
Maximum individual risks	2320.375	1	437
Minimum individual risks	1335.4	1	1247
Minimum weighted sum utility	0.0914942	14	32474

Table 6 presents the selection of suppliers, selection of candidate locations for production plants and warehouses, as well as the value of corresponding weighted sum utility of four selected scenarios: individual minimum costs, individual minimum GHG emissions, individual minimum risks and minimum overall sum weighted utility. It is noted that the maximum value of each individual scenario is not taken into consideration in this comparison, because they are introduced in weighted sum utility method as bench mark parameters to represent the "worst solution" and determine maximum achievable deviation between the "best solution" and the "worst solution" of each scenario, and the independent comparison of the "worst solutions" is therefore meaningless to achieve the optimal solution in this case study. As shown in the table, the individual minimum costs objective has the best weighted sum utility comparing with the other two individual scenarios, and suppliers *s4*, *s5*, *s7*, candidate locations *p1*, *p3*, *p5*, *w3* and w5 are chosen in this scenario. The increase of the overall sum weighted utility are mainly contributed by the individual risk utility which equals to 0.3149, and this is caused by the relatively high risk index in 1st stage transportation of this scenario. When the optimal value of individual GHG emission objective is achieved, suppliers *s3*, *s4*, *s5*, *s7*, candidate locations *p1*, *p2*, *p3*, *p4*, *w2*, *w3* and *w5* are selected. In this scenario, both costs and GHG emissions are increased, the significant increase in cost utility (0.3749) due to more suppliers selected and more facilities opened is the main contributor in the increase of overall weighted sum utility, besides, the individual risk utility is relatively high as well. When the individual risk objective function reaches its optimal value, suppliers *s1*, *s3*, *s6*, and candidate locations *p1*, *p2*, *p3*, *w3* and *w5* are selected. In this scenario, both cost utility and GHG emission utility increase significantly. In order to have higher reliability and lower risks of suppliers, the purchasing costs of raw materials are increased dramatically, and the value of cost utility will then be increased to 0.3831 which takes the largest share of the overall weighted utility. Furthermore, the GHG emission utility increases to 0.3594 due to the increased numbers of transportation of raw materials and finished products in this scenario. If the system performance of optimal overall weighted sum utility is converted to 100%, the system performance of individual minimum costs, individual minimum GHG emissions and individual minimum risks can accordingly be converted to 71.6%, 40.7% and 25.5%, respectively, which is illustrated in Figure 2.

Table 6. The value of weighted sum utility, and selection of suppliers and candidate locations of the four selected scenarios

Scenario	Weighted sum utility	Supplier							Producer					Warehouse				
		s_1	s_2	s_3	s_4	s_5	s_6	s_7	p_1	p_2	p_3	p_4	p_5	w_1	w_2	w_3	w_4	w_5
MinIC[e]	0.127762				■	■		■	■		■		■			■		■
MinIGE[f]	0.223926			■	■	■		■	■	■	■	■			■	■		■
MinIR[g]	0.359397	■		■			■		■	■	■		■	■	■	■		■
MinWSU[h]	0.0914942						■	■	■	■	■					■		■

[e]MinIC=minimum individual costs, the same abbreviation is also applied in subsequent parts of this section.
[f]MinIGE=minimum individual GHG emissions, the same abbreviation is also applied in subsequent parts of this section.
[g]MinIR=minimum individual risks, the same abbreviation is also applied in subsequent parts of this section.
[h]MinWSU=minimum weighted sum utility, the same abbreviation is also applied in subsequent parts of this section.

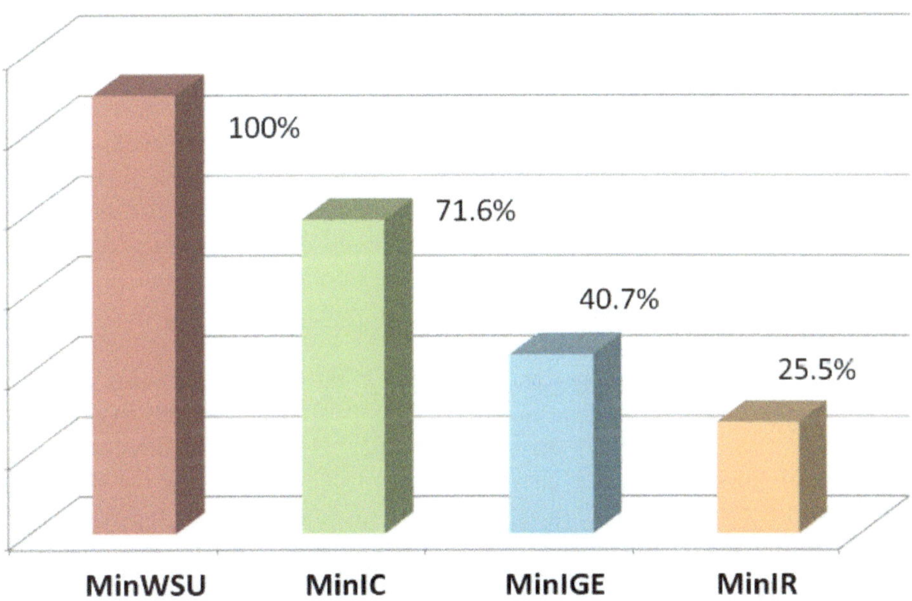

Figure 2. Comparison of the overall system performance of the four selected scenarios

When the overall weighted sum utility objective function achieves its optimal value, suppliers *s6*, *s7*, candidate locations *p1*, *p2*, *p3*, *w3* and *w5* are chosen. Table 7 illustrates the amount of raw materials and/or finished products transported between different facilities at each stage of the supply chain. It is noted that the result achieved in this scenario is mostly influenced by the cost utility due to its relatively large weight, and the result also compromises with the GHG emission utility and risk utility in order to balance the trade-off among those objectives. For instance, the selection of suppliers in optimal overall weighted scenario is achieved through balancing the cost utility and risk utility associated with the purchase and transportation of raw materials. Supplier *s4* has the lowest unit purchasing price and relatively low risk index, however, it is not chosen due to its high unit costs and risk index with respect to the transportation of raw materials, and suppliers *s6* and *s7* are therefore selected to maximize the overall system performance.

The optimal number, locations and inter-facility transportation of raw materials and/or finished products are determined by using the proposed GrSCND model, and the result is achieved through balancing the trade-off among cost utility, GHG emission utility and risk utility. The result is quite convincing, and further sensitive analysis can also be performed, if necessary, to test how and to what extend the optimal supply chain network configuration can be affected by different objectives.

Table 7. The amount of raw materials and/or finished products transported between different facilities at each stage

		s6	*s7*	*p1*	*p2*	*p3*	*w3*	*w5*
AT_{sp} [uw]	*p1*	300						
	p2		250					
	p3	143.75	150					
AT_{pw} [uw]	*w3*					200		
	w5			240	175	35		
AT_{wc} [uw]	*c1*							120
	c2						10	240
	c3						40	
	c4						110	

5. Conclusion

This work has presented a novel research on GrSCND model for companies and supply chains operated in high north arctic regions, where natural and infrastructural challenges bring more complexities in GrSCND than other regions, in order to enhance both competitive competence and sustainability. Compared with previous researches, the formulation and minimization of supply chain risks and reliability are taken into consideration in this study, which is an extremely important influencing factor accounted in the design and planning of supply chains operated in this region. The proposed computational model is formulated based upon multi-objective MIP method which aims to determine the number, locations and inter-facility transportation of raw materials and/or finished products through simultaneously minimizing the overall supply chain costs, GHG emissions and risks, and weighted sum utility method is employed to composite those objectives with different measures of units. A numerical experiment is performed as well to explicitly present the applications of the proposed multi-objective MIP model for GrSCND in high north regions, and Lingo solver is applied in coding and resolving the computational optimization problems.

The model is developed primarily for the design and planning of supply chains operated in high north regions, however, it is also perfectly applicable for the supply chains operated in other regions where the consideration of supply chain risks plays an important role. Besides, the selection of suppliers is also taken into account in this model, which is another crucial influencing factor for GrSCND. For example, in a global supply chain network, the reliability and safety of suppliers in some countries or regions may be significantly affected by some influencing factors, i.e., political stability, tax and tariff, infrastructure, etc., so it is of great importance to account the selection of suppliers in the decisional process of GrSCND so as to minimize the supply chain risks.

The mathematical model is formulated and developed under certain input parameters, however, the design and planning of supply chain network is always a strategic decision which has significant influence on long-term profitability, competitiveness and sustainability of a supply chain, and some input parameters may have great changes within its life span, furthermore, some parameters are stochastic in nature and impossible to be quantified accurately, and this will dramatically increase the level of difficulty in dealing with uncertainties. Therefore, the appropriate treatment of uncertain and stochastic input parameters are suggested for further improvement of the multi-objective MIP model for GrSCND in high north arctic regions.

Acknowledgements

This research was supported by EU-Sustainable Manufacturing and Engineering (SMaE) project (Grand No. 38005) for delivering cross-disciplinary solutions for enhancing competitive competence and sustainability of small and medium sized enterprises (SMEs) in high north arctic regions.

References

[1] Brooks, M.R., Frost, J.D. Providing freight services to remote arctic communities: Are there lessons for practitioners from service to Greenland and Canada's northeast? Research in Transportation Business & Management 2012, 4, 69-78.

[2] Chopra, S., Meindl, P. Supply chain management-Strategy, planning and operation (5th edition). Pearson, 2013.

[3] Solvang, W.D., Hakam, M.H. Sustainable logistics network in sparsely populated areas. Journal of Service Science & Management 2010, 3, 72-77.

[4] Statistical yearbook of Norway. SSB, 2013. Available from: https://www.ssb.no/en/befolkning/artikler-og-publikasjoner/statistical-yearbook-of-norway-2013 (Accessed on: 14. 03. 14).

[5] Handfield, R.B., Walton, S.V., Swwgers, L.K., Melnyk, S.A. "Green" value chain practices in the future industry. Journal of Operations Management 1997, 15 (4), 293-315.

[6] Ali, P., Searcy, C. A comparative literature analysis of definitions for green and sustainable supply chain management. Journal of Cleaner Production 2013, 52, 329-341.

[7] Stivastava, S.K. Green supply-chain management: a state-of-the art literature review. International Journal of Management Reviews 2007, 9 (1), 53-80.

[8] Gonzalez-Benito, J., Gonzalez-Benito, O. The role of stakeholder pressure and managerial values in the implementation of environmental logistics practices. International Journal of Production Research 2006, 44 (7), 1353-1373.

[9] Murphy, P.R., Poist, R.F. Green logistics strategies: an analysis of usage patterns. Transportation Journal 2000, 40 (2), 5-16.

[10] Linton, J.D., Klassen, R., Jayaraman, V. Sustainable supply chains: an introduction. Journal of Operations Management 2007, 25 (6), 1075-7082.

[11] Young, A., Kielkiewicz-Yong, A. Sustainable supply chain network management. Corporate Environmental Strategy 2001, 8 (3), 260-268.

[12] Cruz, Z.M., Matsypura, D. Supply chain networks with corporate social responsibility through integrated environmental decision-making. International Journal of Production Research 2009, 47 (3), 621-648.

[13] Seuring, S., Muller, M. From a literature review to a conceptual framework for sustainable supply chain management. Journal of Cleaner Production 2008, 16 (15), 1699-1710.

[14] Carter, R.C., Rogers, D.S. A framework of sustainable supply chain management: moving toward new theory. International Journal of Physical Distribution & Logistics Management 2008, 38 (5), 360-387.

[15] Sarkis, J., Zhu, Q., Lai, K.-H. An organizational theoretic review of green supply chain management literature. International Journal of Production Economics 2011, 130, 1-15.

[16] Ashby, A., Leat, M., Hudson-Smith, M. Making connections: a review of supply chain management and sustainability literature. Supply Chain Management: An International Journal 2012, 17 (5), 497-516.

[17] Kuo, T.C., Huang, S., Zhang, H. Design for manufacture and design for "X": concepts, applications and perspectives. Computers & Industrial Engineering 2001, 41, 241-260.

[18] Zhu, Q., Sarkis, J., Lai, K.H. Green supply chain management implications for closing the loop. Transportation Research Part E 2008, 44 (1), 1-18.

[19] Fleischmann, M., Beullens, P., Bloemhof-Ruwaard, J.M., Van Wassenhove, L.N. The impact of product recovery on logistics network design. Production & Operations Management 2001, 10, 156-173.

[20] Yu, H., Solvang, W.D. A reverse logistics network design model for sustainable treatment of multi-sourced waste of electrical and electronic equipment (WEEE). Proceeding of the 4th IEEE International Conference on Cognitive Infocommunications. Budapest, Hungary, 2013, pp: 595-600.

[21] Cheng, S., Chan, C.W., Huang, G.H. An integrated multi-criteria decision analysis and inexact mixed integer linear programming approach for solid waste management. Engineering Applications of Artificial Intelligence 2003, 16, 543-554.

[22] Yu, H., Solvang, W.D., Yuan, S. A multi-objective decision support system for simulation and optimization of municipal solid waste management system. Proceeding of the 3rd IEEE International Conference on Cognitive Infocommunications. Kosice, Slovakia, 2012, pp: 193-199.

[23] Elhedhli, S., Merrick, R. Green supply chain network design to reduce carbon emissions. Transportation Research Part D 2012, 17, 370-379.

[24] Farahani, R.Z., Rezapour, S., Drezner, T., Fallah, S. Competitive supply chain network design: An overview of classification, models, solution techniques and applications. Omega 2014, 45, 92-118.

[25] Shen, Z.J. Integrated supply chain models: a survey and future research directions. Journal of Industrial Management and Optimization 2007, 3 (1), 1-27.

[26] Wang, F., Lai, X.F., Shi, N. A multi-objective optimization for green supply chain network design. Decision Support System 2011, 51, 262-269.

[27] Govindan, K., Jafarian, A., Khodaverdi, R., Devika, K. Two-echelon multiple-vehicle location-routing problem with time windows for optimization of sustainable supply chain network of perishable food. International Journal of Production Economics 2014, http://dx.doi.org/10.1016/j.ijpe.2013.12.028.

[28] Quariguasi-Frota-Neto, J., Bloemhof-Ruwaard, J.M., Van Nunen, J.A.E.E., Van Heck, E. Designing and evaluating sustainable logistics networks. International Journal of Production Economics 2008, 111, 195-208.

[29] Harris, I., Naim, M., Palmer, A., Potter, A., Mumford, C. Assessing the impact of cost optimization based on the infrastructure modelling on CO_2 emissions. International Journal of Production Economics 2011, 131, 313-321.

[30] Ubeda, S., Arcelus, F.J., Faulin, J. Green logistics at eroski: a case study. International Journal of Production Economics 2011, 131, 44-51.

[31] Abdallah, T., Farhat, A., Diabat, A., Kennedy, S. Green supply chains with carbon trading and environmental sourcing: formulation and life cycle assessment. Applied Mathematical Modelling 2012, 36, 4271-4285.

[32] Jamshidi, R., FatemiGhomi, S.M.T., Karimi, B. Multi-objective green supply chain optimization with a new hybrid memetic algorithm using the Taguchi method. ScientiaIranica E 2012, 19 (6), 1876-1886.

[33] Latha Shankar, B., Basavarajappa, S., Kadadevaramath, R.S., Chen, J.C.H. A bi-objective optimization of supply chain design and distribution operations using non-dominated sorting algorithm: A case study. Expert Systems with Applications 2013, 40, 5730-5739.

[34] Sheu, J.-B., Lin, A.Y.-S. Hierarchical facility network planning model for global logistics network configuration. Applied Mathematical Modelling 2012, 36, 3053-3066.

[35] Ene, S., Ozturk, N. Open loop reverse supply chain network design. Procedia – Social and Behavioral Sciences 2014, 109, 1110-1115.

[36] Zeballos, L.J., Mendez, C.A., Barbosa-Povoa, A.P., Novais, A.Q. Multi-period design and planning of closed-loop supply chains with uncertain supply and demand. Computers and Chemical Engineering 2014, http://dx.doi.org/10.1016/j.compchemeng.2014.02.027.

[37] Pishvaee, M.S., Razmi, J. Environmental supply chain network design using multi-objective fuzzy mathematical programming. Applied Mathematical Modelling 2012, 36, 3433-3446.

[38] Ramezani, M., Kimiagari, A.M., Karimi, B., Hejazi, T.H. Closed-loop supply chain network design under a fuzzy environment. Knowledge-Based Systems 2014, 59, 108-120.

[39] Amin, S.H., Zhang, G.Q. A multi-objective facility location model for closed-loop supply chain network under uncertain demand and return. Applied Mathematical Modelling 2013, 37, 4165-4176.

[40] Yu, M.-C., Goh, M. A multi-objective approach to supply chain visibility and risk. European Journal of Operational Research 2014, 233, 125-130.

[41] Nema, A.K., Gupta, S.K. Optimization of regional hazardous waste management systems: an improved formulation. Waste Management 1999, 19, 441-451.

[42] Sheu, J.-B. A coordinated reverse logistics system for regional management of multi-source hazardous wastes. Computers & Operations Research 2007, 34, 1442-1462.

2

Improvement the performance of a proton exchange membrane fuel cell by changing the channel geometry

I. Khazaee

Faculty of Mechanical and Energy Engineering, Shahis Beheshti University, A.C., Tehran, Iran.

Abstract
In this study the effect of placing different blocks on the performance of a proton exchange membrane (PEM) fuel cell are investigated numerically for different Aspect Ratios. A complete two-dimensional and single phase model is used to that the proposed model is a full cell model, which includes all the parts of the PEM fuel cell, flow channels, gas diffusion electrodes, catalyst layers and the membrane. Coupled transport and electrochemical kinetics equations are solved in a single domain; therefore no interfacial boundary condition is required at the internal boundaries between cell components. The results show that the predicted polarization curves by using this model are in good agreement with the experimental results. Also the results show that the transverse installation of a rectangular and triangle block in the fuel flow channel can effectively enhance the local cell performance of a PEMFC. The results show that by increasing the aspect ratio of the blocks, the performance of the cell enhances due to enhance the electrochemical reaction at the catalyst layer of the cell.

Keywords: PEM fuel cell; Rectangular block; Triangle block; Numerical modeling; Cell performance.

1. Introduction

A fuel cell is an electro-chemical energy device that converts the chemical energy of fuel directly into electricity and heat, with water as a by-product of the reaction. Based on the types of electrolytes used, they are categorized into polymer electrolyte membrane fuel cells (PEMFCs), solid oxide fuel cells (SOFCs), phosphoric acid fuel cells (PAFCs), molten carbonate fuel cells (MCFCs), and direct methanol fuel cells (DMFCs). The polymer exchange membrane fuel cell (PEMFC) is considered to be the most promising candidate for electric vehicles by virtue of its high power density, zero pollution, low operating temperature, quick start-up capability and long lifetime. PEMFC can also be used in distributed power systems, submarines, and aerospace applications [1].

The single-cell PEMFC consists of a carbon plate, a gas diffusion layer (GDL), a catalyst layer, for each of the anode and the cathode sides, as well as a PEM membrane at the center.

Flow channel geometry is of critical importance for the performance of fuel cells containing proton exchange membranes (PEM) but is of less concern for solid oxide fuel cells (SOFC). The reactants, as well as the products, are transported to and from the cell through flow channels. Flow field configurations, including parallel, serpentine, interdigitated, and other combined versions, have been developed.

The performance of the fuel cell system is characterized by current-voltage curve *(i.e. polarization curve).* The difference between the open circuit potential of the electrochemical reaction and cell voltage

occurs from the losses associated with the operation. The corresponding voltage drop is generally classified in three parts:
i. activation over-potential caused by the electrochemical reactions,
ii. ohmic drop across the polymer electrolyte
iii. mass transfer limitations of reactants
These associated losses dominate over different current density ranges. For low current densities; the activation over-potential is dominant. For high current densities, which are of particular interest for vehicle applications because of higher power density; the mass transfer limitations dominates the losses. For moderate current densities, the ohmic drop across the polymer membrane dominates. Moreover, for high current densities, water starts to exist in liquid form leading to a two-phase transport of reactants to reaction site, which is an additional transport phenomenon of PEM Fuel Cell operation.
West and Fuller [2] proposed a two-dimensional numerical analysis of the rib spacing in PEM electrode assemblies on current and water distribution within the cell. The results indicated that increasing the rib width strongly affected the membrane water content before the catalyst utilization is reduced. Therefore, the two-dimensional effect has a significant influence on water management.
Chiang and Chu [3] investigated the effects of transport components on the transport phenomena and performance of PEM fuel cells by using a three-dimensional model. The impacts of channel aspect ratio (AR) and GDL thickness were examined. It was found that a flat channel with a small AR or a thin GDL generates more current at low cell voltage due to the merits of better reactant gas transport and liquid water delivery.
Wang et al. [4] developed a two-dimensional numerical model to study the two-phase flow transport in the air cathode of a PEMFC. In this paper, the model encompassed both single- and two-phase regimes corresponding to low and high current densities and was capable of predicting the transition between the two regimes.
Jenn-Kun Kuo et al. [5] performed numerical simulations to evaluate the convective heat transfer performance and velocity flow characteristics of the gas flow channel design to enhance the performance of proton exchange membrane fuel cells (PEMFCs). Their study has simulated low Reynolds number laminar flow in the gas flow channel of a PEMFC. The heat transfer performance and enhanced gas flow velocity characteristics of four different channel geometries have been considered, namely a conventional straight gas flow channel and a gas flow channel with the three novel periodic patterns geometries.
The results indicated that, compared to the conventional gas flow channel, the novel gas flow channels proposed in this study provide a significantly improved convective heat transfer performance and a higher gas flow velocity and, hence, an improved catalysis reaction performance in the catalyst layer.
Based on available literature, the present study numerically investigated the effect of using rectangular and triangle baffle blocks and rectangular and triangle blocks in flow fields on cell performance and voltage losses of a PEM fuel cell. The objective of the current work is to show the effect of increasing the aspect ratio of the obstacles in the channels that it may be of interest to engineers attempting to develop the optimization of a PEMFC and to researchers interested in the flow modification aspects of the PEMFC performance enhancement.

2. Numerical model

The cathode electrochemical reactions produce a large amount of liquid water at high electric current. If the liquid water is not properly removed and accumulates in the pores of the porous layers, it restricts the oxygen transport to the gas diffusion layer and the catalyst layer, thereby reducing the electrochemical reaction rate.
The numerical model for the fuel cell used here includes the anode flow channels, anode gas diffusion layer, anode catalyst layer, proton exchange membrane, cathode catalyst layer, cathode gas diffusion layer, and cathode flow channels. Miniature fuel cells with dimensions of 4.731mm×100mm are considered in this investigation. The gas diffusion layer is 0.33mm thick, the catalyst layer is 0.01mm thick, and the proton exchange membrane is 0.051mm thick. The geometrical relations in this study are for two blocks in four aspect ratios as AR=0.2, 0.4, 0.6 and 0.8.
The physical problem considered in this paper is the two dimensional cell model of the PEMFC system as shown in Figure 1 that the dimensionless parameter in this paper named aspect ratio, $AR = \frac{L}{H}$, is defined to characterize the blockage effect of the blocks.

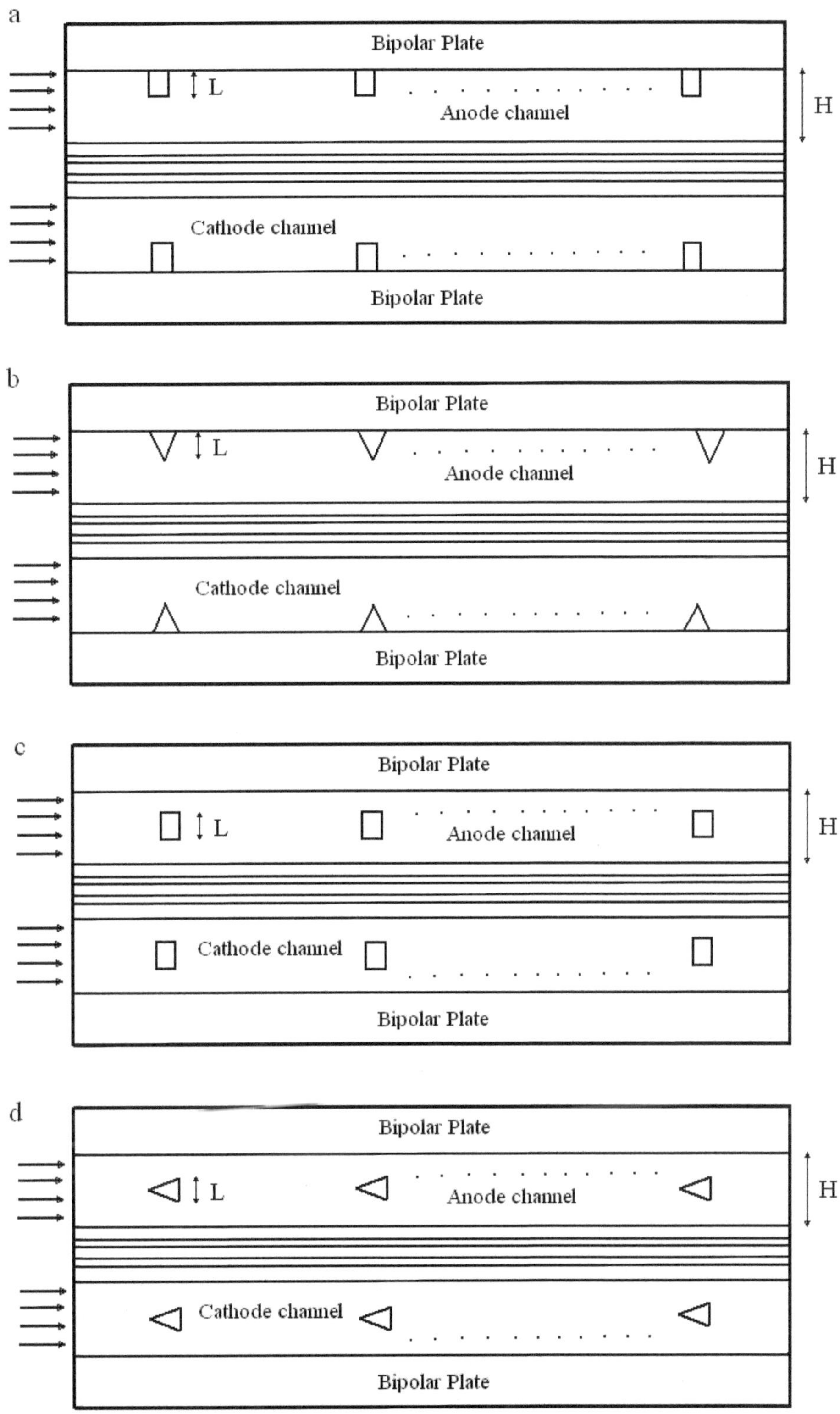

Figure 1. Computational domain of PEMFC with (a) rectangular baffle block, (b) triangle baffle block, (c) rectangular block and (d) triangle block

The proposed model that described in previous work [6] does not require any internal boundary conditions between the components of PEM Fuel Cell system. The different physical properties and transport parameters are incorporated into a single set of governing equations using a single domain formulation. The model aims to study the electrochemical kinetics, current distribution, reactant flow fields and multi-component transport of oxidizer and fuel streams in a multi-dimensional domain. The assumptions made in developing the model are as follows:

- Ideal gas mixtures
- Incompressible and Laminar flow because low flow velocities and low fuel utilization
- Isotropic and homogeneous porous electrodes, catalyst layers and membrane
- Isothermal operation
- Negligible ohmic resistance at porous electrodes and current collectors

The model assumes that the system is steady; the inlet reactants are ideal gases; the flow is laminar; and the porous layers such as the diffusion layer, catalyst layer and PEM are isotropic. The model includes continuity, momentum and species equations for gaseous species, liquid water transport equations in the channels, gas diffusion layers, and catalyst layers, water transport equation in the membrane, electron and proton transport equations. The Bulter–Volumer equation was used to describe electrochemical reactions in the catalyst layers.

The conservation equations of mass, momentum, species and charge are as follows:

$$\frac{\partial(\rho\varepsilon)}{\partial t} + \nabla.(\varepsilon\rho\vec{u}) = 0 \tag{1}$$

$$\frac{\partial(\rho\varepsilon\vec{u})}{\partial t} + \nabla.(\varepsilon\rho\vec{u}\vec{u}) = -\varepsilon\nabla p + \nabla.(\varepsilon\mu\nabla\vec{u}) + S_u \tag{2}$$

$$\frac{\partial(\varepsilon X_k)}{\partial t} + \nabla.(\varepsilon\vec{u}X_k) = \nabla.(D_k^{eff}\nabla X_k) + S_k \tag{3}$$

$$\nabla.(\sigma_g^{eff}\nabla\phi_e) + S_\phi = 0 \tag{4}$$

where $\vec{u}, X_k and \phi_e$ denotes intrinsic velocity vector, molar fraction of kth species and electrolyte phase potential, respectively. The corresponding source terms treating the electrochemical reactions and porous media are presented in Table 1.

Table 1. Source terms

	S_v (Momentum)	S_k (Species)	$S_Ø$ (Phase potential)
Gas channels	0		
Backing layers	$-\frac{\mu}{k}\varepsilon^2\vec{u}$		
Catalyst layers	$-\frac{\mu}{k}\varepsilon^2\vec{u}$	$-\frac{j_a}{2F}\frac{\rho}{\varepsilon_m\varepsilon_{mc}}$: anode H_2 $\frac{j_c}{2F}\frac{\rho}{\varepsilon_m\varepsilon_{mc}}$: cathode O_2 $-\frac{j_c}{F}\left(\frac{1}{2}+n_d\right)\frac{\rho}{\varepsilon_m\varepsilon_{mc}}$: cathode H_2O $-\frac{n_d}{F}j_a\frac{\rho}{\varepsilon_m\varepsilon_{mc}}$: anode H_2O	j
Membrane	$-\frac{\mu}{k}\varepsilon^2\vec{u}$		

It is of benefit to further explain the corresponding diffusivities of the governing equations. The diffusivities for species concentration equations and ionic conductivity for membrane phase potential equation are modified using Bruggman correlation to account for porous electrodes, which can be expressed as:

$$D_k^{eff} = \varepsilon_m^{1.5} D_k \tag{5}$$

$$\sigma_e^{eff} = \varepsilon_m^{1.5} \sigma_e \tag{6}$$

It is worth further explaining the mole fraction of oxygen appearing in Eq. 3 because oxygen is a gaseous species in the cathode flow channel and gas-diffusion electrode but becomes a species dissolved in the electrolyte in the catalyst layer and membrane regions. Our definition is given by

$$X_k = \begin{cases} C_k^g / C_{tot} & in\ gas \\ C_k^e / C_{tot} & in\ electrode \end{cases} \tag{7}$$

where C_k is the molar concentration of species k and superscripts g and e denote the gas and the electrolyte phases, respectively. Thus, X_k is a true mole fraction in the gas phase but is a pseudo mole fraction when species k is in the dissolved form. In addition, there is a discontinuity in the value of X_k at the interface between the gas-diffusion electrode and the catalyst layer due to the following thermodynamic relation

$$C_k^{e,sat} = \frac{RT}{H} C_k^g \tag{8}$$

where H is the Henry's law constant equal to $2\times10^5\ atm\ cm^3/mol$ for oxygen in the membrane.

Either generation or consumption of chemical species k and the creation of electric current occurs only in the active catalyst layers where electrochemical reactions take place. The S_k and S_ϕ terms are therefore related to the transfer current between the solid matrix and the membrane phase inside each of the catalyst layers. These transfer currents at anode and cathode can be expressed as follows

$$j_a = aj_{0,a}^{ref}\left(\frac{X_{H_2}}{X_{H_2,ref}}\right)^{1/2}\left(\frac{\alpha_a + \alpha_c}{RT}.F.\eta\right) \tag{9}$$

$$j_c = -aj_{0,c}^{ref}\left(\frac{X_{O_2}}{X_{|O_2,ref}}\right)^{1/2}\left(-\frac{\alpha_c.F}{RT}.\eta\right) \tag{10}$$

The above kinetics expressions are derived from the general Bulter- Volmer equation based on the facts that the anode exhibits fast electrokinetics and hence a low surface overpotential to justify a linear kinetic rate equation, and that the cathode has relatively slow kinetics to be adequately described by the Tafel equation. In Eq. 9 and 10, the surface overpotential, $\eta(x, y)$, is defined as

$$\eta(x,y) = \phi_s - \phi_e - V_{oc} \tag{11}$$

where ϕ_S and ϕ_ε stand for the potentials of the electronically conductive solid matrix and electrolyte, respectively, at the electrode electrolyte interface. V_{OC} is the reference open-circuit potential of an electrode. It is equal to zero on the anode but is a function of temperature on the cathode namely

$$V_{oc} = 1.17 - 2.756\times10^{-4}(T - 373.15) + 4.308\times10^{-5}\ln\left(\frac{a_{H_2}(a_{O_2})^{1/2}}{a_{H_2O}}\right) \tag{12}$$

where T is in Kelvin and V_{oc} is in volts. Notice that V_{oc} is not the true open-circuit potential of an electrode, which would then depend upon reactant concentrations according to the Nernst equation. This equilibrium potential is calculated from thermodynamic data of reaction enthalpy and entropy changes while the product water is in gaseous phase. The definition of a_{H_2}, a_{O_2} and a_{H_2O} are:

$$a_{H_2O} = \frac{p_{H_2}}{p^0}, \qquad a_{O_2} = \frac{p_{O_2}}{p^0}, \qquad a_{H_2O} = \frac{p_{H_2O}}{p^0_{H_2O}}. \tag{13}$$

The dependence of the cathodic exchange current density on temperature can be fitted as

$$\frac{i_0(T)}{i_0(353K)} = \exp(0.014189(T-353)) \tag{14}$$

The species diffusivity, D_k, varies in different subregions of the PEMFC depending on the specific physical phase of component k. In flow channels and porous electrodes, species k exists in the gaseous phase, and thus the diffusion coefficient takes the value in gas, whereas species k is dissolved in the membrane phase within the catalyst layers and the membrane, and thus takes the value corresponding to dissolved species, which is usually a few orders of magnitude lower than that in gas. In addition, the diffusion coefficient is a function of temperature and pressure.

$$D(T) = D_0\left(\frac{T}{T_0}\right)^{3/2}\left(\frac{p_0}{p}\right) \tag{15}$$

The proton conductivity in the membrane phase has been correlated as

$$\sigma_e(T) = 100\exp\left[1268\left(\frac{1}{303}-\frac{1}{T}\right)\right](0.005139\xi - 0.00326) \tag{16}$$

where the water content in the membrane, ξ, depends on the water activity, a, according to the following fit of the experimental data

$$\xi = \begin{cases} 0.043 + 17.18a - 39.85a^2 + 36a^3 & 0 \prec a \le 1 \\ 14 + 1.4(a-1) & 1 \le a \le 3 \end{cases} \tag{17}$$

The water activity is in turn calculated by

$$a = \frac{X_{H_2O}\,p}{p_{sat}} \tag{18}$$

where the saturation pressure of water vapor can be computed from the following empirical equation

$$\ln(p^{sat}) = 70.43464 - \frac{7362.698}{T} + 0.006952T - 9\ln(T) \tag{19}$$

In a fuel cell system, the inlet flow rates are generally expressed as stoichiometric ratios of inlet streams based on a reference current density. The stoichiometric ratios inlet streams are given by the following equations.

$$S^{anode} = C_{H_2} v^{anode} \frac{2F}{I_{ref} A} \tag{20}$$

$$S^{cathode} = C_{O_2} v^{cathode} \frac{4F}{I_{ref} A} \tag{21}$$

Water transport through the polymer electrolyte membrane has been investigated by several researchers in different aspects. Most interesting studies in this area includes the determination of water diffusion coefficient [7] and water drag coefficient [8] by Zawodzinski et al. and investigating the diffusion of water in Nafion membranes by Motupally et al.[9].

The electro-osmotic drag coefficient is defined as the number of water molecules transported by each hydrogen proton H^+. The electro-osmotic drag coefficient can be expressed with the following equation:

$$n_d = \frac{2.5\xi}{22} \tag{22}$$

The diffusion coefficient of water in Polymer Membrane is also highly dependent on the water content of the membrane. The relation is given as:

$$D_w^m = \begin{Bmatrix} 3.1\times10^{-7}\,\xi(\exp(0.28\xi)-1)\exp(-2346/T) & 0 \prec \xi \prec 3 \\ 4.17\times10^{-8}\,\xi(1+161\exp(-\xi))\exp(-2346/T) & otherwise \end{Bmatrix} \tag{23}$$

Once the electrolyte phase potential is determined in the membrane, the local current density along the axial direction can be calculated as follows

$$I(y) = -\sigma_e^{eff} \left.\frac{\partial \phi_e}{\partial x}\right|_{x = I.F.} \tag{24}$$

where I.F. means the interface between the membrane and cathode catalyst layer. The average current density is then determined by

$$I_{avg} = \frac{1}{L}\int_0^L I(y)dy \tag{25}$$

where L is the cell length.

The boundary conditions prescribed at the inlets of the gas channels are:

$$u_{in}^{anode} = u_a^0 \qquad u_{in}^{cathode} = u_c^0$$

$$C_{H_2}^{anode,in} = C_{H_2}^0 \qquad C_{O_2}^{cathode,in} = C_{O_2}^0$$

$$C_{H_2O}^{anode,in} = C_{H_2O}^{0,a} \qquad C_{H_2O}^{cathode,in} = C_{H_2O}^{0,c}$$

A mesh of 241×74×25 was found to provide required spatial resolution for four-channel geometry. The solution is considered to be converged when the difference between successive iterations is less than 10^{-7} for all variables.

The Electrochemical and Transport Parameters used in these simulations are summarized in Table 2, and the operational parameters are presented in Table 3.

Table 2. Electrochemical and transport properties

Description	Unit	Value
Anode reference exchange current density	A/m^3	1.5e9
Cathode reference exchange current density	A/m^3	4000000
Anode transfer coefficient		2
Cathode transfer coefficient		2
Faraday constant	C/mol	96487
H_2 Diffusivity	m^2/s	3e-5
O_2 Diffusivity	m^2/s	3e-5
H_2O Diffusivity at anode	m^2/s	3e-5
H_2O Diffusivity at cathode	m^2/s	3e-5
Anode backing layer porosity		0.5
Cathode backing layer porosity		0.5
Permeability of anode backing layer	m^2	e-12
Permeability of cathode backing layer	m^2	e-12
Equivalent weight of membrane	kg/mol	1.1

Table 3. Operational parameters

Description	Unit	Value
reference average current density	A/cm^2	1.0
Anode inlet velocity	m/s	3
Cathode inlet velocity	m/s	2
Anode inlet pressure	Atm	1
Cathode inlet pressure	Atm	1
Cell temperature	C°	70
Anode inlet molar concentration		
Hydrogen	mol/m^3	35.667
Oxygen	mol/m^3	0
Water vapor	mol/m^3	16.121
Cathode inlet molar concentration		
Hydrogen	mol/m^3	0
Oxygen	mol/m^3	0 7.51
Water vapor	mol/m^3	16.121

4. Results and discussion

In order to show that the program in this study can handle the cell performance of a PEMFC, we apply the present method to solve the whole of a PEMFC as described in Miansari and others ' paper [10]. The physical parameters and properties of the fuel cell are listed in Table 2. The mesh employed for the comparison with the reference was 241×74×25. The steady-state solution is obtained by the numerical procedure as mentioned in the previous section. As shown in Figure 2, the result of the present predictions of the polarization curve agreeing fairly closely with Miansari and others' predictions [10] gives one confidence in the use of the present program.

Figure 3 (triangle block and triangle baffle block) and Figure 4 (rectangular block and rectangular baffle block) illustrate the polarization curves of the fuel cell performance for different channel geometries with nine blocks (n=9) to investigate the influence of the internal flow modification on the overall fuel cell performance. An overall inspection of Figures 3 and 4 indicates that at the conditions of the higher operating voltage (lower over- potential), the influence of the internal flow modification on the overall fuel cell performance is negligibly small. At lower operating voltage conditions, on the other hand, the effect of the internal flow modification on the polarization curves becomes important. Also Figures 3(a),(b) and 4(a),(b) show that the influence of baffle block on the overall fuel cell performance is small. In addition, for the installation of triangle and rectangular blocks, the overall cell performance decreases when the aspect ratio *AR* decreases. This is due to the fact that a reflection of the more efficient fuel transport and the chemical reaction around the triangle and rectangular blocks reduces the local current density more obviously in the downstream region behind the triangle and rectangular blocks.

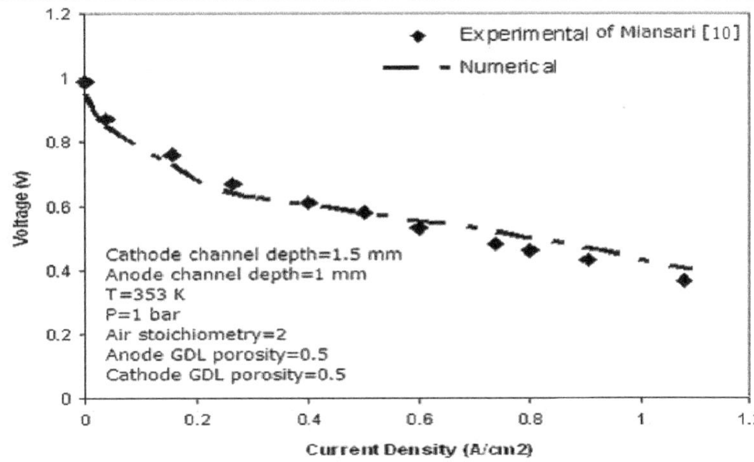

Figure 2. Comparison between results of this paper and results of Miansari et al [10]

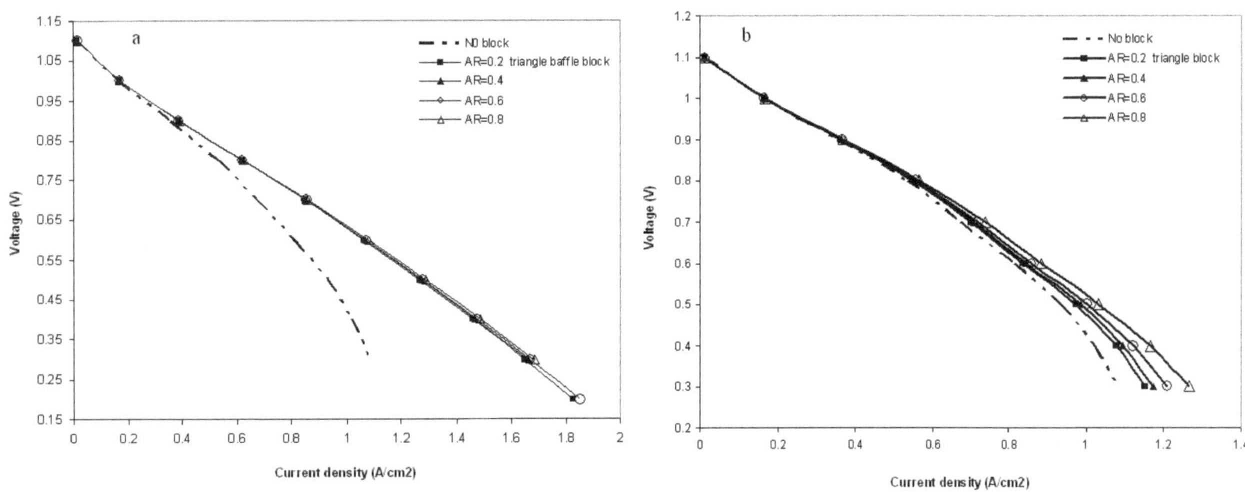

Figure 3. Variation of cell performance at different AR for (a) triangle baffle block and (b) triangle block

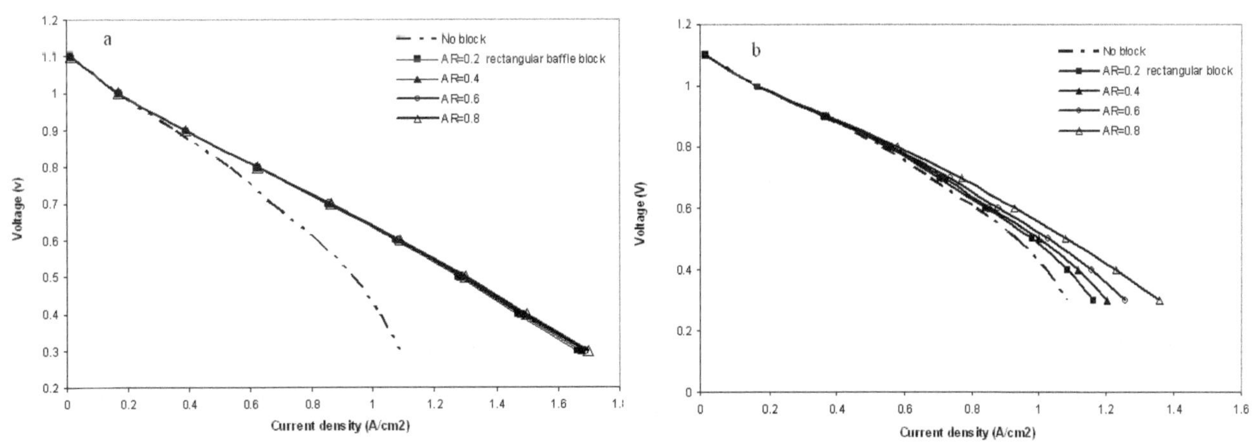

Figure 4. Variation of cell performance at different AR for (a) rectangular baffle block and (b) rectangular block

Figure 5 shows the polarization curves of the fuel cell performance for seven different channel geometries with n=0, n=9, n=13 and n=24 for AR=0.8 to investigate the influence of the internal flow modification on the overall fuel cell performance. It is clear that by increasing the number of the blocks in the channels and installation of rectangular block, the overall cell performance increases. This is due to the fact that a reflection of the more efficient fuel transport and the chemical reaction around the rectangular block reduces the local current density more obviously in the downstream region behind the rectangular block. Also it is clear that the performance of the cell is at higher value by rectangular block with n=9 than triangle block with n=13, therefore there is a best condition for lower capital cost of the fuel cell.

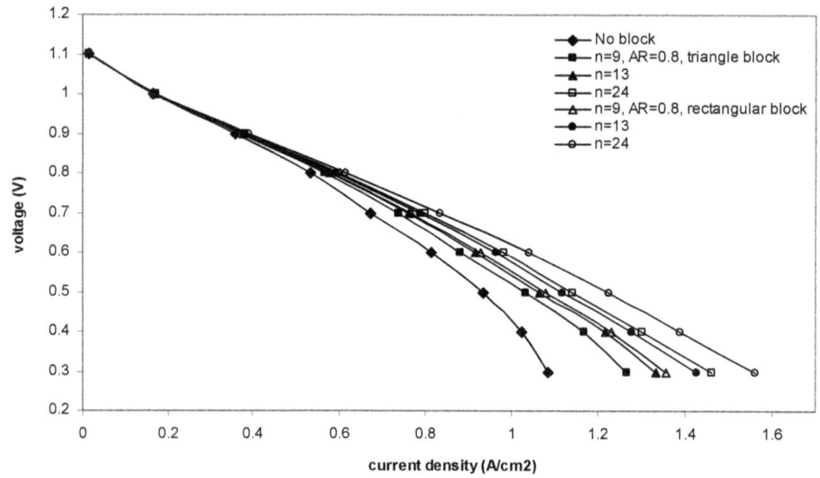

Figure 5. Variation of cell performance for n=0, n=9, n=13 and n=24 and AR=0.8 for triangle and rectangular block

The effect of kind of blocks in the channels presented in Figure 6 for n=9 and AR=0.2, 0.4, 0.6 and 0.8. It is clear that by increasing the aspect ratio of the blocks in the channel, the overall performance of the cell increases but the difference between the performance of the cell at a constant AR for rectangular block than triangle block is very high that this is due to more decreases of local current density in the downstream region behind the rectangular block. Also it is clear that at the conditions of the higher operating voltage (lower over- potential), the influence of the internal flow modification on the overall fuel cell performance is negligibly small and at lower operating voltage conditions the effect of the internal flow modification on the polarization curves becomes important.

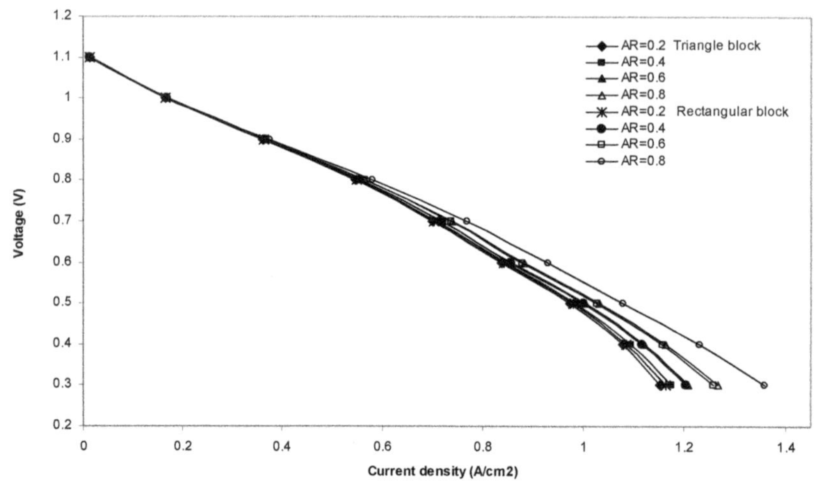

Figure 6. Comparison of cell performance between triangle block and rectangular block for different AR

In Figures 7 and 8, the local current density along the catalyst surface decreases when the axial direction x increases except for the peak regions but for triangle and rectangular baffle blocks it increases in the downstream region behind the baffle blocks. Also it is clear that the local current density is at higher value for baffle blocks that it is due to the more blockage of the fuel gas around the baffle blocks and forcing the more fuel gas into the gas diffusion layer to enhance the chemical reaction at the catalyst layer. The local current density around the baffle blocks and blocks increases with increasing the aspect ratio (AR), and the difference of the peak regions between the baffle blocks and the blocks is very high and also the difference between the triangle block and rectangular block is very high. These peak regions of the distribution of the local current density are caused by the transverse installation of a baffle blocks and blocks. This phenomena means that better cell performance is achieved around the baffle blocks or blocks. This is due to the fact that the fuel gas is blocked by the blocks installed in the flow channel, and more fuel gas is forced into and passes through the gas diffusion layer, which enhances the chemical reaction at the catalyst surface.

In the downstream region behind the baffle blocks and blocks, the local current density decreases as the aspect ratio (AR) decreases. The reduction of the local current density is because of increasing the fuel velocity and decreasing the chemical reaction and less flow into the GDL in the downstream region behind the baffle blocks and blocks. But the difference between current density behind the blocks is very high for triangle and rectangular blocks than the triangle and rectangular baffle blocks.

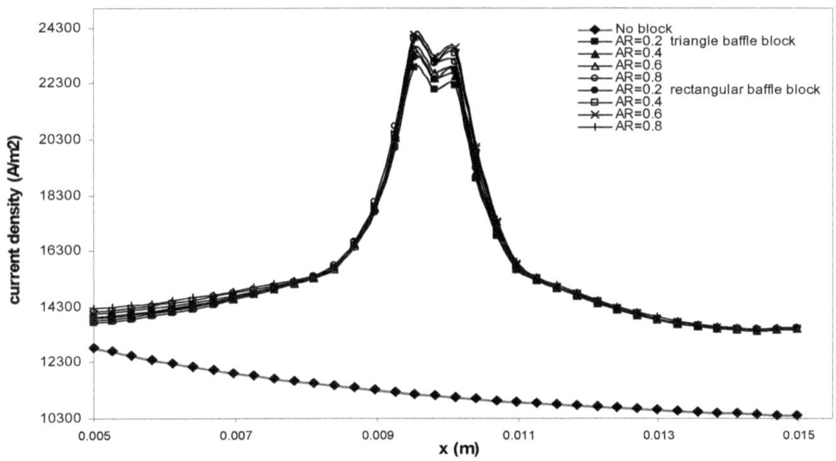

Figure 7. The local current density distributions along the catalyst surface for different AR and triangle and rectangular baffle blocks

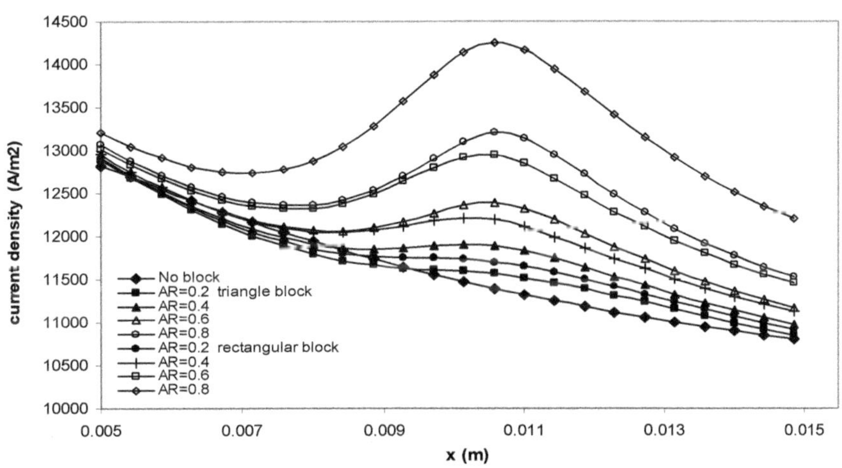

Figure 8. The local current density distributions along the catalyst surface for different AR and triangle and rectangular blocks

5. Conclusion

A complete two-dimensional and single phase model for proton exchange membrane (PEM) fuel cells has been used to investigate the effect of placing triangle and rectangular baffle blocks and triangle and rectangular blocks in the flow channels on the performance for different aspect ratios. The complete two-dimensional model for PEM fuel cells based on the two-fluid method was numerically solved with constant-temperature boundary condition at surfaces of anode and cathode current collectors. The results of this paper are in good agreement with experimental results of miansari et al[10]. The results show that by installing the baffle blocks or blocks into the channels and by increasing the aspect ratio of the blocks the overall cell performance increases. Also the results show that the difference between the performance of the cell at a constant AR for rectangular block than triangle block is very high and by increasing the number of the blocks in the channels and installation of rectangular block, the overall cell performance increases. The local current density is at higher value for baffle blocks than blocks by increasing the AR that it is due to the more blockage of the fuel gas around the baffle blocks and forcing the more fuel gas into the gas diffusion layer. however, the installation of rectangular baffle blocks and rectangular blocks have better overall cell performance than the installation of triangle baffle blocks and triangle blocks and enhances the overall cell performance of a PEMFC system.

Nomenclature

A	Superficial electrode area, m^2
C	Molar concentration, mol/m^3
D	Species diffusivity, m^2/s
I	Current density, A/cm^2
i_0	Reference current density, A/cm^2
U	Inlet velocity, m/s
j	Transfer current density, A/m^3
$\vec{u}$	Velocity vector, m/s
P	pressure, Pa
p^0	Reference pressure, Pa
S	Stoichiometric Ratio
T	Temperature, K

Greek letters

η	Overpotential, V
ρ	Density, kg/m^3
ε	Porosity
σ	Ionic conductivity, S/m
ϕ	Phase potential, V
ν	Volumetric flow rate, m^3/s
ξ	Water content of the membrane
μ	Viscosity, kg m/s
α	Transfer coefficient for the reaction

Acknowledgment

This work was partially supported by Renewable Energy Organization of Iran.

References

[1] J. Larminie, and A. Dicks, Fuel cell system explained., John Wiley and Sons.,2nd edition, 2003.

[2] A.C. West, and T.F. Fuller, "The Influence of Rib Spacing in Proton-Exchange Membrane Electrode Assemblies", Journal of Applied Electrochemistry, 26(1996), pp. 557–565.

[3] M.S. Chiang, and H.S. Chu, "Transient behavior of CO poisoning of the anode catalyst layer of a PEM fuel cell", Journal of Power Sources, 160 (2006), pp.340–352.

[4] Z.H. Wang, C.Y. Wang, and K.S. Chen, "Two-phase flow and transport in the air cathode of proton exchange membrane fuel cells", Journal of Power Sources, 94 (2001), pp.40–50.

[5] J.K. Kuo, T.S. Yen, and C.K. Chen, "Improvement of performance of gas flow channel in PEM fuel cells", Energy Conversion and Management, 49 (2008), pp.2776–2787.

[6] I. Khazaee, and M. Ghazikhani, "Performance improvement of proton exchange membrane fuel cell by using annular shaped geometry", Journal of Power Sources, 196(2011), pp.2661-2668.

[7] T.A. Zawodzinski, M. Neeman, L.O. Sillerud, and S. Gottesfeld, "Determination of water diffusion coefficients in perfluorosulfonate ionomeric membranes", The Journal of Physical Chemistry, 95 (1991), pp.6040-6044.

[8] T.A. Zawodzinski, J. Davey, J. Valerio, and S. Gottesfeld, "The water content dependence of electro-osmotic drag in proton-conducting polymer electrolytes", Electrochimica Acta, 40 (1995), pp.297-302.

[9] S. Motupally, A.J. Becker, and J.W. Weidner, "Diffusion of Water through Nafion® 115 Membranes", Journal of the Electrochemical Society, 147 (2000), pp.3171-3177.

[10] ME. Miansari, K. Sedighi, M. Amidpour, E. Alizadeh, and MO. Miansari, "Experimental and thermodynamic approach on proton exchange membrane fuel cell performance", Journal of Power Sources, 190 (2009), pp.356-361.

Experimental study on the heat transfer enhancement by Dean Vortices in spiral tubes

Hui Zhu[1], Hanqing Wang[2], Guangxiao Kou[2]

[1] School of Energy Science and Engineering, Central South University,NO.932, South Lushan Road, Changsha, Hunan Province, 410083, China.
[2] School of Civil Engineering, Hunan University of Technology, West Taishan Road, Zhuzhou, Hunan Province, 412007, China.

Abstract
Spiral tubes have been reported to be able to improve the heat transfer performance of the fluid flowing through, which is traditionally believed to be caused by the helix structure. In this literature, however, the mechanism of the heat transfer enhancement of spiral tubes was studied experimentally in terms of the control factors of Dean vortices such as Reynolds number, fluid viscosity and curvature ratio. Meanwhile, the torsion of the spiral tube was also taken into consideration. Experiments were carried out under constant wall temperature condition with one straight copper tube and 7 kinds of spiral copper tubes. The results showed that the Nusselt number of the flow in spiral tubes increases with Reynolds number, viscosity and curvature ratio, but changes anomalously with the torsion. And it also clearly revealed that the pressure drop for spiral tubes increases with Reynolds number, curvature ratio and viscosity, while decreases with the torsion. In addition, the heat transfer performance of spiral tubes is apparently better than that of the straight tube. Based on the experiment results, the effect of Dean number on Nusselt number and pressure drop of the spiral tubes was determined, and the correlation equations of heat transfer enhancement by Dean vortices were obtained through multiple regression.

Keywords: Spiral tube; Dean Vortices; Heat transfer enhancement; Nusselt number.

1. Introduction

Spiral tubes have been introduced as an important heat transfer enhancement component of heat exchangers and are widely used in various industrial applications, such as the air conditioning systems, heat recovery processes, chemical reactors. Traditionally, the mechanism of heat transfer enhancement by spiral tubes was attributed to their large heat transfer areas, the compact structure, and, above all, their special geometry. Therefore, studies were carried out in terms of the structure of the spiral tubes to determine the mechanism of the heat transfer enhancement. For example, Heo [1] described the heat transfer augmentation phenomenon of spiral coils under different tube diameter, length, height, pitch, radius, and number of turns. And the natural convection heat transfer from spiral coiled tubes was studied at different inlet temperatures, bath temperatures, coil dimensions, and fluid flow rate by Prabhanjan [2]. There are still other similar works, such as the studies by Yi [3], Ali [4, 5], E. Martínez [6] and so on.

However, the heat transfer characteristics in spiral tubes have rarely been investigated in terms of the Dean vortices, though recent studies have proved that the heat transfer enhancement for the spiral tube strongly depends on the motion of Dean vortices [7]. The Dean vortices, caused by the centrifugal force, is a kind of secondary flow first found by Dean in 1920s. One unique of the Dean vortices is the counter-rotational-vortex structure [8]. Perturbation solutions were obtained by Dean under the curved tube with small curvature ratio and circular cross-section, and the dimensionless Dean number which could be used to weigh the strength of the vortices was presented [9],

$$De = \mathrm{Re}\cdot(r/R_c)^{0.5} = 2r\cdot V\cdot \upsilon^{-1}\cdot(r/R_c)^{0.5} \quad (1)$$

where De denotes the Dean number, Re means the Reynolds number, r and R_c indicate the radius and curvature radius of the tube(m), respectively. And the symbols V and υ are the main flow velocity(m/s) and viscosity of the fluid in the tube, respectively.

Theoretically, the flow pattern of Dean vortices can promote the mixing effect of the fluid in the tube, thus augment of the heat and mass transfer of the fluid. Many researchers have studied, theoretically and experimentally, the Dean vortices for a long time. However, most of the studies were focused on the flow characteristics and mass transfer augmentation. For example, R.Moll and his colleagues studied the mass transfer characteristics experimentally and numerically by measuring the velocity profiles and the local wall shear stress induced by Dean vortices [10,11]. And H. Fellouah presented the studies on Dean instability for Newtonian fluids in laminar secondary flow in 180° curved channels experimentally and numerically [12], and he applied the numerical and experimental methods to analyze the Dean instability in power-law and Bingham fluids in a 180° curved channels of rectangular cross section to understand the effect of rheological fluid behavior on the Dean instability [13]. Except the results stated above, there are still many authors studying on the flow characteristics [14, 15] and mass transfer augmentation characteristics [16-19] of Dean vortices. But few published studies on the heat transfer enhancement by Dean vortices were found, except the works by Ligrani [20]and Tilak [21].

Critical literature review reveals that no explicit correlation between Nusselt number and Dean number is presented to better describe the heat transfer augmentation characteristics in spiral tubes in the present studies available. Also, a lack of consideration exists on the experiments in the literatures: no experiment considered all the controlling factors of Dean vortices. Therefore, it is of necessity to carry out experiments considering the Reynolds number, viscosity, curvature ratio, and torsion of spiral tubes, to better study the heat transfer augmentation by the Dean vortices. Finally the relationship between the Nusselt number and Dean number, and the relationship between pressure drop and Dean number can be determined. The determination of these relationship is an important step in the design of heat exchangers.

2. Experimental work

2.1 Experimental setup

A schematic diagram of the experimental setup is shown in Figure 1. The apparatus needed in the experiments are given in Table 1, and the physical dimensions of test tubes used in the experiment are listed in Table 2.

The inner diameter of the tubes is 19mm each, including the straight tube. And all the tubes have a smooth and plain inner wall surface, therefore the surface roughness has negligible influences. It will not have significant influences on heat transfer phenomenon. The tubes in the experiments were connected to the circulation pipe with straight joints. And the length of the straight part of the circulation pipe connected to the inlet of the test tubes was at least 950mm (10 times of the tube diameter), to minimize to influences of entrance effects. What's more, the circulating pipe was covered by thermal isolating materials to minimize the heat loss. Digital thermometers and pressure transducers were installed at the inlet and outlet of the tubes to acquire pressure and temperature data. For the spiral tubes, the outlet located in the water tank due to the dimensions of the tank. As a result, the thermometer and pressure transducer should be covered with thermal insulation material to ensure the accuracy of the data acquisition. For the purpose of diminishing the influence of radiation, both the surfaces of the water tank and water container were covered by heat insulation material. In order to achieve a constant wall temperature condition, a electric heater with a power of 2500W was employed in the experiments. The electric heater was controlled by a thermal regulator with a precision of 0.1 °C, thus the water in the water tank could be stabilized at a constant temperature approximately. The flow rate in the test tubes was controlled by the pump and the ball valves. When the pump was running, a certain flow rate, which indicated a certain inlet velocity of the tubes, could be obtained by controlling the ball valves in the circulation pipes. The exact flow rate could be determined by measuring the volume of the fluid flowing out of the drain point, with the help of the graduate and timer. Finally, 3 kinds of fluid/solution with different viscosities were introduced to test the influences of viscosity on the heat transfer characteristics of test tubes. Apparatus of the experiments are listed in Table 1.

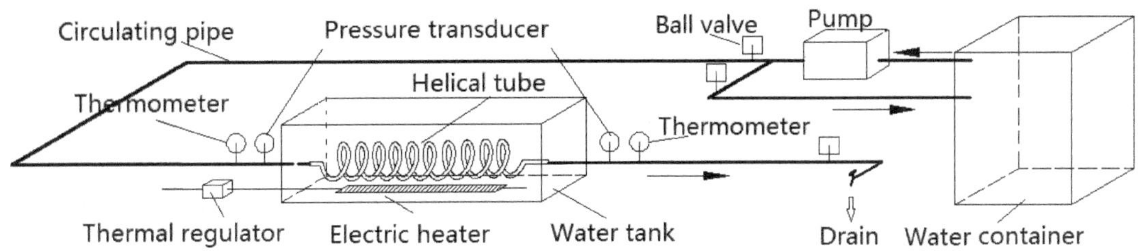

Figure 1. Scheme of the experimental setup

Table 1. Apparatus needed in experiments

Name	Parameter	Name	Parameter
Water container	1000×1000×500mm	Copper tube	L=750mm;r=19mm
Water tank	800×200×200mm	Electric heater	2500W
Pump	1.5m3/h	Thermal regulator	Precision: 0.1°C
Ball valve	DN20	graduate	Max scale: 1000ml
Pressure transducer	Precision: 1Pa	Timer	—
Thermometer	0-100°C, Precision: 0.1°C	Heat insulation material	—
90°elbow	DN20	Kerosene	$0.5m^3$
Straight tee	DN20	glycol solution	Concentration: 35%, $0.5m^3$
Straight joint	DN20	glycol solution	Concentration: 90%, $0.5m^3$

Table 2. Physical dimensions of spiral tubes used in experiment

NO.		r (mm)	R_c (mm)	r/R_c	Torsion τ	Pitch S (mm)
Straight tube	#1	19	$+\infty$	0	0	0
Spiral tube	#2	19	38	0.25	4	38
	#3	19	38	0.25	5.9	57
	#4	19	38	0.25	7.6	76
	#5	19	38	0.25	9	95
	#6	19	57	0.167	4	86
	#7	19	76	0.125	4	162
	#8	19	95	0.1	4	189

Note: The torsion is applied here to evaluate the pitch of a spiral tube according to the correlation $\tau=(S/2\pi)/[R_c^2+(S/2\pi)^2]$, where S and R_c mean the pitch and curvature of a spiral tube, respectively [22].

As for the test tubes used in the experiment, 7 different kinds of spiral tubes and 1 straight tube were included. The parameters of these tubes are listed in Table 2. The straight tube has a length of 750mm, a inner diameter of 19mm, and wall thickness of 0.5mm. And all the spiral tubes are constructed with such a straight tube. The spiral tubes differ from each other in the curvatures and pitches(turns). However, the dimensionless parameter is preferred during the research, therefore the curvature and pitch are replaced by curvature ratio and torsion, respectively. As a result, there are 4 kinds of curvature ratios varying from 0.1 to 0.25 and 4 kinds of torsions changing from 4 to 9. Among them, the curvature ratio is closely relevant to the motion of the Dean vortices. And the torsion, another important parameter of spiral tube, varies with the curvature and pitch of spiral tubes.

2.2 Experimental procedure

The fluids were pumped to the spiral tubes using a centrifugal pump at an initial inlet temperature of 10 °C and different flow rate(that means different velocities or different Reynolds numbers). During the experiment, the fluid was taken from the water container which is made of plexiglass covered by heat insulation material(to minimize the influence of radiation heat loss), and was heated by the hot water in the water tank while passing through spiral tube, then flew out of the system as shown in Figure 1. The inlet and outlet temperatures of the spiral tubes were measured by digital thermometers calibrated to meet limits of error ±0.1°C. Similarly, the inlet and outlet pressure was obtained by the pressure transducers installed at the inlet and outlet of the test tube. In addition, the data acquisition started after the system had run for about 15 minutes, because it had been presumed that it would take about 15 minutes to reach a steady condition for a certain flow rate [23].

The exact procedure is detailed as follows: (1) Water was chosen as the fluid in the experiment with a initial temperature of 10 °C, and was pumped to the test tubes at different flow rate under which the Reynolds number ranges from 760 to 19000. The pressure and temperature data at the inlet and outlet of tubes can be recorded when the pressure and temperature values measured stop fluctuating greatly. The flow rate in the experiment and the corresponding Reynolds numbers are listed in Table 3. (2) Water was chosen again as the test fluid to continue the experiment with spiral tubes at the Reynolds number of 4000 and an initial inlet temperature of 10 °C, which suggests an experiment investigating into the influences of the curvature ratio and torsion on the heat transfer characteristic of the spiral tubes, at constant flow rate and viscosity. (3) Water was replaced by the fluids with higher viscosities to continue the experiment, at a flow rate of 58.8mL/s. This part of experiment was carried out in the straight tube and any one of the spiral tube(the spiral tube with curvature ratio of 0.25 and torsion of 4 was chosen in the experiment). The fluids with higher viscosities used in the experiment include the Kerosene, 35% glycol solution and 90% glycol solution, the dynamic viscosity of which at a temperature of 10 °C are 2.4mPa.s, 13.5mPa.s and 20.2mPa.s, respectively.

Table 3. Flow rate and Reynolds numbers of the experiment

Test tube	#1 to #8										
Flow rate(ml/s)	11.2	22.4	33.6	42.0	58.8	89.6	117.6	162.4	207.2	224	280
Mean velocity(m/s)	0.04	0.08	0.12	0.15	0.21	0.32	0.42	0.58	0.74	0.8	1
Re	760	1520	2280	2850	4000	6000	8000	11000	14000	15200	19000

2.3 Calculation of the experimental data

In the experiment, thermo-physical properties of test fluids flowing through the test tubes are assumed constant along the tube length and evaluated at an average bulk temperature. Heat flux from the hot water in the water tank to the cold water through the tube is calculated from:

$$q = \frac{c_p \cdot m \cdot (T_{out} - T_{in})}{\pi \cdot d \cdot L} \tag{2}$$

The heat flux calculated from above equation was then used to calculate the overall heat transfer coefficient, α, as

$$\alpha = \frac{q}{T_{wall} - T_{fluid}} \tag{3}$$

The symbol T_{wall} denotes the wall temperature(°C) of the test tubes, which is assumed to be equal to the temperature of the water in the water tank, 90 °C. And T_{fluid} indicates the average fluid temperature in the test tubes, which is assumed to be determined by $T_{fluid}=0.5(T_{in}+T_{out})$ due to the limited length of the test tubes.

During the calculation and latter discussion, the dimensionless parameter Nusselt number, rather than the heat transfer coefficient, is chosen to express the heat transfer characteristic. The Nusselt number can be calculated from:

$$Nu = \frac{\alpha \cdot d}{\lambda} \tag{4}$$

λ is the thermal conductivity(W/m. °C) of the fluid flowing through the test tubes.

3. Results and discussion

3.1 Effects of Reynolds number on heat transfer for test tubes

Plenty of research has revealed that the heat transfer performance can be improved by increasing the Reynolds number, for both straight and curved tubes. In this study, experiment on straight tube and spiral tubes was carried out at a series of Reynolds numbers, and finally the results were compared.

Figure 2 depicts the heat transfer characteristic of test tubes under different Reynolds numbers. It is clear that the Nusselt number increases with Reynolds number, for all the test tubes. Another result that can be achieved is that the Nusselt number for spiral tubes is higher than that for straight tube under the same Reynolds number. For example, the Nusselt number for the spiral tube, with a curvature ratio of 0.25 and torsion of 4, is 39.4% higher than that for straight tube at the Reynolds number of 19000 (Table 4). In addition, the Nusselt number in spiral tubes increases faster than that of the straight tube for Reynolds number larger than 15000. Figure 3 demonstrates the changes of pressure drop of test tubes at different Reynolds numbers. It can be observed that the pressure drop for all the test tubes increases with the Reynolds number, but the pressure drop for spiral tubes is higher than that for straight tube. What's more, the spiral tube with different curvature ratios and different torsions exhibits different pressure drop, which will be discussed in 3.2 and 3.3.

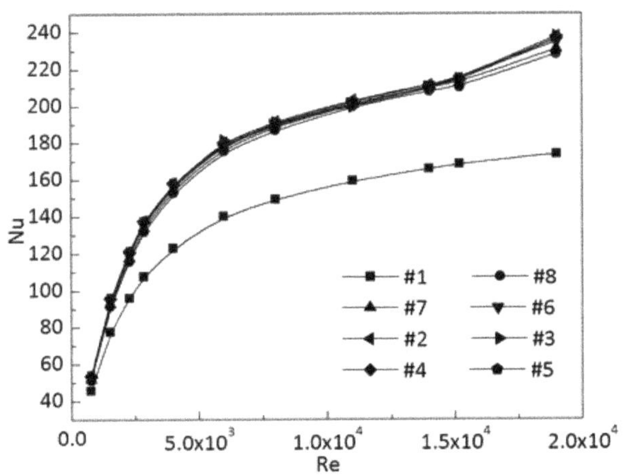

Figure 2. Nu as a function of Re

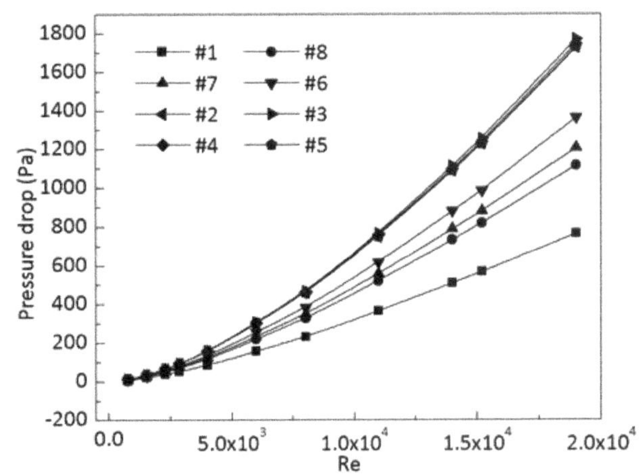

Figure 3. Pressure drop as a function of Re

Table 4. Percentage of increment in Nu compared with #1 at Re=19000

Test tube	#1	#8	#7	#6	#2	#3	#4	#5
Nu	170.7	228.2	231	235.1	237.9	238	237.4	237.5
increment	-	33.7%	35.3%	37.7%	39.4%	39.4%	39.1%	39.1%

For the straight tube, the Nusselt number and pressure drop increase with Reynolds number because the turbulence intensity of the fluid in the tube increases at high Reynolds number, thus improves the mixing effect and causes a penalty in pressure drop. However, the fluid field in the spiral tubes will not go into "chaos" due to the existence of the "subsidiary flow with certain flow pattern"(the Dean vortices), although there is an increase in Reynolds number. It is implied that an increase in Reynolds number under experiment conditions can only increase the intensity of the Dean vortices, while not destroy the vortices(However, perhaps there will be no difference in Nusselt number when the Reynolds number increases to a certain value). As result, the forced convection in the spiral tubes is improved. As for the pressure drop of the spiral tubes, the profile drag dominates. Therefore the pressure drop is closely related to the geometry factors such as curvature ratio.

3.2 Effects of curvature ratio on heat transfer of spiral tubes

Experiment was performed to investigate the influence of the curvature ratio at a certain Reynolds number. Figure 4 illustrates the variation of Nusselt number versus curvature ratio for spiral tubes (#2, #6, #7 and #8) with curvature ratios ranging from 0.1 to 0.25 and a torsion of 4. From this figure, it can be seen that the Nusselt number increases when the curvature ratio varies from 0.1 to 0.25. There is a similar tendency for the pressure drop as is shown in Figure 5.

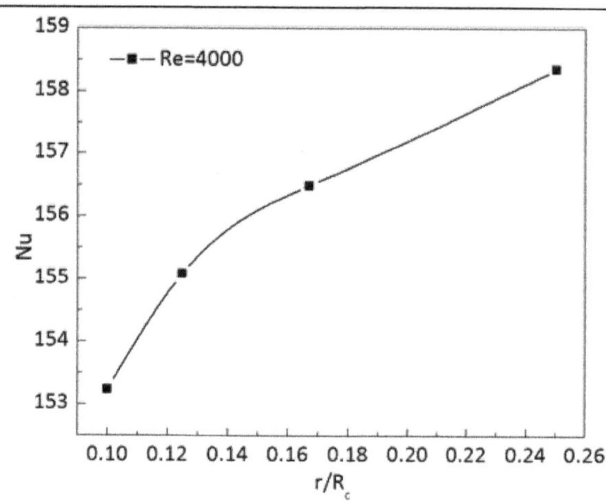

Figure 4. Nu as a function of r/R_c

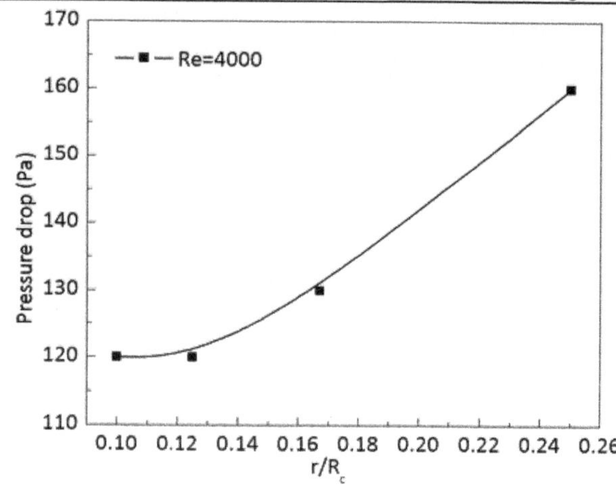

Figure 5. Pressure drop as function of r/R_c

For a spiral tube with a certain diameter, a greater curvature ratio means that the curvature is small, therefore the spiral tube has more turns than spiral tubes with smaller curvature ratios. As a result, fluid flowing through the tube has to change its main flow direction more times. Therefore the forced convection provided by Dean vortices works longer, but it inevitably increases the pressure drop.

3.3 Effects of torsion on heat transfer for spiral tubes

The spiral tubes with torsions ranging from 4 to 9 and a curvature ratio of 0.25 are used in this part of experiment. The effect of torsion on heat transfer enhancement characteristic of the spiral tubes is depicted in Figure 6 and Figure 7.

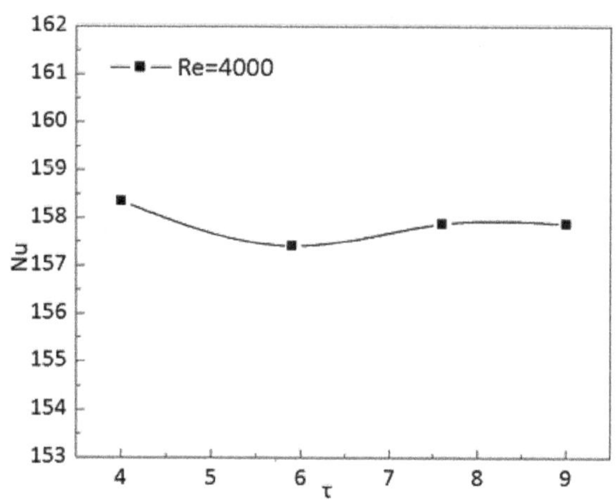

Figure 6. Nu as a function of torsion

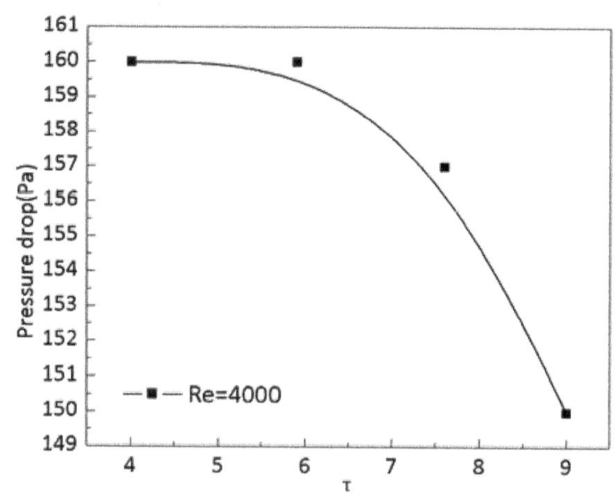

Figure 7. Pressure drop as function of torsion

The figure shows that the Nusselt numbers are fluctuating between 157 and 159, which suggests that the torsion has an insignificant influence on Nusselt number for spiral tubes. As for the influence of torsion on pressure drop, Figure 7 clearly illustrates that the pressure drop decreases from 160Pa to 150Pa as the torsion increases.

3.4 Effects of viscosity on heat transfer for test tubes

In this part of experiment, Kerosene, 35% glycol solution and 90% glycol solution are used, the dynamic viscosities of which are 2.4mPa.s, 13.5mPa.s and 20.2mPa.s respectively, at a temperature of 10 °C. However, the dimensionless Prandtl number is employed instead of viscosity to indicate the fluid viscosity in the discussion. The Prandtl numbers of Kerosene, 35% glycol solution (35% glycol and 65% water, in volume) and 90% glycol solution(90% glycol and 10% water, in volume) are presumed to be 34, 66 and 150 respectively at 10 °C.

Figure 8 shows the variation of Nusselt number with Prandtl number in test tube #1 and #2, at flow rate of 58.8ml/s. As indicated in the figure, the Nusselt number tends to increase with the rise of Prandtl number for all cases. Meanwhile, the Nusselt number for #2 is higher than that for #1 under the same Prandtl number. It can be concluded that the spiral tube exhibits a better heat transfer performance than straight tube when the fluid has a high viscosity. The influence of the Prandtl number on pressure drop of the tube is shown in Figure 9. As indicated in the figure, the pressure drop of #1 and #2 increases with the rise of Prandtl number. It is also found that the pressure drop for #2 is higher than that for #1 under the same Prandtl number.

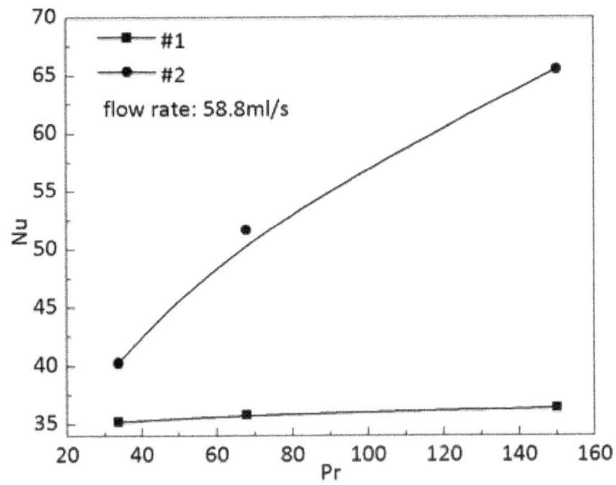

Figure 8. Nu as a function of Pr

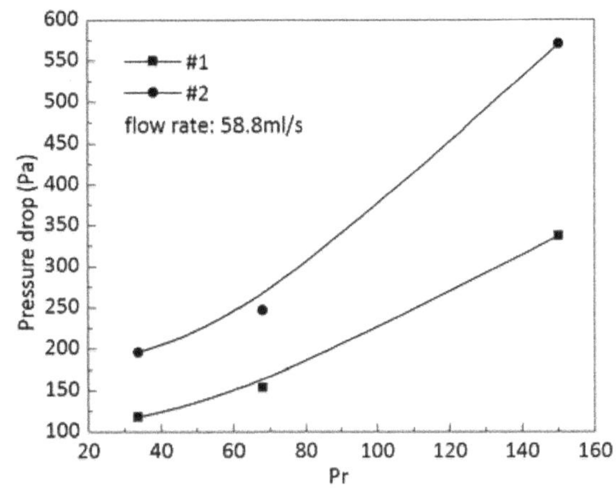

Figure 9. Pressure drop as a function of Pr

The spiral tubes in this part of experiment are characteristic of a higher Nusselt number due to the inertial force(centrifugal force) that drives the Dean vortices. In the spiral tube, the inertial force dominates, while in the straight force the reverse is true(which means that the viscous force dominates). Therefore, the heat transfer between the center and the boundary layer continues in the spiral tube even with a high viscosity fluid. Whereas in the straight tube, the heat exchange between the fluid in boundary layer and the core of the tube is inadequate. As for the pressure drop, the profile drag plays an important role, therefore the spiral tube exhibits higher pressure drop than straight ones.

4. Correlation equation of heat transfer enhancement by Dean vortices

4.1 Effects on Dean number on heat transfer of spiral tubes

The results above revealed that the heat transfer performance of spiral tubes is influenced by Reynolds number, curvature ratio and viscosity of the fluid. According to Eq.(1), the Dean number is a function of the Reynolds number and curvature ratio of the spiral tube. As a result, it can be predicted that the heat transfer performance is influenced by Dean numbers.

Based on the results above, the Dean numbers are calculated with different Reynolds numbers and curvature ratios. Then the Nusselt numbers and pressure drop obtained in the experiment are matched with the Dean numbers. Figure 10 demonstrates the variation of Nusselt number under different Dean numbers. From the figure it can be seen that the Nusselt numbers is approximately in parabolic increase with the Dean numbers, which means an improved heat transfer performance. As to the changes in pressure drop under different Dean numbers, Figure 11 provides a clear depiction. As is shown in the figure, the pressure drop increases with the Dean numbers.

However, there is a lack of evidence to judge whether a spiral tube with Dean vortices has a better heat transfer performance than straight tube, by comparing only the Nusselt numbers and pressure drop. Therefore, the thermal enhancement factor, η, is introduced to better evaluate the heat transfer performance of the spiral tubes($\eta=(Nu/Nu_{ref})(f/f_{ref})^{-1/3}$) [24]. In this literature, the straight test tube is regarded as the reference tube. A value of η greater than 1 indicates better performance of heat transfer. Figure 12 illustrates the variation of η versus Dean number for all the spiral tubes in the experiment.

The figure shows that spiral tubes perform better than straight tubes. The comprehensive heat transfer performance of spiral tube improves with increasing Dean number before the Dean number reaches approximately 2000. When Dean numbers are greater than 2000, the value of η starts to decrease, but still greater than 1. It implies that the spiral tube with Dean vortices has a better heat transfer enhancement performance than straight tubes, especially when the Dean number is approaching 2000. (The pressure drop increases much faster when Dean number is great than 2000, which decrease the comprehensive heat transfer performance, although the Nusselt number keeps increasing).

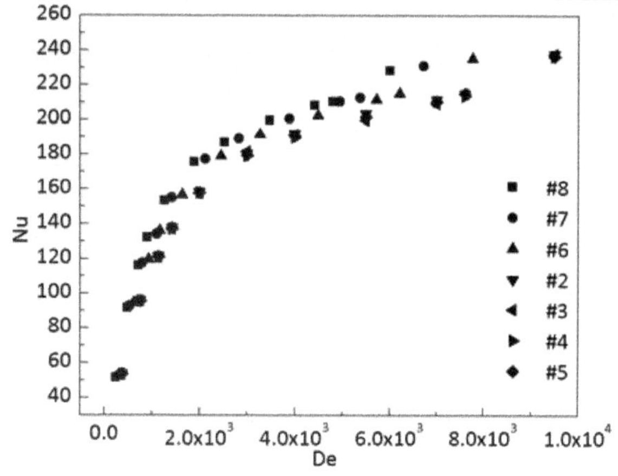

Figure 10. Nu as a function of De

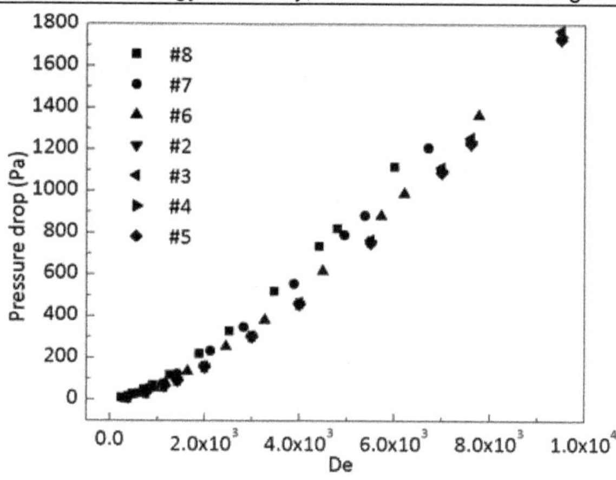

Figure 11. Pressure drop as function of De

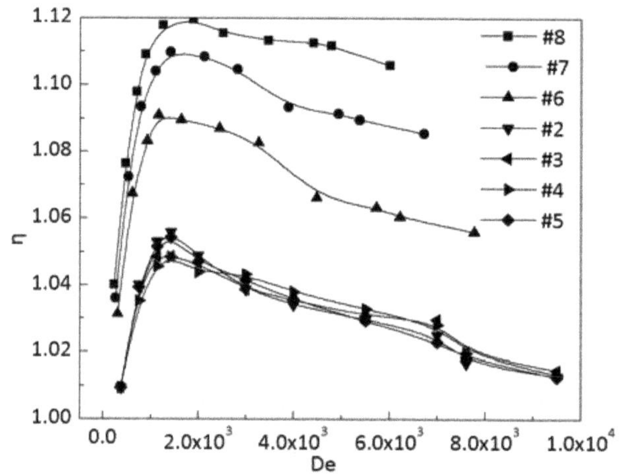

Figure 12. The value of η as a function of De

4.2 Deduction of the correlation equation

As mentioned in 4.1, the heat transfer enhancement performance is closely relevant to the Dean number and the viscosity that is indicated by dimensionless Prandtl number. As a result, the Nusselt number can be expressed as:

$$Nu \propto f(De \cdot \Pr) \tag{5}$$

Meanwhile, the heat transfer enhancement is achieved by forced convection caused by Dean vortices. Therefore, the mathematic model of forced convection in tubes can be assumed to be:

$$Nu = C \cdot De^{m} \cdot \Pr^{n} \tag{6}$$

where C, m and n are the undetermined constants which is influenced by the flow and heat transfer characteristics.

According to Eq.(6), multiple regression was performed using STATISTICA with the data obtained from the experiment. The results can be expressed as the following equations:

$$Nu = 5.3611 \cdot De^{0.553637} \cdot \Pr^{0.497215} \tag{7}$$

$$\Delta P = 0.003 \cdot De^{1.503735} \cdot \Pr^{0.118736} \tag{8}$$

Eq.(7) and Eq.(8) are believed to be effective when Dean numbers ranges from 240 to 9500, and Prandtl number ranges from 9 to 150, and the correlation coefficient for Eq.(7) and Eq.(8) are 0.9737 and 0.9825, respectively (9 denotes the Prandtl number of water at 10 °C).

5. Conclusion

The heat transfer enhancement characteristics for spiral tubes were experimentally investigated in terms of Dean vortices. From the data obtained from the experiment, it is found that the Dean vortices induced by centrifugal force in the spiral tube plays an significant role in the heat transfer enhancement. A few important conclusions are stated below.

Results depict that both the Nu and pressure drop increase with Re, and pressure drop in all spiral tubes is higher than that of the straight tube. In addition, Nu and pressure drop for spiral tubes increase with the curvature ratio at constant flow rate, viscosity and torsion. And the torsion has been found to be ineffective to heat transfer performance, but it reduces the pressure drop of the flow in spiral tubes. What's more, it was observed that the Nu for both spiral tube and straight tube increase with the Prandtl number and it was also found that the Nu for the spiral tube was much greater, which suggests that the spiral tube has a better heat transfer performance even under high viscosity fluid. Finally, considering the influence of Re, curvature ratio and Pr simultaneously, the effect of Dean number on heat transfer was investigated. The result reveals that both Nu and pressure drop increase with De. And what deserves attention is that η is greater than 1 for all the spiral tubes and reaches its top value when De approximates 2000. Based on the results above, correlation equations were obtained via multiple regression, which are believed to be effective when Dean number ranges from 240 to 9500 and Prandtl number ranges from 9 to 150.

Acknowledgements

The authors would like to express the appreciation to the National Natural Science foundation of China(Grant No. 51276057) and Hunan Provincial Innovation Foundation For Postgraduate (2014) for the financial support for this study.

Nomenclature

De	Dean number
Re	Reynolds number
r	inner radius of the tube, m
d	inner diameter of the tube, m
L	length of the tube, m
R_c	curvature of spiral tube, m
S	pitch of spiral tube, m
V	main velocity of the fluid, m/s
q	heat flux, W/m2
c_p	specific heat capacity, J/(kg. °C)
m	mass flow rate, kg/s
T	temperature, °C
Nu	Nusselt number
Pr	Prandtl number
f	drag coefficient
ΔP	pressure drop, Pa

Greek symbols

υ	kinetic viscosity, m^2/s
α	heat transfer coefficient ,W/(m^2. °C)
λ	thermal conductivity, W/(m. °C)
η	comprehensive evaluation factor
τ	torsion of spiral tube

Subscripts

in	inlet
out	outlet
wall	tube wall
fluid	fluid
ref	reference tube

References

[1] Jeong-Hwan Heo, Bum-Jin Chung. Influence of helical tube dimensions on open channel natural convection heat transfer, Int. J. Heat Mass Transfer. 55(2012)2829-2834.

[2] D.G.Prabhanjan, T.J.Rennie, G.S.Vijaya Raghavan. Natural convection heat transfer from helical coiled tubes, Int.J.Therm.Sci.43(2004)359-365.

[3] Jie Yi, Zhen-Hua Liu, Jing Wang. Heat transfer characteristics of the evaporator section using small helical coiled pipes in a looped heat pipe, Appl. Therm. Eng. 23(2003)89-99.

[4] M.E.Ali. Laminar natural convection from constant heat flux helical coiled tubes, Int. J. Heat Mass Transfer. 41(1998)2175-2182.

[5] M.E.Ali. Experimental investigation of natural convection from vertical helical coiled tubes, Int. J. Heat Mass Transfer. 37(1994)665-671.

[6] E.Martínez, W.Vicente, G.Sotoa, et al. Methodology for determining the optimal fin dimensions in helically segmented finned tubes, Appl. Therm. Eng. 31(2011)1744-1750.

[7] M.M.Abo Elazm, A.M.Ragheb, A.F.Elsafty, et al. Numerical investigation for the heat transfer enhancement in helical cone coils over ordinary helical coils, J.Eng.Sci Tech. 8(2013)1-15.

[8] W. R. Dean. The stream-line motion of fluid in a curved pipe, Philos. Mag. 5(1928)673-695.
[9] W. R.Dean. Note on the motion of fluid in a curved pipe, Philos. Mag. 20(1927)208-223.
[10] R. Moll, D. Veyret, F. Charbit, et al. Dean vortices applied to membrane process: Part I. Experimental approach, J.Memb.Sci.288(2007)307-320.
[11] R. Moll, D. Veyret, F. Charbit, et al. Dean vortices applied to membrane process: Part II: Numerical approach, J.Memb.Sci.288(2007)321-335.
[12] H. Fellouah, C. Castelain, A. Ould El Moctar. A criterion for detection of the onset of Dean instability in Newtonian fluids, Eur.J.Mech B-Fluid. 25(2006)505-531.
[13] H. Fellouah, C. Castelain,A. Ould-El-Moctar,et al. The Dean instability in power-law and Bingham fluids in a curved rectangular duct, J.Non-Newton.Fluid Mech.165(2010)163-173.
[14] R Moll, Ph Moulin,D Veyret,et al. Numerical simulation of Dean vortices: fluid trajectories, J.Memb.Sci. 197(2002)157-172.
[15] K. Raghavan, M.M. Bernitsas. Experimental investigation of Reynolds number effect on vortex induced vibration of rigid circular cylinder on elastic supports, Ocean.Eng.38(2011) 719-731.
[16] D. Kaufhold, F. Kopf, C. Wolff,et al. Generation of Dean vortices and enhancement of oxygen transfer rates in membrane contactors for different hollow fiber geometries, J. Memb. Sci. 423 (2012)342-347.
[17] J.Kaur, G.P Agarwal. Studies on protein transmission in thin channel flow module: the role of dean vortices for improving mass transfer, J.Memb.Sci.196(2002)1–11.
[18] D.N. Kuakuvi, P. Moulin, F. Charbit.Dean vortices: a comparison of woven versus helical and straight hollow fiber membrane modules, J.Memb.Sci.171(2000)59–65.
[19] Jin Xuana, Michael K.H. Leungb, Dennis Y.C. Leung, et al. Density-induced asymmetric pair of Dean vortices and its effects on mass transfer in a curved micro-channel with two-layer laminar stream, Chem. Eng. J. 171(2011)216-223.
[20] P.M. Ligrani, S. Choi. Effects of Dean vortex pairs on surface heat transfer in curved channel flow, Int. J. Heat Mass Transfer. 39(1996)27-37.
[21] Tilak T.Chandratilleke, Nima Nadim,Ramesh Narayanaswamy. Vortex structure-based analysis of laminar flow behavior and thermal characteristics in curved ducts, Int.J.Therm.Sci.59(2012) 75-86.
[22] Germano, M. On the effect of torsion on a helical pipe flow, J.Fluid Mech. 125(1982)1-8.
[23] S.S. Pawar, Vivek K. Sunnapwar. Experimental studies on heat transfer to Newtonian and non-Newtonian fluids in helical coils with laminar and turbulent flow, Exp. Therm Fluid Sci.44 (2013) 792-804.
[24] Umesh Potdar, Nilesh Shinde, Manoj Hambarde. Study of heat transfer coefficient & friction factor of stationary square channel with V shaped & 45°angled arc of circle ribs with different blokage ratio, Int.J.Appl. Sci Eng Res.1(2012)47-56.

4

Development a new equation of polarization curve for a proton exchange membrane fuel cell at different channel geometry

I. Khazaee

Department of Mechanical Engineering, Torbat-e-jam branch, Islamic Azad University, Torbat-e-jam, Iran.

Abstract
The polarization curve of a proton exchange membrane fuel cell is an important parameter that is used to investigate the performance of it that is expressed with the Nernst equation with the equation of losses the voltage such as activation loss, ohmic loss and concentration loss that they are a function of temperature of the cell and the current density. In this study a new correlation for polarization curve is obtained that it is a function of temperature, current density and a new parameter of cross-section geometry of channels. For this purpose three PEM fuel cells with different channels geometry of rectangular, elliptical and triangular have constructed. The active area of each cell is $25cm^2$ that its weight is $1300gr$. The material of the gas diffusion layer is Carbon clothes, the membrane is nafion112 and the catalyst layer is a plane with 0.004 gr/cm^2 Platinum. Also a test bench designed and constructed for testing the cell and a series of experiments are carried out to investigate the influence of the geometry of the cell on performance of the cell. The results show that when the geometry of channel is rectangular the performance of the cell is better than the triangular and elliptical channel.

Keywords: Fuel cell; Curve fitting; Channel geometry; Performance.

1. Introduction

A fuel cell is an electro-chemical energy device that converts the chemical energy of fuel directly into electricity and heat, with water as a by-product of the reaction. Based on the types of electrolytes used, they are categorized into polymer electrolyte membrane fuel cells (PEMFCs), solid oxide fuel cells (SOFCs), phosphoric acid fuel cells (PAFCs), molten carbonate fuel cells (MCFCs), and direct methanol fuel cells (DMFCs). The polymer exchange membrane fuel cell (PEMFC) is considered to be the most promising candidate for electric vehicles by virtue of its high power density, zero pollution, low operating temperature, quick start-up capability and long lifetime.

Furthermore, the PEM fuel cell is being investigated as an alternate power generation system especially for distributed generation and transportation. The PEM fuel cell is providing reliable power at steady state; however, it is not able to respond promptly to a load step change. Since the fuel cell is an electrochemical energy conversion device that converts fuel into electricity, its dynamic behavior depends both on chemical and thermodynamic processes [1]. The polymer electrolytes work at low temperature, which brings this further advantage that a PEM fuel cell can start quickly. PEM fuel cells are being actively developed for use in cars and buses, as well as for a very wide range of portable applications, and also for combined heat and power systems. It could be argued that PEM fuel cells exceed all other electrical energy generating technologies in the breadth of scope of their possible applications. Scrivano et al. [2] presented the results of an experimental analysis performed on an Exchange miniaturized, 6W Proton Membrane Fuel Cell (PEMFC) system, integrated with on-site hydrogen production by electrolysis; in particular, they investigated the effects of environmental parameters such as the external temperature and the humidity on the performance of fuel cells. Also they proposed a simple semi-empirical mathematical model capable to perform rough prediction on the behavior of such systems when exposed at different ambient temperatures. The model threats the stacks as black boxes, not investigating singularly the inner phenomena which occur in the cell. Amphlett et al. [3, 4] investigated a theoretical model which was employed to provide the structure of the equations, and then, the parameters of these equations were found by using the regression techniques to fit the experimental results. Also they studied a semi-empirical model with a theoretical background that takes into account the main variables of the fuel cell operation such as the operating temperature, the partial pressures at the electrodes and the fuel cell current. Del Real et al. [5] investigated a simple empirical equation to model the fuel cell voltage with considering the variations of the main process variables. The model equation has 11 parameters: one parameter related to the mass of liquid water at the anode channel must be estimated due to technical constraints, and the other parameters are obtained from experimental data. Although the model proposed by them, fitted well with the experimental data, the equation of the fuel cell voltage does not have a theoretical basis, and, therefore, it is based on assumptions relating to the effects of temperature and partial pressures that are not proven to be general for fuel cells other than those used in [5]. Khazaee [6] investigated numerically the effect of using different obstacles on the performances, current density and gas concentration for different Aspect Ratios and compared the results with experimental results of a triangular channel geometry PEM fuel cell. He found that by increasing the hydrogen flow rate from 0.3 L/min to 0.7 L/min and the oxygen flow rate from 0.5 L/min to 0.9 L/min the cell performance enhances. Also he found that by increasing the number of the blocks in the channels and installation of rectangular block, the overall cell performance increases. Ferng et al. [7] performed analytical and experimental work to investigate a single PEM fuel cell. In their paper, they presented a study of the cell performance covering the effects of operating temperature and pressure on performance and the flow characteristics within the cell. Their paper shows the positive effects of temperature and pressure on the performance of a single PEM fuel cell. Hussain et al. [8] investigated a thermodynamic model of a polymer electrolyte membrane (PEM) fuel cell power system for transportation applications. Their analysis includes the operation of all the components in the system, which consists of two major modules: PEM fuel cell stack module and system module and a cooling pump. System module includes air compressor, heat exchanger, humidifier and a cooling loop. They found that with the increase of external load (current density), the difference between the gross stack power and net system power increases and the largest irreversibility rate occurs in the fuel cell stack. Park and Xianguo [9] investigated a numerical and experimental study to investigate the cross flow in a PEM fuel cell. Experimental measurements revealed that the pressure drop in a PEM fuel cell is significantly lower than that without cross flow and three-dimensional numerical simulation has been performed for wide ranges of flow rate, permeability and thickness of gas diffusion layer to analyze the effects of those parameters on the resultant cross flow and the pressure drop of the reactant streams. Xianguo et al. [10] investigated the characteristics of liquid water removal from GDL experimentally, through measuring unsteady pressure drop in a cell which has the GDL initially wet with liquid water. They controlled the thickness of GDL carefully by inserting various thicknesses of metal shims between the plates. They found that severe compression of GDL could result in excessive pressure drop from channel inlet to channel outlet. Xianguo et al. [11] developed a non-isothermal stack model to analyze the effects of flow variance and temperature distribution on the performance of a polymer electrolyte membrane fuel cell stack. Their stack model consists of the flow network solver for pressure and mass flow distributions for the reactant gas streams and cooling water, and the heat transfer solver for temperature distribution among the cells in the stack, as well as the fuel cell model for individual cell performance. They found that the effect of temperature is dominant on the cell voltage variance when the flow variance is small for sufficiently uniform distribution of reactant flow among the cells in the stack. In this study the effect of changing the cross-section geometry of channels has been investigated experimentally and a new equation for polarization curve obtained that is a function of temperature, current density and the geometry of channels. Therefore The main objective of this paper contrary to another works is to analyze the influence of different geometries of the channels or flow fields on the voltage (and consequently the electric power) supplied to a PEM fuel cell at different levels of cell current.

2. Description of the experiments and method of the measurements

For experimental investigation of the performance of the fuel cell a setup has been fabricated. A schematic flow of the test bench is shown in Figure 1. It allows controlling several physical parameters, and the measurement of many output data. In fact, the polymeric membrane has permeability to hydrogen and oxygen; due to the high-pressure gradient from cathode to anode, this driving force could push hydrogen from cathode to anode across the membrane and a dangerous mix with oxygen could occur; this concentration must always be kept below a safety level. The test bench is made up of four main subsystems. First, the gases supply system, which sends the oxygen and hydrogen flow into the system for electrochemical reaction. Second, there is two humidifier that humidify the oxygen and hydrogen before going into the cell for complete transferring of proton from the membrane to the cathode side. Third, the nitrogen supply system is applied to inert any flammable mix inside the ducts and to purge the system before activation. Finally, there is the electrical power supply, regulated from an AC/DC voltage regulator driven from the control panel.

The specifications of the test system for this study are:

- The humidifier system is membranous.
- The test bench has the system of announcement the leakage of hydrogen.
- The system can control and show the temperature of the oxygen and hydrogen.
- The system can control and show the temperature of the cell.
- The system can control and show the flow rate of the oxygen and hydrogen.
- The system can control and show the inlet pressure of the oxygen and hydrogen.
- The system can show the voltage of the cell.
- The system can show the current of the cell.

Table 1 shows the environs of operation of the experimental setup in this study.

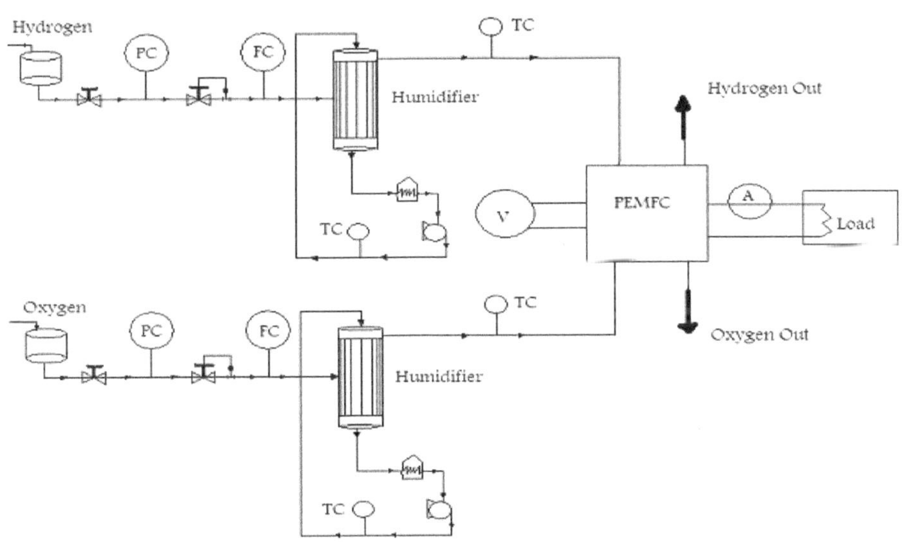

Figure 1. Schematic of the PEMFC system

Table 1. Operational characteristics of the test bench

Voltage	0-2 V
Current	0-20 A
Power	0-22 W
Moisture	100%
Flow rate	0-2 L/min
Gas temperature	Up to 75 °C
Cell temperature	Up to 75 °C

The PEM fuel cell considered in this study is a single cell with the size of 45x95x101 mm^2 and an active area of 25 cm^2 and serpentine and triangular flow field geometries of channels with the weight of 1300 gr. The width, land width and depth of the channel were selected to be 1, 0.8 and 2 mm respectively. For a bipolar plate, non-porous graphite is selected. A Nafion 117 membrane with 0.004 gr/cm^2 Platinum for the anode and cathode was employed as a membrane electrode assembly. On both sides of the MEA, there were 0.33 mm thick carbon papers that acted as diffusion layers. The thickness of the catalyst layer and the proton exchange membrane is about 0.01 mm and 0.051 mm. The geometry of the channels of the cell in the experimental setup is shown in Figure 2. The examined prototype can operate at a maximum 5 bar absolute pressure; a pressure regulator valve is included, to make possible to vary the operating pressure of the FC system and the accuracy of monitoring the pressure is ±2%. Two flow meters is used to measure the flow rate of the oxygen and hydrogen that the accuracy of them is ±0.1L/min.

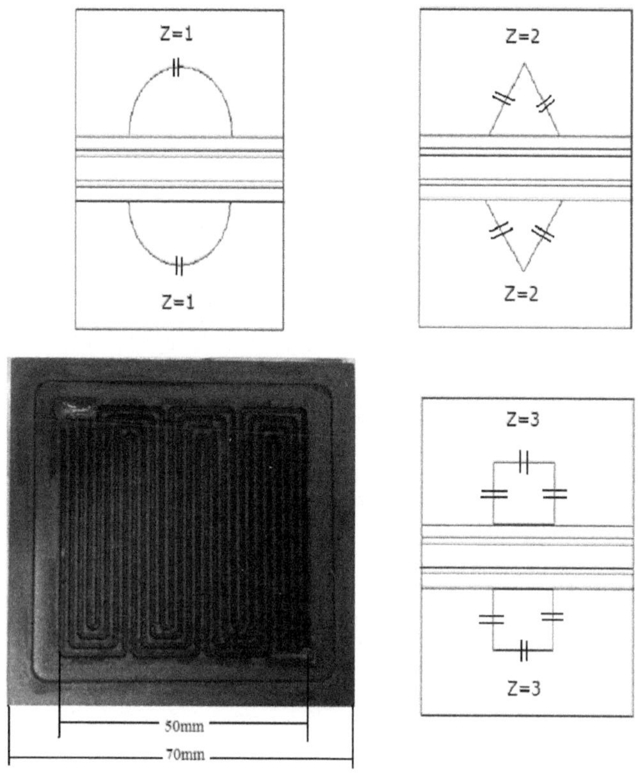

Figure 2. The channels of the PEM fuel cell

In order to plot the polarization curve and simulate a variable load, a resistors box was used that the accuracy of monitoring the voltage and ampere is ±1%. The resistors box, located outside the test chamber, is manually operated; the box and the cables do not introduce relevant errors because they are shielded from external magnetic fields (due to the very low current values). In order to operate in equilibrium conditions, current and voltage values corresponding to each particular value of the total resistance were measured after a sufficient time period to ensure stationary conditions to have been reached as concerns both fuel cell performance and the values of humidity and temperature in the test chamber. The temperature of the inlet gases was measured by digital thermometer with ±0.1°C accuracy. The changed parameters are: input oxygen temperature (T_{O2}), input hydrogen temperature (T_{H2}), cell temperature (T_{cell}), input pressure (P), oxygen flow rate ($\dot{Q}_{O2}$) and hydrogen flow rate ($\dot{Q}_{H2}$) and the measured parameters are voltage and current of the cell. The Range of changing the parameters in this study is shown in Table 2.

At first, we perform the experiments by humidifying the membrane of the fuel cell by saturation water vapor and then change the input oxygen temperature, input hydrogen temperature, cell temperature, input pressure, oxygen flow rate and hydrogen flow rate and measure the pointed parameters and the voltage and the current of the cell after steady state condition. Figure 3 shows the experimental setup.

Table 2. Range of changing the parameters in this study

($\dot{m}_{H2}$) Hydrogen flow rate	0.3-0.9 L/min
($\dot{m}_{O2}$) Oxygen flow rate	0.5-1.3 L/min
(P_{O_2}) Oxygen pressure	2-5 bar
(P_{H2}) Hydrogen pressure	2-5 bar
(T_{H2}) Hydrogen temperature	40-60 °C
(T_{O_2}) Oxygen temperature	45-65 °C
(T_{cell}) Cell temperature	40-60 °C

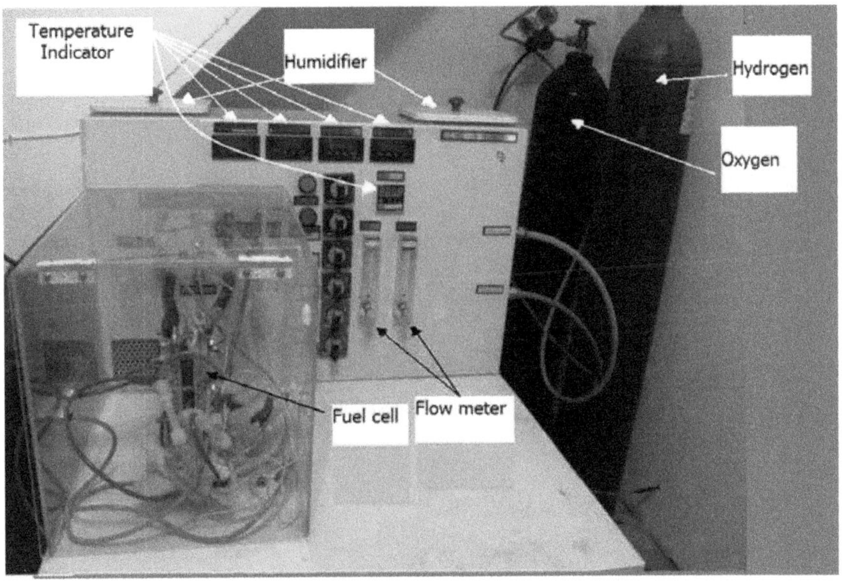

Figure 3. A photo of the experimental setup

3. Results and discussion

The main goal of this study is to investigate the effects of some parameters on the polarization curve of a PEM fuel cell and the performance of it. The Range of changing the parameters in this study is shown in Table 2 and the experiments for each of the parameters done and repeated while the steady condition occurred. Figures 4-6 show the effect of cell temperature on the performance of the fuel cell at P=2.905 bar, T_{O_2}=55 °C, T_{H2}=55 °C, $\dot{Q}_{O2}$=0.5 L/min and $\dot{Q}_{H2}$=0.3 L/min for elliptical, rectangular and triangular channel geometry. It is clear that increasing in the cell temperature leads to the increase in the performance of the cell which is due to the decreasing of activation overpotential and increase in the electrochemical reaction. This is because of the exchange current density of the oxygen reduction reaction increases rapidly with temperature due to the enhanced reaction kinetics, which reduces activation losses. A higher temperature leads also to a higher diffusivity of the hydrogen protons in the electrolyte membrane, thereby reducing the membrane resistance and this leads to reducing the potential loss in the membrane. Also they indicate that at the conditions of the higher operating voltage (lower over- potential), the influence of the internal flow modification on the overall fuel cell performance is negligibly small. At lower operating voltage conditions, on the other hand, the effect of the internal flow modification on the polarization curves becomes important.

In Figure 7 the effect of channel geometry on the overall cell performance of three PEM fuel cells for T_{cell}=60 °C, T_{O_2}=55 °C, T_{H2}=55 °C, $\dot{m}_{O2}$=0.5 L/min, $\dot{m}_{H2}$=0.3 L/min and P=2.905 bar is shown. It is clear that when the geometry of the cell is rectangular, the overall cell performance is better than the elliptical geometry and triangular geometry that this is due to decrease in penetration of the gases into the gas diffusion layer and decreasing the chemical reaction at the surface of catalyst layers. At these conditions the maximum difference between the performance of rectangular and elliptical geometry is about 12% and between the rectangular and triangular geometry is about 18%.

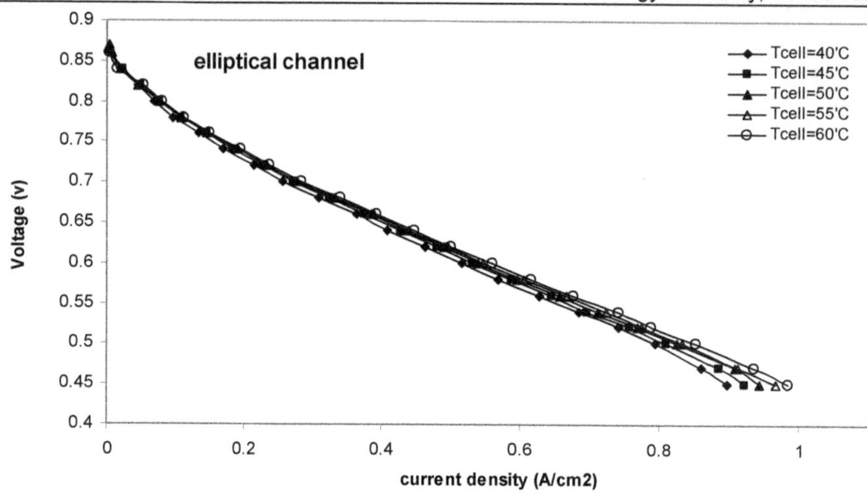

Figure 4. Variation of cell performance of elliptical channel geometry at different cell temperatures for P=2.905 bar, $\dot{m}_{H2}$=0.3 L/min, $\dot{m}_{O2}$=0.5 L/min, T_{O_2}=55 °C, T_{H2}=55 °C

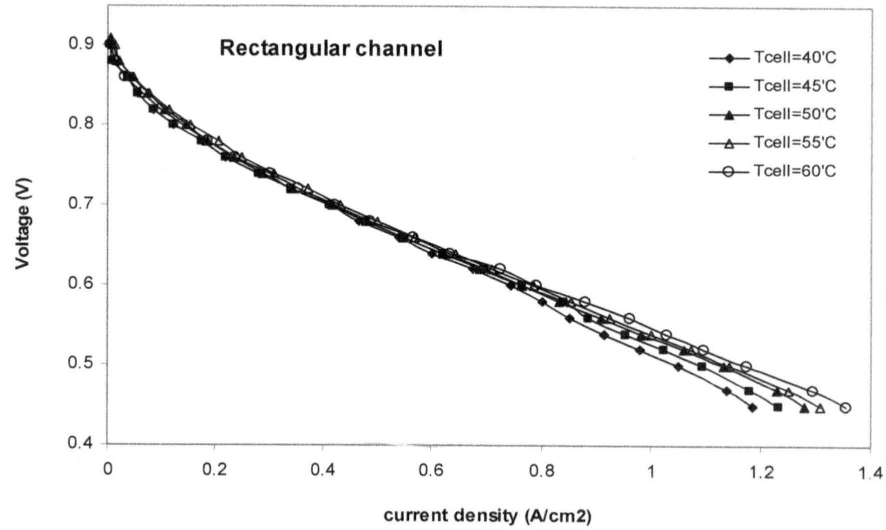

Figure 5. Variation of cell performance of rectangular channel geometry at different cell temperatures for P=2.905 bar, $\dot{m}_{H2}$=0.3 L/min, $\dot{m}_{O2}$=0.5 L/min, T_{O_2}=55 °C, T_{H2}=55 °C

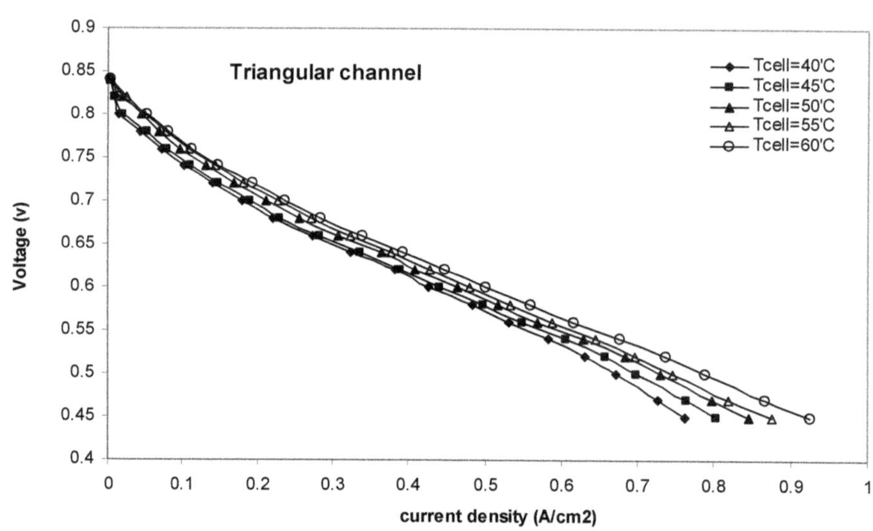

Figure 6. Variation of cell performance of triangular channel geometry at different cell temperatures for P=2.905 bar, $\dot{m}_{H2}$=0.3 L/min, $\dot{m}_{O2}$=0.5 L/min, T_{O_2}=55 °C, T_{H2}=55 °C

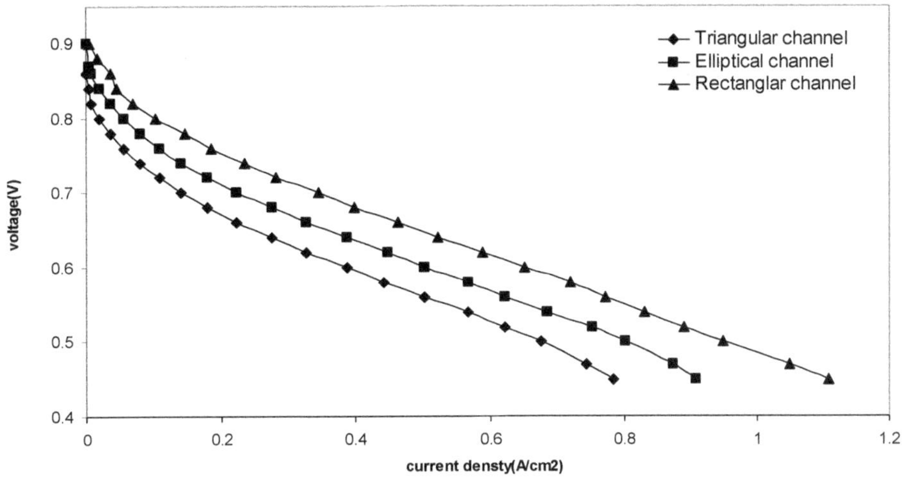

Figure 7. Variation of cell performance at different channel geometry for T_{cell}=60 °C, $\dot{m}_{O2}$=0.5 L/min, $\dot{m}_{H2}$=0.3 L/min, P=2.905 bar, T_{O_2}=55 °C, T_{H2}=55 °C

3.1 A new equation for polarization of the PEM fuel cell

By doing the experiments, it is clear that some parameters such as cell temperature (T_{cell}) and channel geometry specification (Z) affect the performance of the cell. The main reason of changing the polarization curve by changing these parameters is the electrochemical reaction at the catalyst surfaces but the more details described in previous sections. Therefore the main goal of this paper is to obtain a semi-empirical equation for polarization curve that the method of fitting used in this paper is the least square method. This method fits a set of data points (x_i, y_i) to a function that is a combination of any number of functions of the independent variable x. The goal of nonlinear regression is to determine the best-fit parameters for a model by minimizing a chosen merit function. Where nonlinear regression differs is that the model has a nonlinear dependence on the unknown parameters, and the process of merit function minimization is an iterative approach. The process is to start with some initial estimates and incorporates algorithms to improve the estimates iteratively. The new estimates then become a starting point for the next iteration. These iterations continue until the merit function effectively stops decreasing. The nonlinear model to be fitted can be represented by:

$$y = y(x;a) \tag{1}$$

The merit function minimized in performing nonlinear regression the following:

$$\chi^2(a) = \sum_{i=1}^{N}\left\{\frac{y_i - y(x_i;a)}{\sigma_i}\right\}^2 \tag{2}$$

where σ_i is the measurement error, or standard deviation of the *i*th data point. For understanding how the results calculated we have:

The *ith* predicted, or fitted value of the dependent variable Y, is denoted by $\hat{Y}_i$. This value is obtained by evaluating the regression model $\hat{Y} = f(X, \hat{\beta}_j)$, where $\hat{\beta}_j$ are the regression parameters, or variables. Then the residuals $(Y_i - \hat{Y}_i)$ and sum of the residuals $\sum_{i=1}^{n}(Y_i - \hat{Y}_i)$ calculated and then, the average of residuals and residual of sum of squares calculated.

SSE = Residual or Error Sum of Squares (Absolute) = $\sum_{i=1}^{n}(Y_i - \hat{Y}_i)^2$

SSE_R = Residual or Error Sum of Squares (Relative) $=\sum_{i=1}^{n}[(Y_i-\hat{Y}_i)^2 * W_i]$ where $W_i = 1/\sigma_i^2$ normalized so that $\sum_{i=1}^{n} W_i = n$, σ_i =the standard deviation of the i_{th} data point Y_i and n is the number of data points, or observations.

The principle behind nonlinear regression is to minimize the residual sum of squares by adjusting the parameters $\hat{\beta}_j$ in the regression model to bring the curve close to the data points. This parameter is also referred to as the error sum of squares, or *SSE*. If the residual sum of squares is equal to 0.0, the curve passes through every data point.

Thus, the model that is used in this paper is an equation that is similar to Nernst equation [1]:

$$V = E_{rev}(T) - \eta_{act} - \eta_{ohm} - \eta_{conc} \tag{3}$$

The $E_{rev}(T)$ is a function of temperature of the cell that is:

$$E_{rev}(T) = 1.229 + aT + bT^2 \tag{4}$$

η_{act}, η_{ohm} *and* η_{conc} are the activation loss, ohmic loss and the concentration loss that in this paper they have selected similar to equations that have described in reference [1] as:

$$\eta_{act} = T[c + dLn(i)] \tag{5}$$

$$\eta_{ohm} = (e + fT + gT^2)i \tag{6}$$

$$\eta_{conc} = (h + jT + kT^2)(l + mZ + nZ^2)Ln\left[1 - \frac{i}{o + pT + qT^2}\right] \tag{7}$$

that in this equation current density is in A/cm^2, temperatures are in °C and Z=1 for elliptical channel and Z=2 for triangular channel and Z=3 for rectangular channel that this parameter is only in concentration loss that it can be seen in Figure 2.

In equations (4), (5), (6) and (7) the constants a, b, c, d, e, f, g, h, j, k, l, m, n, o, p and q are undefined and by using a software as Datafit which fits the results of experiment from one to more independent variables that in Table 3 the value of these constants was shown.

Figure 8 shows the comparison between the experimental results of Miansari et al. [12] at Z=3 and the 1mm depth of the channels and the correlated equation at T_{O_2}=70 °C, T_{H2}=70 °C, T_{cell}=70 °C, and P=1.905 bar. It is clear that there are significant agreements with them. Figure 9 shows the comparison between the experimental results and the correlated equation of the polarization curve for T_{O_2}=55 °C, T_{H2}=55 °C, T_{cell}=60 °C, P=3.905 bar, $\dot{m}_{O2}$=0.5 L/min, $\dot{m}_{H2}$=0.3 L/min and three different channels geometry. It is clear that there are significant agreements with them.

Table 3. Value and limits of constants of correlated equation

constant	value	constant	value
a	9921.0	j	37177.-586
b	00013478.0	k	73922.9
c	00789.1	l	56617.647
d	00033137.0	m	75213.0
e	25095.-139	n	21267.-0
f	78474.-17	o	72615.36237
g	25433.0	p	55053.-210
h	60255.-12736	q	39646.1

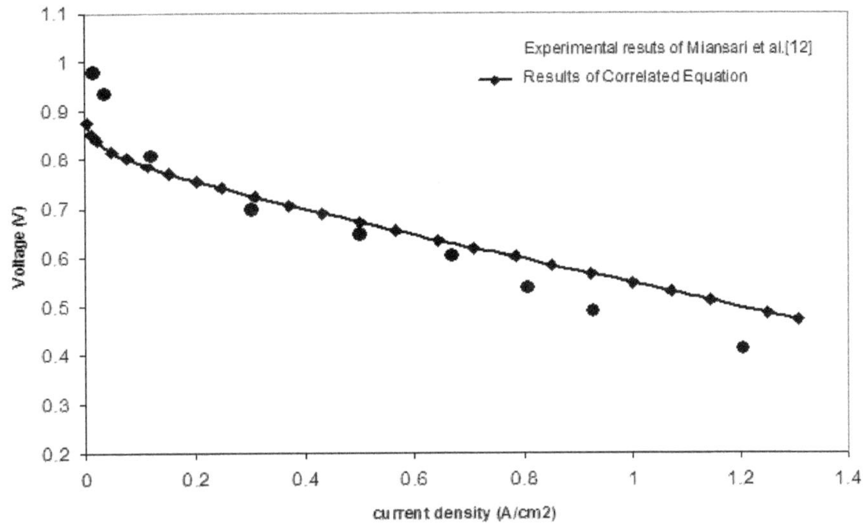

Figure 8. Comparison between the experimental results of Miansari et al. [12] at Z=3 and the correlated equation at To_2=70 °C, T_{H2}=70 °C, T_{cell}=70 °C, P=1.905 bar

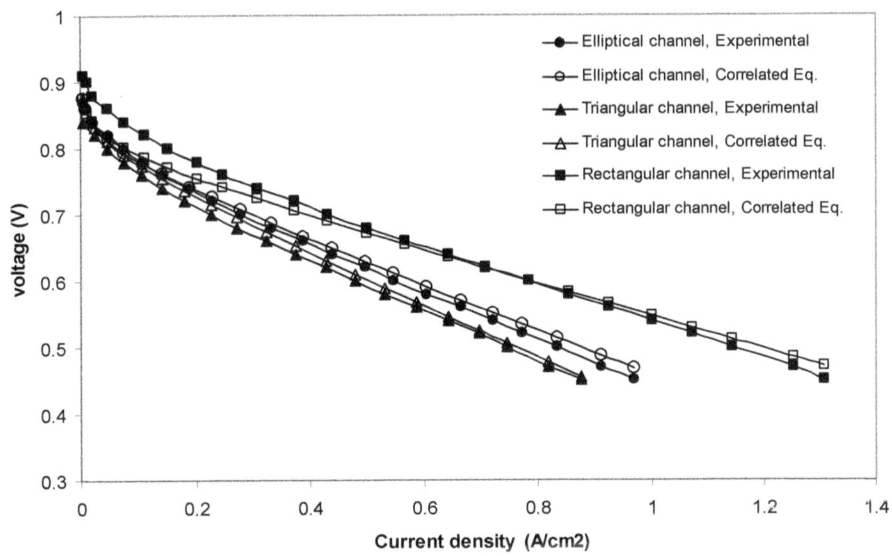

Figure 9. Comparison between the experimental results and the correlated equation of the polarization curve for To_2=55 °C, T_{H2}=55 °C, T_{cell}=60 °C, P=3.905 bar, $\dot{m}_{O2}$=0.5 L/min, $\dot{m}_{H2}$=0.3 L/min and three different channels geometry

4. Conclusion

In this study a new correlation for polarization curve is obtained that it is a function of temperature, current density and a new parameter of cross-section geometry of channels. We have found out that:

- Increasing in the cell temperature from 40 °C to 60 °C leads to the increase in the performance of the cell about 17%, 11% and 9% for triangular, rectangular and elliptical channel which is due to the decreasing of activation overpotential and increase in the electrochemical reaction.
- When the geometry of the cell is rectangular, the overall cell performance is better than the elliptical geometry and triangular geometry that this is due to decrease in penetration of the gases into the gas diffusion layer and decreasing the chemical reaction at the surface of catalyst layers.
- A new correlation for predicting the polarization curve of a PEM fuel cell according to cell temperature and the parameter of cross-section channel geometry was proposed and there was a good agreement between its results and the experimental results of Miansari et al. [12].

Acknowledgment
This work was partially supported by Renewable Energy Organization of Iran.

References

[1] J. Larminie, and A. Dicks, Fuel cell system explained., John Wiley and Sons.,2nd edition, 2003.
[2] G. Scrivano, A. Piacentino, and F. Cardona, "Experimental characterization of PEM fuel cells by micro-models for the prediction of on-site performance", Renewable energy. 34(2009), 634-639.
[3] J.C. Amphlett, R.M. Baumert, R.F. Mann, B.A. Peppley, and P.R. Roberge, "Performance modeling of the Ballard Mark IV solid polymer electrolyte fuel cell. I - mechanistic model development", Electrochem. Sc. Tech.142 (1995), pp.1-8.
[4] R.F. Mann, J.C. Amphlett, M.A.I. Hooper, H.M. Jensen, B.A. Peppley, and P.R. Roberge, "Development and application of a generalised steady-state electrochemical model for a PEM fuel cell", Journal of Power Sources. 86(2000), pp.173-180.
[5] A.J. Del Real, A. Arce, and C. Bordons, "Development and experimental validation of a PEM fuel cell dynamic model", Journal of Power Sources. 173(2007), 310-324.
[6] Khazaee, I. "Effect of placing different obstacles in flow fields on performance of a PEM fuel cell: Numerical investigation and experimental comparison" Heat and Mass transfer, 49(9), pp. 1287-1298, (2013).
[7] Y.M. Ferng, Y.C. Tzang, B.S. Pei, C.C. Sun, and A. Su, "Analytical and experimental investigations of a proton exchange membrane fuel cell", International Journal of Hydrogen Energy, 29, (2004), 381-91.
[8] M.M. Hussain, J.J. Baschuk, X. Li, I. Dincer, "Thermodynamic analysis of a PEM fuel cell power system", International Journal of Thermal Sciences, 44, (2005), 903-911.
[9] J. Park and L. Xianguo, "An experimental and numerical investigation on the cross flow through gas diffusion layer in a PEM fuel cell with a serpentine flow channel", Journal of Power Source, 163, (2007), 853-863.
[10] K. Jiao, J. Park, and L. Xianguo, "Experimental investigations on liquid water removal from the gas diffusion layer by reactant flow in a PEM fuel cell", Applied Energy, 87, (2010), 2770-2777.
[11] J. Park and L. Xianguo, "Effect of flow and temperature distribution on the performance of a PEM fuel cell stack", Journal of Power Source, 162, (2006), 444-459.
[12] M. Miansari, K. Sedighi, M. Amidpour, E. Alizadeh, and M.O. Miansari, "Experimental and thermodynamic approach on proton exchange membrane fuel cell performance", Journal of Power Source, 190, (2009), 356-361.

5

Thermodynamic analysis for a regenerative gas turbine cycle in coking process

Zelong Zhang[1, 2, 3], Lingen Chen[1, 2, 3], Yanlin Ge[1, 2, 3], Fengrui Sun[1, 2, 3]

[1] Institute of Thermal Science and Power Engineering, Naval University of Engineering, Wuhan 430033, China.
[2] Military Key Laboratory for Naval Ship Power Engineering, Naval University of Engineering, Wuhan 430033, China.
[3] College of Power Engineering, Naval University of Engineering, Wuhan 430033, China.

Abstract
A regenerative gas turbine cycle driven by residual coke oven gas is proposed in this paper. The thermal efficiency and the work output (per ton of coke) of the system are analyzed based on thermodynamics and the theory of gas turbine cycle. The influences of the gas release rate, the residual gas rate and the effectiveness of regenerator on the performance of the cycle are analyzed by using numerical examples. It is found that the work output increases with the increase of the residual gas rate while decreases with the increase of the gas release rate. The cycle with regenerator can reach higher thermal efficiency and bigger work output, which means that the coke oven gas is used more effectively. Moreover, there exist two optimal pressure ratios of compressor which lead the maximum thermal efficiency and the maximum specific work, respectively.

Keywords: Coke oven gas; Regenerative Brayton cycle; Thermodynamic analysis; Thermal efficiency; Work output.

1. Introduction

Steel is a crucial material in human technological development, and is widely used in our lives. China's iron and steel industry is the basis of the national economy, and has made considerable progress in the past two decades [1]. With technological progress and proper redistribution, China's iron and steel industry has made tremendous achievements in energy conservation [2-11]. However, in comparison with developed countries, there is a big potential to save energy in China's iron and steel industry.

Coke is one of the most important materials consumed by the steelmaking process. It performs several functions in the blast furnace. In the coking process, bituminous coals are carbonized, and form coke, tar and COG (coke oven gas). After residual heat recovery and a series of cleaning treatments, COG consists mainly of H2 and CH4. COG is a prime fuel which has high heating value. It is an important fuel in steelmaking process. In the process of steelmaking, in order to maintain the mass flow rate in a constant, the energy input must be residual. As for gas system of an integrated iron and steel plant, the COG is always residual. Furthermore, with the progress in steelmaking process and metallurgy technology, the amount of residual COG will be more and more. Make good use of the residual COG can decrease the energy consumption in steelmaking process. Sun et al [12] designed a cogeneration system for producing coke and generating electricity based on the principle of comprehensive and stepped utilization of energy. In the system, clean power generation with high efficiency is realized by decreasing the amount of air required for coking and supplying the combined cycle with enriched COG as much as possible. Hu et al [13] analyzed material flow, energy flow and sulfur flow of coking process in integrated iron and steel plants, and indicated that the utilization of COG should take into account various factors including the optimization of iron and steel manufacturing process and gas balance of plant, etc.. Villar et al [14] analyzed new waste-to-energy technologies in continuous industries in terms of conversion, energy saving, heat recovery, electricity generation, transportation fuel, storing energy and fuel, environmental emission, and recycling management. Mert et al [15] provided exergoeconomic analysis of an electricity and thermal energy cogeneration plant in an iron and steel factory.

Considering energy saving in steelmaking process, this paper proposes a regenerative gas turbine cycle by using residual COG as fuel. This system can reach higher thermal efficiency because the regeneration can recovery a part of waste heat in exhausted gas, and the work output of the system can be used for generating electricity and driving blowers, etc. Furthermore, the thermal efficiency and work output of the system are analyzed based on classical thermodynamics and the theory of gas turbine cycles [16-31]. According to the thermodynamic analysis, one can estimate the system's work output in different amount of residual COG. This paper can provide a basis for the optimization of residual COG utilization in further steps.

2. System description

In the coking process, a large amount of COG is produced as a by-product. After residual heat recovery and a series of cleaning treatments, COG can be used as good fuel which consists mainly of H_2 and CH_4. Part of COG is used in other process of ironmaking, and the residual coke oven gas is used as fuel in the regenerative gas turbine cycle proposed in this paper in order to save energy. COG is compressed in gas compressor, then is mixed with compressed air and burned in combustion chamber. The working fluid flows through the turbine and generates power, and the regenerator recovers waste heat in exhaust gas. The system layout of the regenerative gas turbine plant in coking process is shown in Figure 1. Figure 2 is the corresponding T-s diagram of the system.

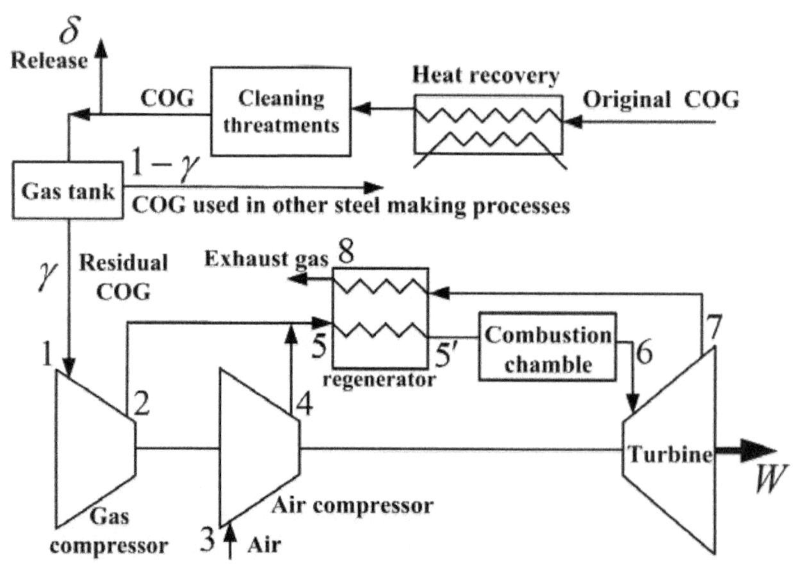

Figure 1. The system layout of a the regenerative gas turbine in coking process

3. Performance analyses

It is set that the COG's density is ρ_g=0.49 kg/m^3 and heating value is $H = 16.706\ \text{MJ/m}^3 = 34.0939\ \text{MJ/kg}$. Producing one ton coke can generate COG of 430 m^3 or 210.7kg. The mass of residual COG is

$$m_g\ (\text{kg}) = 210.7(1-\delta)\gamma \tag{1}$$

where δ is gas release rate, and γ is residual gas rate.

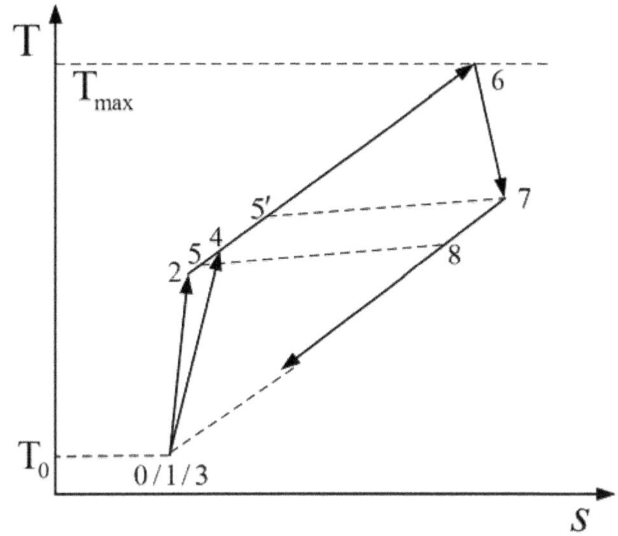

Figure 2. The T-s diagram of the system

Air consists 21% O_2 and 79% N_2 and its density is ρ_a=1.169 kg/m^3. According to COG's components listed in Table 1 and air's components, one can obtain theoretical air quantity for complete combustion of 1 m^3 COG:

$$L_0 = \frac{100}{21}\left[56.83\%/2 + 22.49\% \times 2 + 5.49\%/2 + 2.54 \times (\mathrm{n} + \mathrm{m}/2)\right] \approx 3.8638 \text{ m}^3 \tag{2}$$

Table 1. Proportions of COG's component (T_0, p_0)

Component	H_2	CH_4	CO	$N_2 + Ar$	CO_2	C_nH_m	Other components
Volume ratio	56.83%	22.49%	5.49%	10.26%	2.04%	2.54%	0.35%

The air-fuel ratio is

$$m_g/m_a = \frac{\rho_g}{\rho_a \lambda L_0} \tag{3}$$

where λ is excess air ratio, and m_a is mass of air.
The gas compression process is adiabatic and irreversible, the efficiency of gas compressor is defined as $\eta_{cg} = (T_{2s} - T_1)/(T_2 - T_1)$, the isentropic temperature ratio across the gas compressor is $T_{2s}/T_1 = (P_2/P_2)^{\frac{k_g-1}{k_g}}$, and the temperature ratio of compression process is

$$\frac{T_2}{T_1} = \frac{T_2 - T_1}{T_{2s} - T_1}\left(\frac{T_{2s}}{T_1} - 1\right) + 1 = 1 + \left[(P_2/P_2)^{\frac{k_g-1}{k_g}} - 1\right]\Big/\eta_{cg} = 1 + \left(\varphi_{cg}{}^{n_g} - 1\right)/\eta_{cg} \tag{4}$$

where T is temperature, P is pressure, k_g is specific heat ratio of COG, $n_g = (k_g - 1)/k_g$, and $\varphi_{cg} = P_2/P_1$ is pressure ratio of gas compressor.
The work required for gas compressor is

$$W_{cg} = m_g(h_2 - h_1) = m_g c_{pg} T_1 (T_2/T_1 - 1) = m_g\, c_{pg} T_1 \left(\varphi_{cg}{}^{n_g} - 1\right)/\eta_{cg} \tag{5}$$

where c_{pg} is isobaric specific heat of COG and T_1 is ambient temperature.

The air compression process is adiabatic and irreversible, the efficiency of air compressor is defined as $\eta_{ca} = (T_{4s} - T_3)/(T_4 - T_3)$, the isentropic temperature ratio across the air compressor is $T_{4s}/T_3 = (P_4/P_3)^{\frac{k_a-1}{k_a}}$, and the temperature ratio of compression process is

$$\frac{T_4}{T_3} = \frac{T_4 - T_3}{T_{4s} - T_3}\left(\frac{T_{4s}}{T_3} - 1\right) + 1 = 1 + \left[(P_4/P_3)^{\frac{k_a-1}{k_a}} - 1\right]\Big/\eta_{ca} = 1 + \left(\varphi_{ca}^{\ n_a} - 1\right)/\eta_{ca} \tag{6}$$

where k_a is specific heat ratio of air, $n_a = (k_a - 1)/k_a$, and $\varphi_{ca} = P_4/P_3$ is pressure ratio of air compressor. The work required for air compressor is

$$W_{ca} = m_a(h_4 - h_3) = m_g \frac{\rho_a \lambda L_0}{\rho_g} c_{pa} T_3 (T_4/T_3 - 1) = m_g \frac{\rho_a \lambda L_0}{\rho_g} c_{pa} T_3 \left(\varphi_{ca}^{\ n_a} - 1\right)\Big/\eta_{ca} \tag{7}$$

The inlet temperature of regenerator depends on the temperature of compressed gas and compressed air. The mixing process of gas and air is defined as ideal gas mixing process. According to the temperature and quality of compressed gas and compressed air, one can derive the inlet temperature of regenerator:

$$\begin{aligned} T_5 &= \frac{m_g T_2 + m_a T_4}{m_g + m_a} = T_2\left(\frac{1}{1 + \rho_a \lambda L_0/\rho_g}\right) + T_4\left(\frac{1}{1 + \rho_g/(\rho_a \lambda L_0)}\right) \\ &= T_1\left\{\left[1 + \frac{\left(\varphi_{cg}^{\ n_g} - 1\right)}{\eta_{cg}}\right]\left(\frac{1}{1 + \rho_a \lambda L_0/\rho_g}\right) + \left[1 + \frac{\left(\varphi_{ca}^{\ n_a} - 1\right)}{\eta_{ca}}\right]\left(\frac{1}{1 + \rho_g/(\rho_a \lambda L_0)}\right)\right\} \end{aligned} \tag{8}$$

The turbine expansion process is adiabatic and irreversible, the efficiency of turbine is defined as $\eta_t = (T_6 - T_7)/(T_6 - T_{7s})$, the isentropic temperature ratio across the turbine is $T_{7s}/T_6 = (P_7/P_6)^{\frac{k_w-1}{k_w}}$, and the temperature ratio of expansion process is

$$\frac{T_7}{T_6} = 1 - \frac{T_6 - T_7}{T_6 - T_{7s}}\left(1 - \frac{T_{7s}}{T_6}\right) = 1 - \eta_t\left[1 - (P_7/P_6)^{\frac{k_w-1}{k_w}}\right] = 1 - \eta_t\left[1 - \left(1 - 1/\varphi_t^{\ n_w}\right)\eta_t\right] \tag{9}$$

where $n_w = (k_w - 1)/k_w$ is specific heat ratio of working fluid (after air and gas being mixed and burned), and $\varphi_t = P_6/P_7$ is pressure ratio of turbine.
The work output of turbine is

$$\begin{aligned} W_t &= m_w(h_6 - h_7) = m_g\left(1 + \rho_a \lambda L_0/\rho_g\right) c_{pw} T_6 (1 - T_7/T_6)\eta_t \\ &= c_{pw} T_1 \tau m_g \left(1 + \rho_a \lambda L_0/\rho_g\right)\left(1 - 1/\varphi_t^{\ n_w}\right)\eta_t \end{aligned} \tag{10}$$

where c_{pw} is isobaric specific heat of working fluid, and $\tau = T_6/T_1$ is temperature ratio of the system. According to the balance of heat, $c_{pw}(T_7 - T_8) = c_{px}(T_{5'} - T_5)$, and the effectiveness of regenerator, $E_R = (T_7 - T_8)/(T_7 - T_5)$, the outlet temperature of regenerator is

$$T_{5'} = \frac{E_R \tau T_1\left[1 - \left(1 - 1/\varphi_t^{\ n_w}\right)\eta_t\right] + \left(c_{px}/c_{pw} - E_R\right)T_5}{c_{px}/c_{pw}} \tag{11}$$

where c_{px} is isobaric specific heat of mixed gas (air and gas are mixed before combustion).
According to the balance of heat in combustion chamber, the amount of heat added to the system is

$$Q_{in} = c_{px}\left(m_a + m_g\right)\left(T_6 - T_{5'}\right) = c_{px} m_g \left(1 + \rho_a \lambda L_0 / \rho_g\right)\left(T_6 - T_{5'}\right) = m_g \eta_B H \tag{12}$$

Combing Eq. (2) with Eq. (11), the excess air ratio can be obtained

$$\lambda = \frac{\eta_B \rho_g H / \left[c_{px}\left(T_6 - T_{5'}\right)\right] - \rho_g}{\rho_a L_0} \tag{13}$$

The work output of the system is

$$W = W_t - W_{cg} - W_{ca}$$

$$= m_g T_1 \left[c_{pw} \tau \left(1 + \frac{\rho_a \lambda L_0}{\rho_g}\right)\left(1 - \frac{1}{\varphi_t^{n_w}}\right)\eta_t - \frac{c_{pg}\left(\varphi_{cg}^{n_g} - 1\right)}{\eta_{cg}} - \frac{\rho_a \lambda L_0}{\rho_g \eta_{ca}} c_{pa}\left(\varphi_{ca}^{n_a} - 1\right)\right] \tag{14}$$

The thermal efficiency of the system is

$$\eta = W/Q_{in} = \frac{c_{pw}\tau\eta_t(1+\rho_a\lambda L_0/\rho_g)(1-1/\varphi_t^{n_w}) - c_{pg}(\varphi_{cg}^{n_g}-1)/\eta_{cg} - \rho_a\lambda L_0 c_{pa}(\varphi_{ca}^{n_a}-1)/(\rho_g\eta_{ca})}{c_{pw}(1+\rho_a\lambda L_0/\rho_g)\{\tau - \{E_R\tau[1-\eta_t(1-1/\varphi_t^{n_w})] + (c_{px}/c_{pw} - E_R)\{[1+(\varphi_{cg}^{n_g}-1)/\eta_{cg}][1/(1+\rho_a\lambda L_0/\rho_g)] + [1+(\varphi_{ca}^{n_a}-1)/\eta_{ca}][1/(1+\rho_g/\rho_a\lambda L_0)]\}\}c_{pw}/c_{px}\}} \tag{15}$$

From Eq. (15), one can see that the residual gas rate and the gas release rate have no influence on the thermal efficiency of the system.

4. Numerical examples

To see how the various parameters influence thermal efficiency and work output of the system, numerical examples are provided. In the calculations, it is set that the isobaric specific heat of air is $c_{pa} = 1.004\ \mathrm{kJ/(kg \cdot K)}$, isobaric specific heat of COG is $c_{pg} = 2.69\ \mathrm{kJ/(kg \cdot K)}$, the isobaric specific heat of mixed gas is $c_{px} = 1.006\ \mathrm{kJ/(kg \cdot K)}$, the isobaric specific heat of working fluid is $c_{pw} = 1.013\ \mathrm{kJ/(kg \cdot K)}$, the specific heat ratio of air is $k_a = 1.400$, the specific heat ratio of COG is $k_g = 1.351$, the specific heat ratio of working fluid is $k_w = 1.392$, the efficiencies of gas compressor and air compressor are $\eta_{ca} = \eta_{cg} = 0.9$, the efficiency of turbine is $\eta_t = 0.85$, the pressure ratios of gas compressor, air compressor and turbine are the same ($\varphi_{cg} = \varphi_{ca} = \varphi_t = \varphi_c$), the efficiency of combustion chamber is $\eta_B = 0.95$, the ambient temperature is $T_1 = T_3 = 300\mathrm{K}$, the ambient pressure is $P_1 = P_3 = 0.1013\mathrm{MPa}$, the temperature ratio of the cycle is $\tau = 5$, the effectiveness of regenerator is $E_R = 0.9$, the gas release rate is $\delta = 15\%$, and the residual gas rate is $\gamma = 20\%$.

Figures 3 and 4 show the influences of the gas release rate δ and the residual gas rate γ on the work output versus pressure ratio ($W - \varphi_c$) characteristic, respectively. One can see that the work output (per ton of coke) increases with increase in the gas release rate δ and decreases with the increase in the residual gas rate γ. In the practical steelmaking process, because of the heating value of coke oven gas is large, the residual coke oven gas should be used in generating electricity as much as possible. The amount of heat demanded in other procedures of steelmaking process can use blast furnace gas and basic oxygen furnace gas instead of coke oven gas. In this way, the amount of residual coke oven gas can be increased. Furthermore, the proper redistribution of the gas buffers such as gas tank and gas-fired boiler can decrease the gas release rate and make coke oven gas use effectively.

Figures 5 and 6 show the influences of the effectiveness of regenerator on the thermal efficiency versus pressure ratio (η-φ_c) and W-φ_c characteristics. When $E_R = 1$, both thermal efficiency η and work output W decrease with the increase in pressure ratio of compressor φ_c. When $E_R < 1$, there exist two optimal pressure ratios of compressor which lead to maximum thermal efficiency and work output, respectively. Furthermore, there is a critical pressure ratio of compressor, when φ_c is smaller than it, both thermal efficiency η and work output W increase with the increase in E_R; when φ_c is larger than it, both thermal efficiency η and work output W decrease with the increase in E_R.

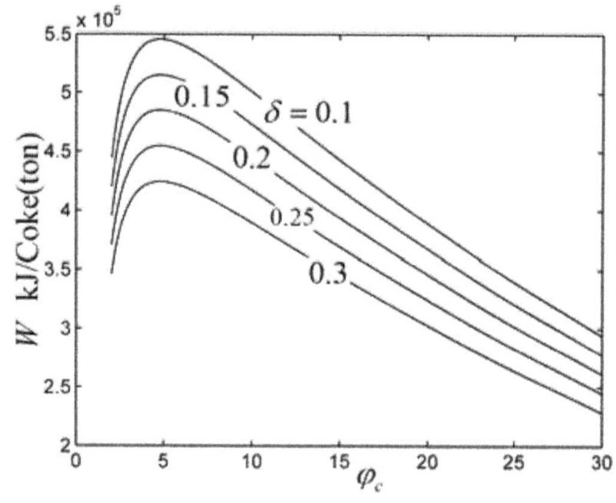

Figure 3. The influence of δ on the $W-\varphi_c$ characteristic

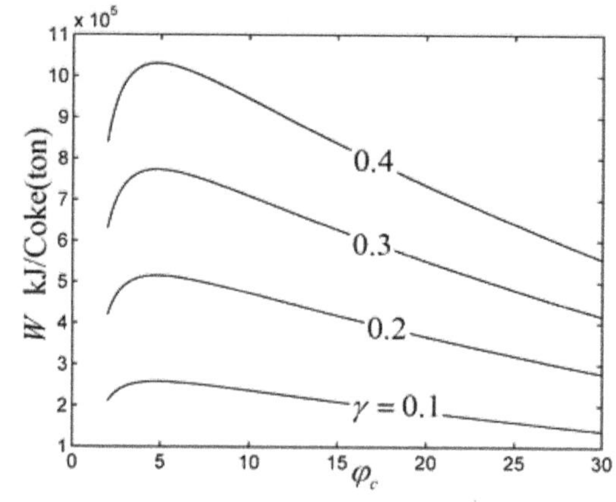

Figure 4. The influence of γ on the $W-\varphi_c$ characteristic

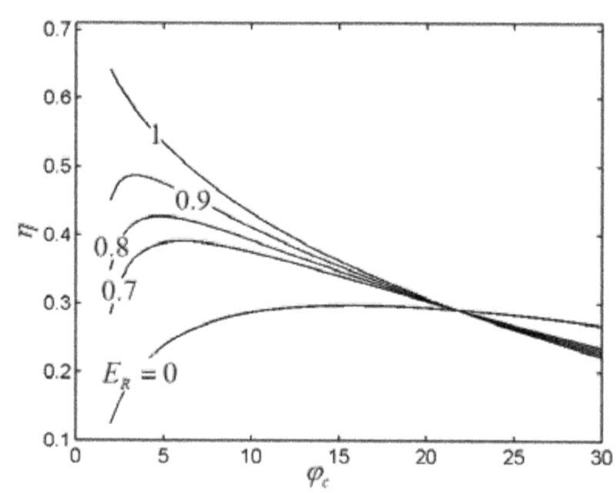

Figure 5. The influence of E_R on the $\eta-\varphi_c$ characteristic

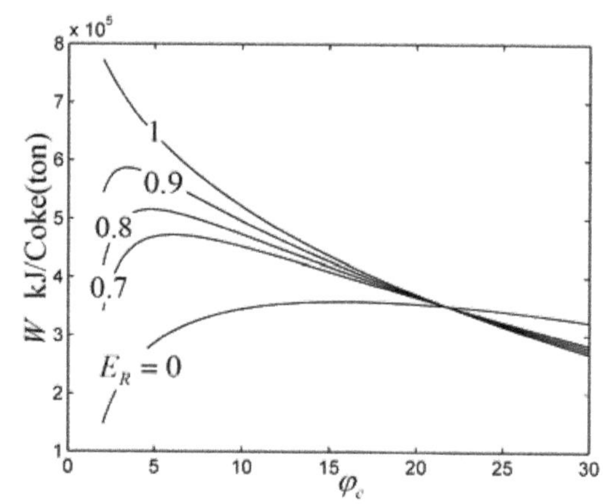

Figure 6. The influence of E_R on the $W-\varphi_c$ characteristic

5. Conclusion

In order to meet the needs in energy conservation of steelmaking industry, a regenerative gas turbine cycle by using residual COG as fuel is proposed in this paper. Work output of the system (the regenerative gas turbine cycle) can be used for generating electricity and driving blowers, etc. Furthermore, the thermal efficiency and work output of the system are analyzed based on classical thermodynamics and the theory of gas turbine cycle. Using numerical calculations, the effects of the gas release rate, the residual gas rate and the effectiveness of regenerator on the performance of the system are analyzed. One can see that the work output (per ton of coke) increases with increase in the gas release rate and decreases with the increase in the residual gas rate. In the practical steelmaking process, the proper redistribution of the gas buffers such as gas tank or gas-fired boiler can decrease the gas release rate and make COG be used effectively. In the system, when the pressure ratio is smaller than the critical one, both thermal efficiency and work output increase with the increase in effectiveness of regenerator, so the system with regenerator uses COG effectively. This paper analyzes the thermodynamic processes of the system, estimates the system's work output in different amount of residual COG and can provides a basis for the optimization of residual COG utilization in further steps.

Acknowledgments
This paper is supported by the National Key Basic Research and Development Program of China ('973' Program) (Grant No. 2012CB720405) and the National Natural Science Foundation of P. R. China (Project No. 10905093).

References

[1] Y. Wei, H. Liao, Y. Fan, An empirical analysis of energy efficiency in China's iron and steel sector, Energy 32(2007) 2262-2270.
[2] National Bureau of Statistics of China, China Statistical Yearbook 2006, China Statistics Press, Beijing, 2006.
[3] OECD/ International Energy Agency, World Energy Outlook 2006, Paris, 2006.
[4] Z. C. Guo, Z. X. Fu, Current situation of energy consumption and measures taken for energy saving in the iron and steel industry in China, Energy 35(2010) 4356-4360.
[5] G. Ma, J. Cai, W. Zeng, H. Dong, Analytical research on waste heat recovery and utilization of China's iron & steel industry, Energy Proce. 14(2012) 1022-1028.
[6] W. Sun, J. Cai, Z. Ye, Advances in energy conservation of China steel industry, Sci. Word J (2013) Article ID 247035.
[7] H. Zhang, L. Dong, H. Q. Li, B. Chen, Q. Tang, T. Fujita, Investigation of the residual heat recovery and carbon emission mitigation potential in Chinese steelmaking plant: A hybrid material/energy flow analysis case study, Sust. Energy Tech. Assess. 2(2013) 67-80.
[8] M. Z. Stijepovic, P. Linke, Optimal waste heat recovery and reuse in industrial zones, Energy, 36(2011) 4019-4031.
[9] Y. Ammar, S. Joyce, R. Norman, Y. Wang, A. P. Roskilly, Low grade thermal energy sources and uses from the process industry in the UK, Appl. Energy 89(2012) 3-20.
[10] C. Gao, D. Wang, H. Dong, J. Cai, W. Zhu, T. Du, Optimization and evaluation of steel industry's water-use system, J. Cleaner Produ. 19(2011) 64-69.
[11] J. Gao, S. Li, Y. Zhang, P. Chen, P. Shen, Process of re-resourcing of converter slag, J. Iron & Steel Res., Inter. 18(2011) 32-39.
[12] S. Sun, H. Jin, L. Gao, W. Han, Study on a multifunctional energy system producing coking heat, methanol and electricity, Fuel 89(2010) 1353-1360.
[13] C. Hu, C. Zhang, X. Zhang, Y. Qi, R. Yin, Material and energy flow analysis of coking process in integrated steel plants, J. Iron Steel Res. 19(2007) 16-20 (in Chinese) .
[14] A. Villar, J. J. Arribas, J. Parrondo, Waste-to-energy technologies in continous process industries, Clean Tech. Environ. Policy 14(2012) 29-39.
[15] M. S. Mert, O. F. Dilmac, S. Ozkan, F. Karaca, E. Bolat, Exergoeconomic analysis of a cogeneration plant in an iron and steel factory, Energy 46(2012) 78-84.
[16] H. Sato, Cycle Theory of Gas Turbine, Sankaido, Tokyo, 1972.
[17] W. A. Woods, P. J. Bevan, D. I. Bevan, Output and efficiency of the closed-cycle gas turbine, Proc. IMechE, Part A: J. Power Energy 205(1991) 59-66.
[18] W. A. Woods, On the role of the harmonic mean isentropic exponent in the analysis of the closed-cycle gas turbine, Proc. IMechE, Part A: J. Power Energy 205(1991) 287-291.
[19] T. H. Frost, B. Agnew, A. Anderson, Optimization for Brayton-Joule gas turbine cycles, Proc. IMechE, Part A: J. Power Energy 206(1992) 283-288.
[20] P. Vadasz, D. Weiner, The optimal intercooling of compressor by a finite number of intercoolers, Trans. ASME, J. Energy Res. Tech. 114(1992) 255-260.
[21] C. A. Frangopoulos, G. G. Dimopoulos, Effects of gas-properties evaluation method on the optimal point of gas turbine cycles, Int. J. Thermodyn. 8(2005) 95-102.
[22] A. Calvo Hernandez, A. Medina, J. M. M. Roco Power and efficiency in regenerative gas turbine, J. Phys. D: Appl. Phys. 28(1995) 2020-2023.
[23] A. Calvo Hernandez, J. M. M. Roco, A. Medina, Power and efficiency in a regenerative gas-turbine cycle with multiple reheating and intercooling stages, J. Phys. D: Appl. Phys. 29(1996) 1462-1468.
[24] M. J. Naser, Exergy analysis and second law efficiency of a regenerative Brayton cycle with isothermal heat addition, Entropy 7(2005) 172-187.
[25] H. Zhao, P. F. Peterson, Multiple reheat helium Brayton cycles for sodium cooled fast reactors, Nucl. Eng. Des. 238(2008) 1535-1546.

[26] R. Lugo, J. M. Zamora, M. Salazar, Optimum pressure ratio for complex gas turbine cycles, Inform. Tecn. 20(2009) 137-151.

[27] H. Chandra, A. Arora, S. C. Kaushik, Thermodynamic analysis and parametric study of a closed cycle gas turbine with intercooler and reheater on the basis of a new isentropic exponent, Int. J. Sust. Energy 30(2011) 82-97.

[28] J. H. Horlock, Advance Gas Turbine Cycles, 1st edition, Elsevier Science Publishers, London, 2003.

[29] H. Canière, A. Willockx, E. Dick, Raising cycle efficiency by intercooling in air-cooled gas turbines, Appl. Thermal Eng. 26(2006) 1780-1787.

[30] O. S. Sanjay, B. N. Prasad, Comparative evaluation of gas turbine power plant performance for different blade cooling means, Proc. IMechE, Part A: J. Power Energy 223(2009) 71-82.

[31] O. S. Sanjay, B. N. Prasad, Comparative performance analysis of cogeneration gas turbine cycle for different blade cooling means, Int. J. Thermal Sci. 48(2009) 1432-1440.

6

Analysis of transport phenomena and electrochemical reactions in a micro PEM fuel cell with nature inspired flow field design

Maher A.R. Sadiq Al-Baghdadi

Fuel Cell Research Center, International Energy and Environment Foundation, Najaf, P.O.Box 39, Iraq.

Abstract

Micro-fuel cells are considered as promising electrochemical power sources in portable electronic devices. The presence of microelectromechanical system (MEMS) technology makes it possible to manufacture the miniaturized fuel cell systems. The majority of research on micro-scale fuel cells is aimed at micro-power applications. Performance of micro-fuel cells are closely related to many factors, such as designs and operating conditions. CFD modeling and simulation for heat and mass transport in micro PEM fuel cells are being used extensively in researches and industrial applications to gain better understanding of the fundamental processes and to optimize the micro fuel cell designs before building a prototype for engineering application. In this research, full three-dimensional, non-isothermal computational fluid dynamics model of a micro proton exchange membrane (PEM) fuel cell with nature inspired flow field designs has been developed. The design inspired from the existed biological fluid flow patterns in the leaf. This comprehensive model accounts for the major transport phenomena such as convective and diffusive heat and mass transfer, electrode kinetics, transport and phase-change mechanism of water, and potential fields in a micro PEM fuel cell. The model explains many interacting, complex electrochemical, and transport phenomena that cannot be studied experimentally.

Keywords: PEM fuel cells; Micro fuel cell; Nature inspired flow channels; Electrochemical; Heat transfer; CFD.

1. Introduction

Fuel cell system is an advanced power system for the future that is sustainable, clean and environmental friendly. Small fuel cells have provided significant advantages in portable electronic applications over conventional battery systems. Competitive costs, instant recharge, and high energy density make fuel cells ideal for supplanting batteries in portable electronic devices. Micro fuel cells belong to the category of portable fuel cells, but are used in the miniaturized electronic devices with low power, such as micro sensors. The presence of microelectromechanical system (MEMS) technology makes it possible to manufacture the miniaturized fuel cell systems for application in portable electronic devices. The majority of research on micro-scale fuel cells is aimed at micro-power applications. There are many new miniaturized applications which can only be realized if a higher energy density power source is available compared to button cells and other small batteries.

The development of physically representative models that allow reliable simulation of the processes under realistic conditions is essential to the development and optimization of fuel cells, the introduction of cheaper materials and fabrication techniques, and the design and development of novel architectures. The difficult experimental environment of fuel cell systems has stimulated efforts to develop models that could simulate and predict multi-dimensional coupled transport of reactants, heat and charged species using computational fluid dynamic (CFD) methods. These models can generally be characterized by the scope of the model. In many cases, modeling efforts focus on a specific part or parts of the fuel cell, like the cathode catalyst layer [1], the cathode electrode (gas diffusion layer plus catalyst layer) [2-4], or the membrane electrode assembly (MEA) [5]. These models are very useful in that they may include a large portion of the relevant fuel cell physics while at the same time having relatively short solution times. However, these narrowly focused models neglect important parts of the fuel cell making it impossible to get a complete picture of the phenomena governing fuel cell behavior. Models that include all parts of a fuel cell are typically two- or three-dimensional and reflect many of the physical processes occurring within the fuel cell [6–16]. In a real PEM fuel cell geometry, the gas diffusion layers are used to enhance the reaction area accessible by the reactants. The effect of using these diffusion layers is to allow a spatial distribution in the current density on the membrane in both the direction of bulk flow and the direction orthogonal to the flow but parallel to the membrane. This two-dimensional distribution cannot be modeled with the well-used two-dimensional models where the mass-transport limitation is absent in the third direction. Berning et al. [13] used the unified approach to develop a three-dimensional model of a straight gas flow channels fuel cell. The model studied reactant concentrations, current density distributions and temperature gradients within the cell as well as water flux and species transport.

Nguyen et al. [14] developed a three-dimensional model, which accounts for mass and heat transfer, current and potential distribution within a cell using a one turn serpentine flow field. Their results show that oxygen concentration along the gas channels decrease in the direction of flow. Also, in the gas diffusion layer, the oxygen concentration is a minimum under the land area. At high current densities the oxygen is almost completely depleted under the land areas. The result is an uneven distribution of oxygen concentration along the catalyst layer resulting in local overpotentials, which vary spatially.

Um and Wang [15] used a three-dimensional model to study the effects an interdigitated flow field. The model accounted for mass transport, electrochemical kinetics, species profiles and current density distribution within the cell. Interdigitated flow fields result in forced convection of gases, which aids in liquid water removal at the cathode. This would help improve performance at high current densities when transport limitations due to excessive water production are expected. The model shows that there is little to no difference at low to medium current densities between an interdigitated flow field and a conventional flow field. However, at higher current densities, a fuel cell with an interdigitated flow field has a limiting current, which is nearly 50% greater than an equivalent cell with a conventional flow field.

Al-baghdadi [16] developed a three-dimensional model to investigate the performance of tubular-shaped PEM fuel cell. The results show that higher power densities are achieved with the tubular design mainly because of lower activation and diffusion overpotentials. Better gas replenishment at the catalyst sites in tubular design results in lower and quite uniform distribution of activation and diffusion potentials.

In this work, a multi-phase, non-isothermal, fully three-dimensional CFD model of a micro PEM fuel cell with nature inspired flow field designs has been developed. The design inspired from the existed biological fluid flow patterns in the leaf. The model accounts for detailed species mass transport, heat transfer in the solids as well as in the gases, potential losses in the gas diffusion layers and membrane, electrochemical kinetics, and the transport of water through the membrane.

2.1. Computational domain

The full computational domains for the micro PEM fuel cell consist of cathode and anode gas flow fields, and the MEA are shown in Figure 1.

2.2. Model equations

2.2.1. Gas flow channels

In the fuel cell channels, the gas-flow field is obtained by solving the steady-state Navier-Stokes equations, i.e. the continuity equation, the mass conservation equation for each phase yields the volume fraction (r) and along with the momentum equations the pressure distribution inside the channels. The continuity equation for the gas phase inside the channel is given by;

(a)

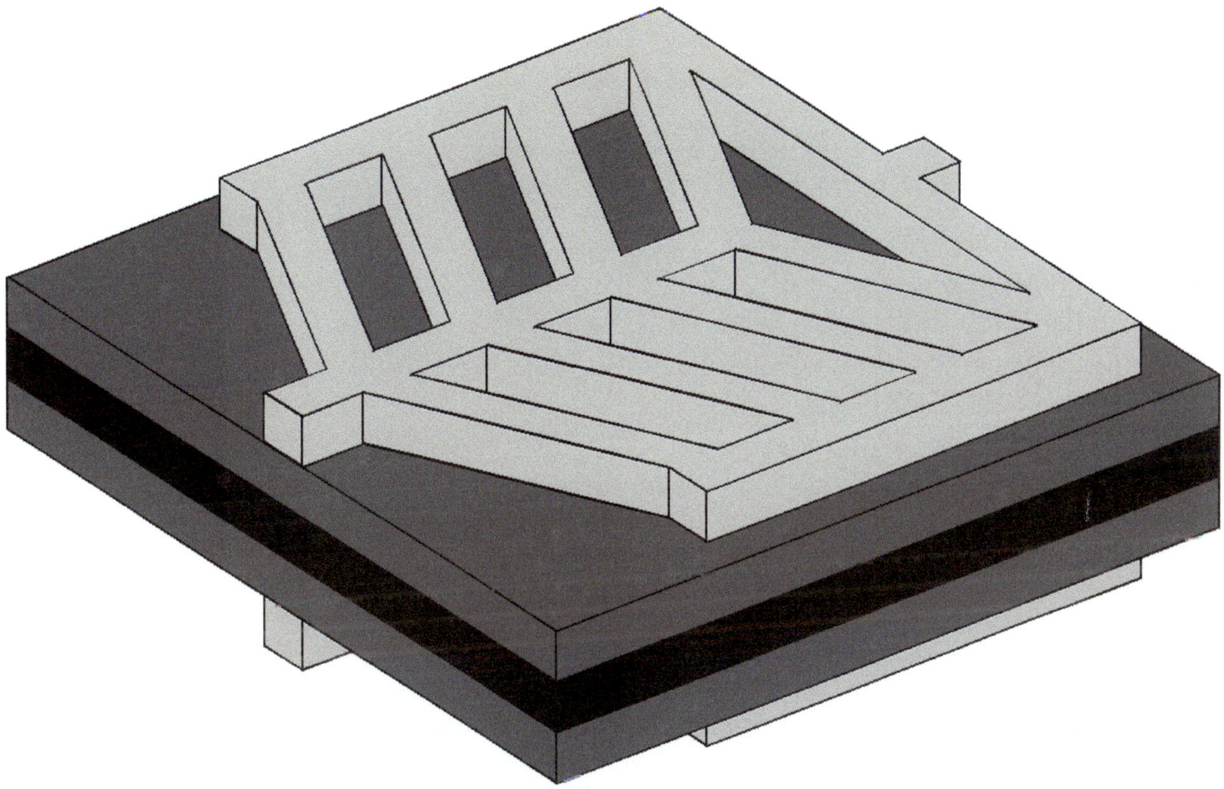

(b)

Figure 1. (a) Fluid distribution patterns in nature (leaf), and (b) three-dimensional computational domain of the micro PEM fuel cell with nature inspired gas flow channels.

$$\nabla \cdot \left(r_g \rho_g \mathbf{u}_g \right) = 0 \tag{1}$$

and for the liquid phase inside the channel becomes;

$$\nabla \cdot \left(r_l \rho_l \mathbf{u}_l \right) = 0 \tag{2}$$

where **u** is velocity vector [m/s], ρ is density [kg/m^3].

Two sets of momentum equations are solved in the channels, and they share the same pressure field. Under these conditions, it can be shown that the momentum equations becomes;

$$\nabla\cdot\left(\rho_g\mathbf{u}_g\otimes\mathbf{u}_g-\mu_g\nabla\mathbf{u}_g\right)=-\nabla r_g\left(P+\frac{2}{3}\mu_g\nabla\cdot\mathbf{u}_g\right)+\nabla\cdot\left[\mu_g\left(\nabla\mathbf{u}_g\right)^T\right] \tag{3}$$

$$\nabla\cdot\left(\rho_l\mathbf{u}_l\otimes\mathbf{u}_l-\mu_l\nabla\mathbf{u}_l\right)=-\nabla r_l\left(P+\frac{2}{3}\mu_l\nabla\cdot\mathbf{u}_l\right)+\nabla\cdot\left[\mu_l\left(\nabla\mathbf{u}_l\right)^T\right] \tag{4}$$

where P is pressure (Pa), μ is viscosity [kg/(m·s)].

The mass balance is described by the divergence of the mass flux through diffusion and convection. Multiple species are considered in the gas phase only, and the species conservation equation in multi-component, multi-phase flow can be written in the following expression for species i;

$$\nabla\cdot\begin{bmatrix}-r_g\rho_g y_i\sum_{j=1}^{N}D_{ij}\dfrac{M}{M_j}\left[\left(\nabla y_j+y_j\dfrac{\nabla M}{M}\right)+\left(x_j-y_j\right)\dfrac{\nabla P}{P}\right]+\\ r_g\rho_g y_i\cdot\mathbf{u}_g+D_i^T\dfrac{\nabla T}{T}\end{bmatrix}=0 \tag{5}$$

where T is temperature (K), y is mass fraction, x is mole fraction, D is diffusion coefficient [m^2/s]. Subscript i denotes oxygen at the cathode side and hydrogen at the anode side, and j is water vapour in both cases. Nitrogen is the third species at the cathode side.

The Maxwell-Stefan diffusion coefficients of any two species are dependent on temperature and pressure. They can be calculated according to the empirical relation based on kinetic gas theory [17];

$$D_{ij}=\frac{T^{1.75}\times10^{-3}}{P\left[\left(\sum_k V_{ki}\right)^{1/3}+\left(\sum_k V_{kj}\right)^{1/3}\right]^2}\left[\frac{1}{M_i}+\frac{1}{M_j}\right]^{1/2} \tag{6}$$

In this equation, the pressure is in atm and the binary diffusion coefficient Dij is in [cm^2/s].
The values for $\left(\sum V_{ki}\right)$ are given by Fuller et al. [17].

The temperature field is obtained by solving the convective energy equation;

$$\nabla\cdot\left(r_g\left(\rho_g Cp_g\mathbf{u}_g T-k_g\nabla T\right)\right)=0 \tag{7}$$

where Cp is specific heat capacity [J/(kg.K)], k is gas thermal conductivity [W/(m.K)].

The gas phase and the liquid phase are assumed to be in thermodynamic equilibrium; hence, the temperature of the liquid water is the same as the gas phase temperature.

2.2.2. Gas diffusion layers

The physics of multiple phases through a porous medium is further complicated here with phase change and the sources and sinks associated with the electrochemical reaction. The equations used to describe transport in the gas diffusion layers are given below. Mass transfer in the form of evaporation $\left(\dot{m}_{phase}>0\right)$ and condensation $\left(\dot{m}_{phase}<0\right)$ is assumed, so that the mass balance equations for both phases are;

$$\nabla\cdot\left(\left(1-sat\right)\rho_g\varepsilon\mathbf{u}_g\right)=\dot{m}_{phase} \tag{8}$$

$$\nabla\cdot\left(sat.\rho_l\varepsilon\mathbf{u}_l\right)=\dot{m}_{phase} \tag{9}$$

where sat is saturation, ε is porosity

The momentum equation for the gas phase reduces to Darcy's law, which is, however, based on the relative permeability for the gas phase (KP). The relative permeability accounts for the reduction in pore space available for one phase due to the existence of the second phase [18].

The momentum equation for the gas phase inside the gas diffusion layer becomes;

$$\mathbf{u}_g=-\left(1-sat\right)Kp.\nabla P/\mu_g \tag{10}$$

where KP is hydraulic permeability [m^2].

Two liquid water transport mechanisms are considered; shear, which drags the liquid phase along with the gas phase in the direction of the pressure gradient, and capillary forces, which drive liquid water from high to low saturation regions [18]. Therefore, the momentum equation for the liquid phase inside the gas diffusion layer becomes;

$$\mathbf{u}_l = -\frac{KP_l}{\mu_l}\nabla P + \frac{KP_l}{\mu_l}\frac{\partial P_c}{\partial sat}\nabla sat \tag{11}$$

where Pc is capillary pressure [Pa].

The functional variation of capillary pressure with saturation is prescribed following Leverett [18] who has shown that;

$$P_c = \tau\left(\frac{\varepsilon}{KP}\right)^{1/2}\left(1.417(1-sat) - 2.12(1-sat)^2 + 1.263(1-sat)^3\right) \tag{12}$$

where τ is surface tension [N/m].

The liquid phase consists of pure water, while the gas phase has multi components. The transport of each species in the gas phase is governed by a general convection-diffusion equation in conjunction which the Stefan-Maxwell equations to account for multi species diffusion;

$$\nabla\cdot\begin{bmatrix} -(1-sat)\rho_g \varepsilon y_i \sum_{j=1}^{N} D_{ij}\frac{M}{M_j}\left[\left(\nabla y_j + y_j\frac{\nabla M}{M}\right) + (x_j - y_j)\frac{\nabla P}{P}\right] + \\ (1-sat)\rho_g \varepsilon y_i \cdot \mathbf{u}_g + \varepsilon D_i^T\frac{\nabla T}{T} \end{bmatrix} = \dot{m}_{phase} \tag{13}$$

In order to account for geometric constraints of the porous media, the diffusivities are corrected using the Bruggemann correction formula [14];

$$D_{ij}^{eff} = D_{ij} \times \varepsilon^{1.5} \tag{14}$$

The heat transfer in the gas diffusion layers is governed by the energy equation as follows;

$$\nabla\cdot\left((1-sat)\left(\rho_g \varepsilon Cp_g \mathbf{u}_g T - k_{eff,g}\varepsilon\nabla T\right)\right) = \varepsilon\beta(T_{solid} - T) - \varepsilon\dot{m}_{phase}\Delta H_{evap} \tag{15}$$

where $keff$ is effective electrode thermal conductivity [W/(m·K)]; the term [εβ(Tsolid−T)], on the right hand side, accounts for the heat exchange to and from the solid matrix of the GDL. The gas phase and the liquid phase are assumed to be in thermodynamic equilibrium, i.e., the liquid water and the gas phase are at the same temperature.

The potential distribution in the gas diffusion layers is governed by;

$$\nabla\cdot(\lambda_e\nabla\phi) = 0 \tag{16}$$

where λe is electrode electronic conductivity [S/m].

In order to account for the magnitude of phase change inside the GDL, expressions are required to relate the level of over- and undersaturation as well as the amount of liquid water present to the amount of water undergoing phase change. In the present work, the procedure of Berning and Djilali [18] was used to account for the magnitude of phase change inside the GDL.

2.2.3. Catalyst layers

The catalyst layer is treated as a thin interface, where sink and source terms for the reactants are implemented. Due to the infinitesimal thickness, the source terms are actually implemented in the last grid cell of the porous medium. At the cathode side, the sink term for oxygen is given by;

$$S_{O_2} = -\frac{M_{O_2}}{4F}i_c \tag{17}$$

where F is Faraday's constant (96487 [C/mole]), i_c is cathode local current density [A/m^2], M is molecular weight [kg/mole].

Whereas the sink term for hydrogen is specified as;

$$S_{H_2} = -\frac{M_{H_2}}{2F} i_a \tag{18}$$

where i_a is anode local current density [A/m^2]

The production of water is modelled as a source terms, and hence can be written as;

$$S_{H_2O} = \frac{M_{H_2O}}{2F} i_c \tag{19}$$

The generation of heat in the cell is due to entropy changes as well as irreversibility's due to the activation overpotential [19];

$$\dot{q} = \left[\frac{T(-\Delta s)}{n_e F} + \eta_{act} \right] i \tag{20}$$

where $\dot{q}$ is heat generation [W/m^2], ne is number of electrons transfer, s is specific entropy [J/(mole.K)], ηact is activation overpotential (V).

The local current density distribution in the catalyst layers is modelled by the Butler-Volmer equation [18, 19];

$$i_c = i_{o,c}^{ref} \left(\frac{C_{O_2}}{C_{O_2}^{ref}} \right) \left[\exp\left(\frac{\alpha_a F}{RT} \eta_{act,c} \right) + \exp\left(-\frac{\alpha_c F}{RT} \eta_{act,c} \right) \right] \tag{21}$$

$$i_a = i_{o,a}^{ref} \left(\frac{C_{H_2}}{C_{H_2}^{ref}} \right)^{1/2} \left[\exp\left(\frac{\alpha_a F}{RT} \eta_{act,a} \right) + \exp\left(-\frac{\alpha_c F}{RT} \eta_{act,a} \right) \right] \tag{22}$$

where C_{H_2} is local hydrogen concentration [mole/m^3], $C_{H_2}^{ref}$ is reference hydrogen concentration [mole/m^3], C_{O_2} is local oxygen concentration [mole/m^3], $C_{O_2}^{ref}$ is reference oxygen concentration [mole/m^3], $i_{o,a}^{ref}$ is anode reference exchange current density, $i_{o,c}^{ref}$ is cathode reference exchange current density, R is universal gas constant (8.314 [J/(mole·K)]), αa is charge transfer coefficient, anode side, and αc is charge transfer coefficient, cathode side.

2.2.4. Membrane

The balance between the electro-osmotic drag of water from anode to cathode and back diffusion from cathode to anode yields the net water flux through the membrane;

$$N_W = n_d M_{H_2O} \frac{i}{F} - \nabla \cdot (\rho D_W \nabla c_W) \tag{23}$$

where Nw is net water flux across the membrane [kg/(m^2·s)], nd is electro-osmotic drag coefficient.

The water diffusivity in the polymer can be calculated as follow [12];

$$D_W = 1.3 \times 10^{-10} \exp\left[2416 \left(\frac{1}{303} - \frac{1}{T} \right) \right] \tag{24}$$

The variable c_W represents the number of water molecules per sulfonic acid group (i.e. mol H_2O/equivalent SO_3^{-1}). The water content in the electrolyte phase is related to water vapour activity via [6, 7];

$$\begin{aligned} c_W &= 0.043 + 17.81a - 39.85a^2 + 36.0a^3 && (0 < a \leq 1) \\ c_W &= 14.0 + 1.4(a - 1) && (1 < a \leq 3) \\ c_W &= 16.8 && (a \geq 3) \end{aligned} \tag{25}$$

The water vapour activity a given by;

$$a = x_W P / P_{sat} \tag{26}$$

Heat transfer in the membrane is governed by;

$$\nabla \cdot \left(k_{mem} \cdot \nabla T\right) = 0 \tag{27}$$

where k_{mem} is membrane thermal conductivity [W/(m·K)].

The potential loss in the membrane is due to resistance to proton transport across membrane, and is governed by;

$$\nabla \cdot \left(\lambda_m \nabla \phi\right) = 0 \tag{28}$$

where λ_m is membrane ionic conductivity [S/m].

2.2.5. Potential drop across the cell

Useful work (electrical energy) is obtained from a fuel cell only when a current is drawn, but the actual cell potential, (E_{cell}), is decreased from its equilibrium thermodynamic potential, (E), because of irreversible losses. The various irreversible loss mechanisms which are often called overpotentials, (η), are defined as the deviation of the cell potential, (E_{cell}), from the equilibrium thermodynamic potential (E). The cell potential (E_{cell}) is obtained by subtracting all overpotentials (η) (losses) from the equilibrium thermodynamic potential (E) as the following expression;

$$E_{cell} = E - \eta_{act} - \eta_{ohm} - \eta_{mem} - \eta_{Diff} \tag{29}$$

The equilibrium potential is determined using the Nernst equation [16];

$$E = 1.229 - 0.83 \times 10^{-3}\left(T - 298.15\right) + 4.3085 \times 10^{-5} T\left[\ln\left(P_{H_2}\right) + \frac{1}{2}\ln\left(P_{O_2}\right)\right] \tag{30}$$

2.2.5.1. Activation Overpotential

Activation overpotential (η_{act}) arises from the kinetics of charge transfer reaction across the electrode-electrolyte interface. In other words, a portion of the electrode potential is lost in driving the electron transfer reaction. Activation overpotential is directly related to the nature of the electrochemical reactions and represents the magnitude of activation energy, when the reaction propagates at the rate demanded by the current. The activation overpotential can be divided into the anode and cathode overpotentials. The anode and cathode activation overpotentials are calculated from Butler-Volmer equation (21 and 22).

2.2.5.2. Ohmic Overpotential in Gas Diffusion Layers

The Ohmic overpotential (η_{ohm}) is the potential loss due to current conduction through the anode and cathode gas diffusion layers, and can be modeled by equation (16).

2.2.5.3. Membrane Overpotential

The membrane overpotential (η_{mem}) is related to the fact that an electric field is necessary in order to maintain the motion of the hydrogen protons through the membrane. This field is provided by the existence of a potential gradient across the cell, which is directed in the opposite direction from the outer field that gives us the cell potential, and thus has to be subtracted. The overpotential in membrane is calculated from the potential equation (28).

2.2.5.4. Diffusion Overpotential

Diffusion overpotential (η_{Diff}) is caused by mass transfer limitations on the availability of the reactants near the electrodes. The electrode reactions require a constant supply of reactants in order to sustain the current flow. When the diffusion limitations reduce the availability of a reactant, part of the available reaction energy is used to drive the mass transfer, thus creating a corresponding loss in output voltage. Similar problems can develop if a reaction product accumulates near the electrode surface and obstructs the diffusion paths or dilutes the reactants. Mass transport loss becomes significant when the fuel cell is operated at high current density. This is created by the concentration gradient due to the consumption of oxygen or fuel at the electrodes. The mass transport loss at the anode is negligible compared to that at the cathode. At the limiting current density, oxygen at the catalyst layer is depleted and no more current increase can be obtained from the fuel cell. This is responsible for the sharp decline in potential at high current densities. To reduce mass transport loss, the cathode is usually run at high pressure. The anode and cathode diffusion overpotentials are calculated from the following equations [19];

$$\eta_{Diff,c} = \frac{RT}{2F}\ln\left(1-\frac{i_c}{i_{L,c}}\right) \tag{31}$$

$$\eta_{Diff,a} = \frac{RT}{2F}\ln\left(1-\frac{i_a}{i_{L,a}}\right) \tag{32}$$

$$i_{L,c} = \frac{2FD_{O_2}C_{O_2}}{\delta_{GDL}} \tag{33}$$

$$i_{L,a} = \frac{2FD_{H_2}C_{H_2}}{\delta_{GDL}} \tag{34}$$

where $i_{L,c}$ and $i_{L,a}$ are cathode and anode local limiting current density [A/m^2], δ_{GDL} is the gas diffusion layer thickness [m], D is the diffusion coefficients [m^2/s].
The diffusivity of oxygen and hydrogen are calculated from the following equations [19];

$$D_{O_2} = 3.2\times10^{-5}\left(\frac{T}{353}\right)^{3/2}\left(\frac{101325}{P}\right) \tag{35}$$

$$D_{H_2} = 1.1\times10^{-4}\left(\frac{T}{353}\right)^{3/2}\left(\frac{101325}{P}\right) \tag{36}$$

2.2.6. Cell power and efficiency

Once the cell potential is determined for a given current density, the output power density is found as;

$$W_{cell} = I.E_{cell} \tag{37}$$

where I is the cell operating (nominal) current density (A/m^2).
The thermodynamic efficiency of the cell can be determined as [19];

$$E_{fc} = \frac{2E_{cell}F}{M_{H_2}.LHV_{H_2}} \tag{38}$$

where LHV_{H2} is the lower heating value of hydrogen [J/kg]

2.3. Boundary conditions

Boundary conditions are specified at all external boundaries of the computational domain as well as boundaries for various mass and scalar equations inside the computational domain.

2.3.1. Inlets

The inlet values at the anode and cathode are prescribed for the velocity, temperature and species concentrations (Dirichlet boundary conditions). The inlet velocity is a function of the desired current density, the geometrical area of the membrane, the channel cross-section area, and stoichiometric flow ratio. The inlet velocities of air and fuel are calculated according to [20]:

$$u_{in,c} = \zeta_c \frac{I}{4F} A_{MEA} \frac{1}{x_{O_2,in}} \frac{RT_{in,c}}{P_c} \frac{1}{A_{ch}} \tag{39}$$

$$u_{in,a} = \zeta_a \frac{I}{2F} A_{MEA} \frac{1}{x_{H_2,in}} \frac{RT_{in,a}}{P_a} \frac{1}{A_{ch}} \tag{40}$$

where A_{MEA} is the area of the MEA [m^2], A_{ch} is the cross sectional area of flow channel [m^2], and ξ is the stoichiometric flow ratio.

2.3.2. Outlets

At the outlets of the gas-flow channels, only the pressure is being prescribed as the desired electrode pressure; for all other variables, the gradient in the flow direction is assumed to be zero (Neumann boundary conditions).

2.3.3. External surfaces

At the external surfaces of the cell, temperature is specified and zero heat flux is applied at the surface of the conducting boundary surfaces.

2.3.4. Interfaces inside the computational domain

Combinations of Dirichlet and Neumann boundary conditions are used to solve the electronic and protonic potential equations. Dirichlet boundary conditions are applied at the land area (interface between the bipolar plates and the gas diffusion layers). Neumann boundary conditions are applied at the interface between the gas channels and the gas diffusion layers to give zero potential flux into the gas channels. Similarly, the protonic potential field requires a set of potential boundary condition and zero flux boundary condition at the anode catalyst layer interface and cathode catalyst layer interface respectively.

2.4. Computational procedure

2.4.1. Computational grid

The governing equations were discretized using a finite volume method and solved using a multi-physics general-purpose computational fluid dynamic (CFD) code. The computational domains are divided into a finite number of control volumes (cells). All variables are stored at the centroid of each cell. Interpolation is used to express variable values at the control volume surface in terms of the control volume center values. Stringent numerical tests were performed to ensure that the solutions were independent of the grid size. A computational quadratic mesh consisting of a total of 11556 nodes and 59814 meshes was found to provide sufficient spatial resolution (Figure 2). The coupled set of equations was solved iteratively, and the solution was considered to be convergent when the relative error was less than 1.0×10^{-6} in each field between two consecutive iterations. The geometric and the base case operating conditions are listed in Table 1. Values of the electrochemical transport parameters for the base case operating conditions are taken from reference [20] and are listed in Table 2.

2.4.2. Solution algorithm

The solution begins by specifying a desired current density of the cell to be used for calculating the inlet flow rates at the anode and cathode sides. An initial guess of the activation overpotential is obtained from the desired current density using the Butler-Volmer equation. Then follows by computing the flow fields for velocities u,v,w, and pressure P. Once the flow field is obtained, the mass fraction equations are solved for the mass fractions of oxygen, hydrogen, water vapor, and nitrogen. Scalar equations are solved last in the sequence of the transport equations for the temperature field in the cell and potential fields in the gas diffusion layers and the membrane. The local current densities are solved based on the Butler-Volmer equation. After the local current densities are obtained, the local activation overpotentials can be readily calculated from the Butler-Volmer equation. The local activation overpotentials are updated after each global iterative loop. Convergence criteria are then performed on each variable and the procedure is repeated until convergence. The properties and then source terms are updated after each global iterative loop based on the new local gas composition and temperature. The strength of the current model is clearly to perform parametric studies and explore the impact of various parameters on the transport mechanisms and on fuel cell performance. The new feature of the algorithm developed in this work is its capability for accurate calculation of the local activation overpotentials, which in turn results in improved prediction of the local current density distribution. The flow diagram of the algorithm is shown in Figure 3.

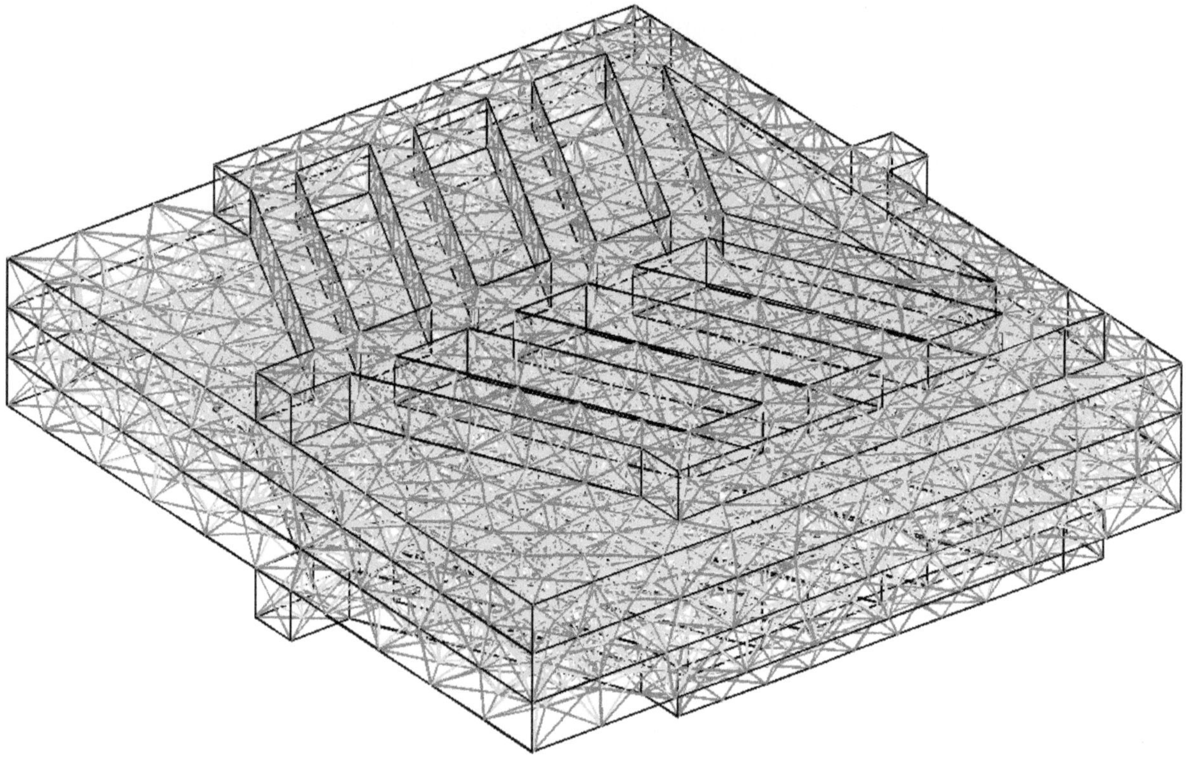

Figure 2. Computational mesh for micro PEM fuel cell with nature inspired gas flow channels.

Table 1. Geometrical and operational parameters for base case conditions.

Parameter	Symbol	Value
Cell length	L_{cell}	5000 μm
Cell width	W_{cell}	5000 μm
Gas channel width	W	250 μm
Gas channel height	H	250 μm
Gas diffusion layer thickness	δ_{GDL}	260 μm
Wet membrane thickness (Nafion® 117)	δ_{mem}	50 μm
Catalyst layer thickness	δ_{CL}	28.7 μm
Hydrogen reference mole fraction	$x_{H_2}^{ref}$	0.84639
Oxygen reference mole fraction	$x_{O_2}^{ref}$	0.17774
Anode pressure	P_a	1e5 Pa
Cathode pressure	P_c	1e5 Pa
Inlet fuel and air temperature	T_{cell}	323.15 K
Relative humidity of inlet fuel and air	ψ	100 %
Air stoichiometric flow ratio	ξ_c	2
Fuel stoichiometric flow ratio	ξ_a	2

Table 2. Electrode and membrane parameters for base case operating conditions.

Parameter	Symbol	Value
Electrode porosity	ε	0.4
Electrode electronic conductivity	λ_e	100 S/m
Membrane ionic conductivity (Nafion®117)	λ_m	17.1223 S/m
Transfer coefficient, anode side	α_a	0.5
Transfer coefficient, cathode side	α_c	1
Cathode reference exchange current density	$i_{o,c}^{ref}$	1.8081e-3 A/m^2
Anode reference exchange current density	$i_{o,a}^{ref}$	2465.598 A/m^2
Electrode thermal conductivity	k_{eff}	1.3 W/m.K
Membrane thermal conductivity	k_{mem}	0.455 W/m.K
Electrode hydraulic permeability	kp	1.76e-11 m^2
Entropy change of cathode side reaction	ΔS	-326.36 J/mol.K
Heat transfer coef. between solid and gas phase	β	4e6 W/m^3
Protonic diffusion coefficient	D_{H^+}	4.5e-9 m^2/s
Fixed-charge concentration	c_f	1200 mol/m^3
Fixed-site charge	z_f	-1
Electro-osmotic drag coefficient	n_d	2.5
Droplet diameter	D_{drop}	1e-8 m
Condensation constant	C	1e-5
Scaling parameter for evaporation	ϖ	0.01

3. Results and discussion

In order to gain some insight into the micro-structured PEM fuel cell, the oxygen and hydrogen distribution, local current densities, and temperature distribution are plotted in Figures 4-8, respectively, at 0.3 A/cm^2 nominal current density.

The velocity profiles in the mid-plane of the cathode (red) and anode (green) gas flow channels are shown in Figure 4. The velocity exhibits a laminar fully developed profile where the highest velocity is located at the centre of the channels and this profile is maintained until the end of the channels. However, the velocity scale in the anode gas flow channel is several orders of magnitude smaller than that in the cathode gas flow channel. This is because the molar fraction of the hydrogen of the incoming humid gas is greater than that of oxygen, and hence, the inlet velocity at the anode gas flow channel is smaller than that at cathode gas flow channel.

The detailed distribution of oxygen weight fraction is shown in Figures 5 and 6. In the GDL, oxygen concentration under the land area is smaller than that under the air inlet area. The concentration of oxygen at the catalyst layer is balanced by the oxygen that is being consumed and the amount of oxygen that diffuses towards the catalyst layer driven by the concentration gradient. The lower diffusivity of the oxygen along with the low concentration of oxygen in ambient air results in noticeable oxygen depletion under the land areas. However, the decrease in molar concentration of the hydrogen under the land areas is smaller than for the oxygen in cathode side due to the higher diffusivity of the hydrogen.

Due to the relatively low diffusivity of the oxygen compared with that of the hydrogen, the cathode operation conditions usually determine the limiting current density. This is because an increase in current density corresponds to an increase in oxygen consumption.

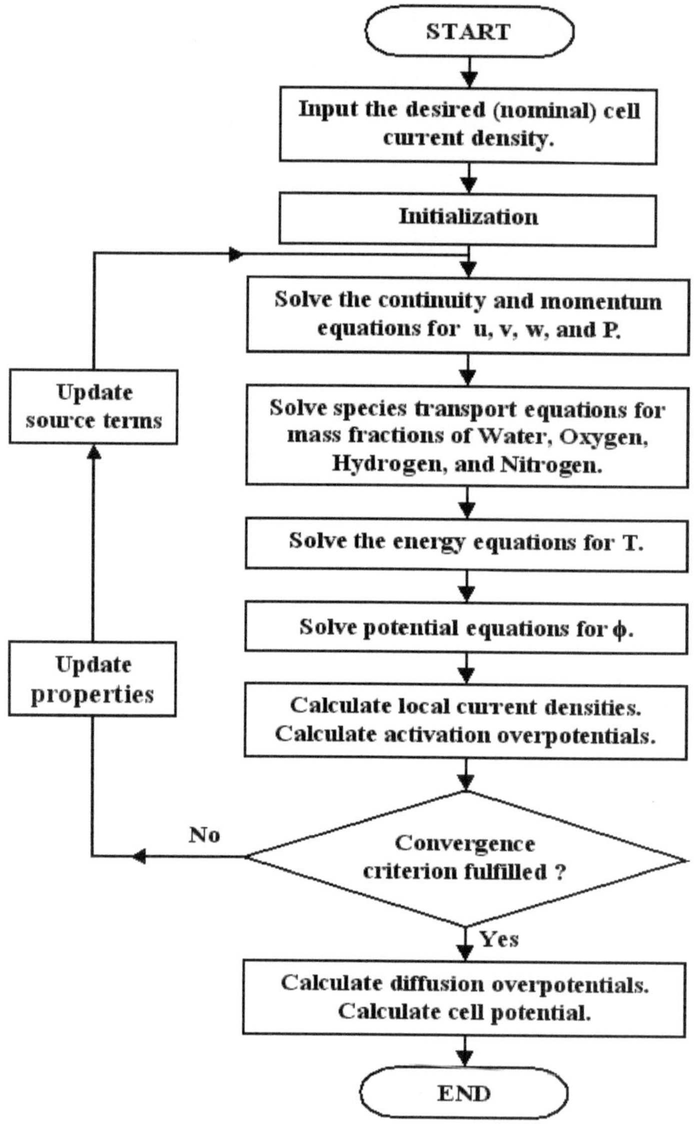

Figure 3. Flow diagram of the solution procedure used.

Figure 7 shows the local current density distribution at the cathode side catalyst layer. The local current densities have been normalized by the nominal current density in each case (i.e. ic/I). The local current density of the cathode side reaction depends directly on the oxygen concentration. The diffusion of the oxygen towards the catalyst layer is the main impediment for reaching high current densities. Therefore, it can be seen that the local current density under the land areas a noticeable decrease takes place. A high fraction of the current is generated at the catalyst layer that lies beneath the air inlet area, leading to under-utilization of the catalyst under the land areas. This can lead to local hot spots inside the membrane electrode assembly. These hot spots can lead to a further drying out of the membrane, thus increasing the electric resistance, which in turn leads to more heat generation and can lead to a failure of the membrane. Thus, it is important to keep the current density relatively even throughout the cell. For optimal fuel cell performance, a uniform current density generation is desirable, and this could only be achieved with a non-uniform catalyst distribution, possibly in conjunction with non-homogeneous gas diffusion layers.

The temperature distribution inside the fuel cell has important effects on nearly all transport phenomena, and knowledge of the magnitude of temperature increases due to irreversibilities might help prevent failure. Figure 8 shows the distribution of the temperature at the cathode side catalyst layer. The temperature peak appears in the cathode catalyst layer near the inlet area, implying that major heat generation takes place in the region. For an optimum fuel cell performance, and in order to avoid large temperature gradients inside the fuel cell, it is desirable to achieve a uniform current density distribution inside the cell.

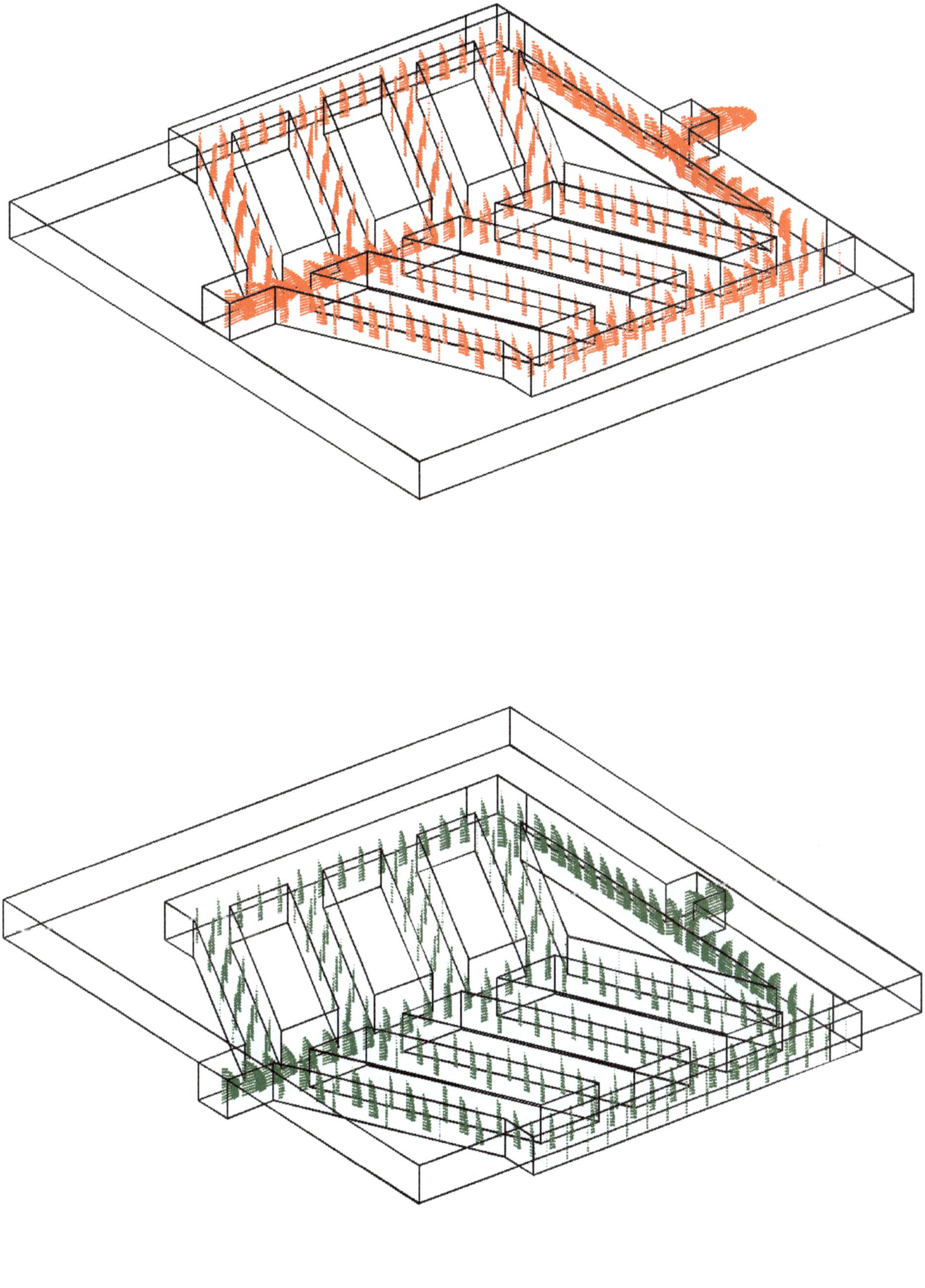

Figure 4. Velocity profiles in the mid-plane of the cathode (red) and anode (green) gas flow channels.

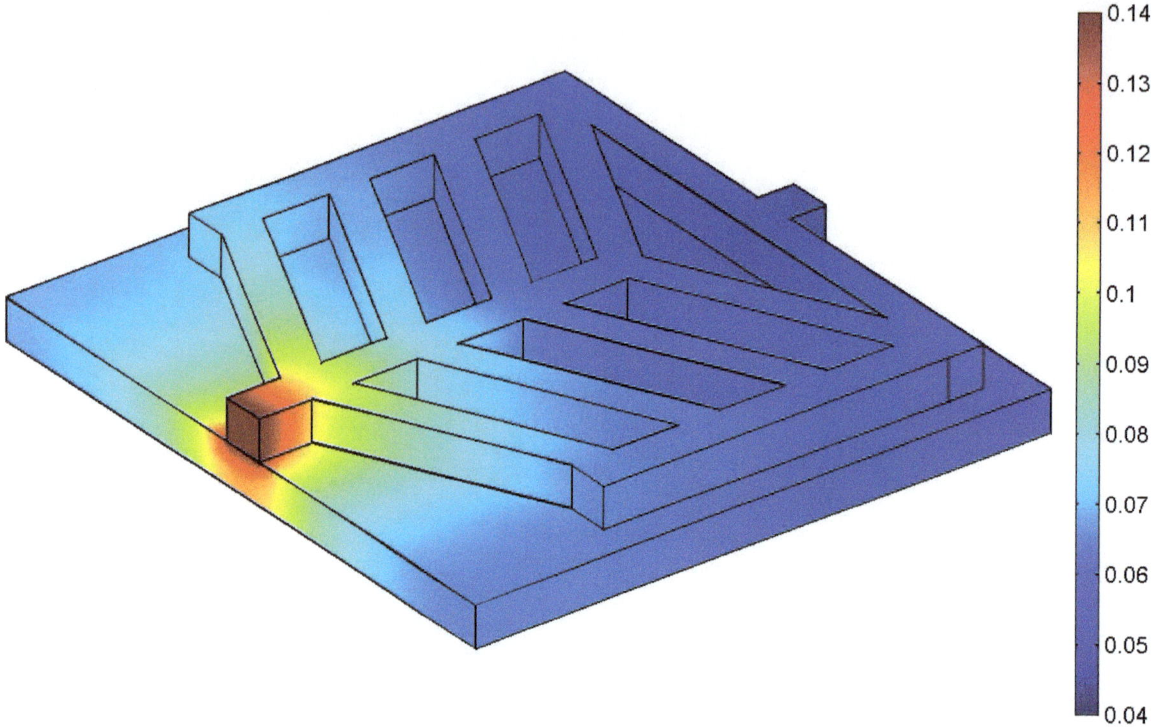

Figure 5. Oxygen weight fraction distribution.

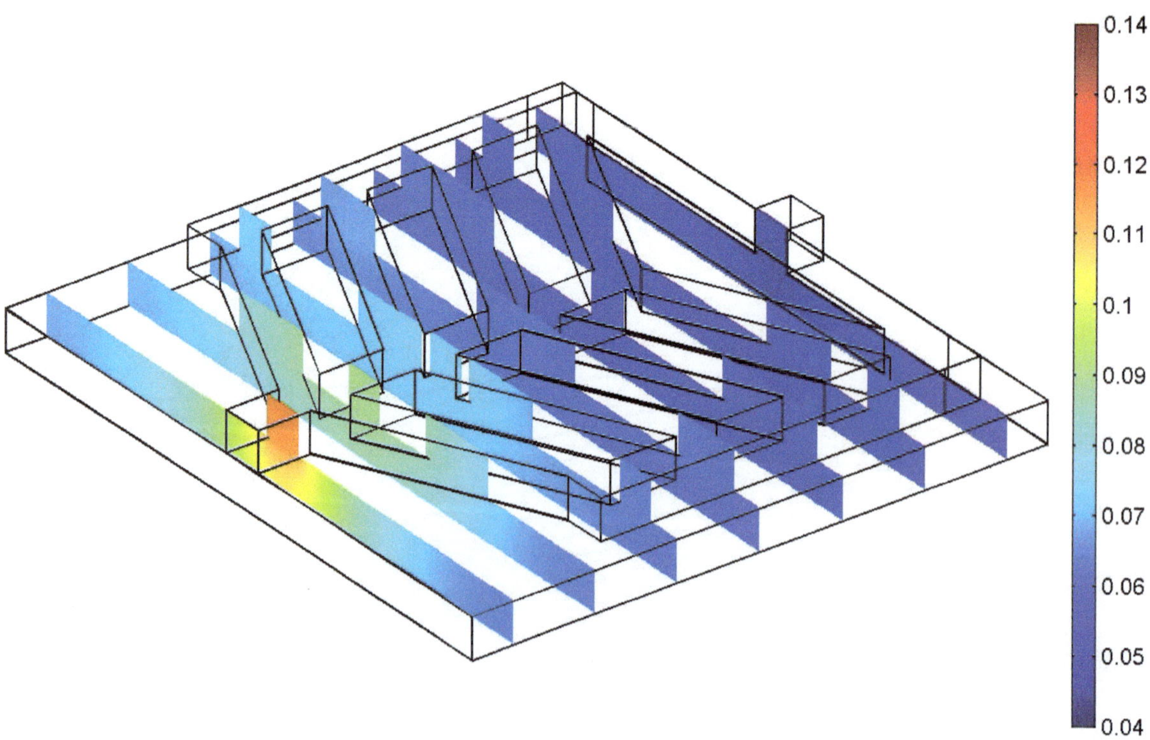

Figure 6. Oxygen weight fraction distribution..

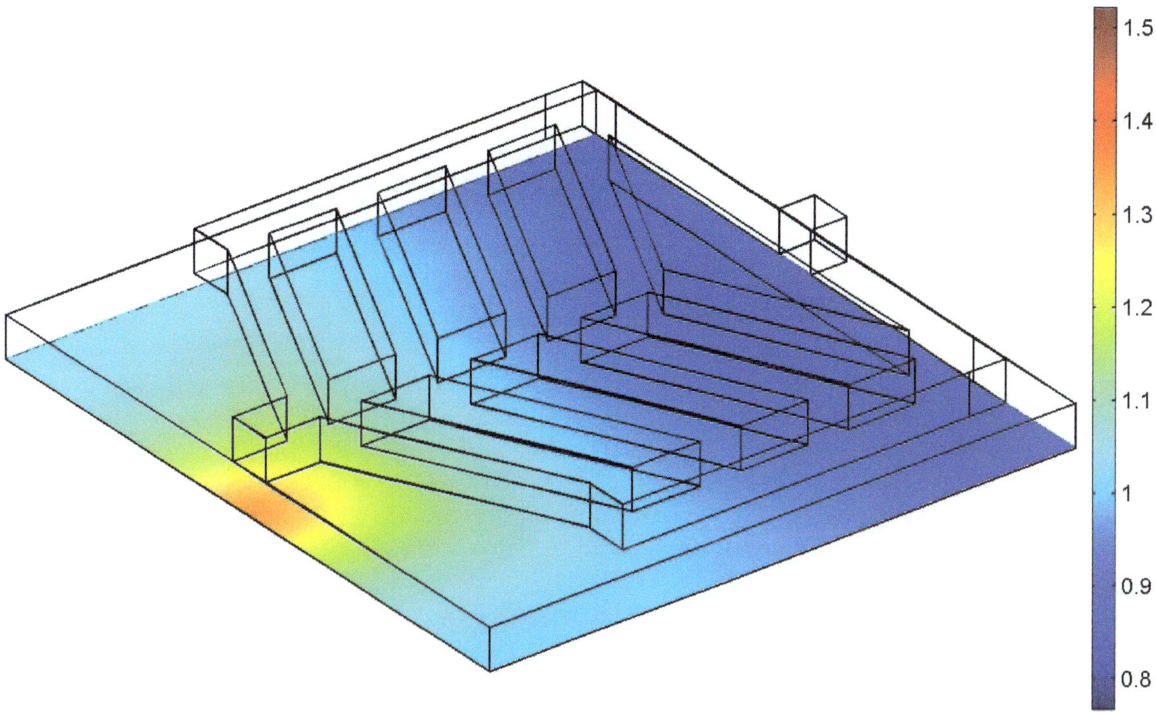

Figure 7. Dimensionless local current density distribution (ic/I) at cathode side catalyst layer.

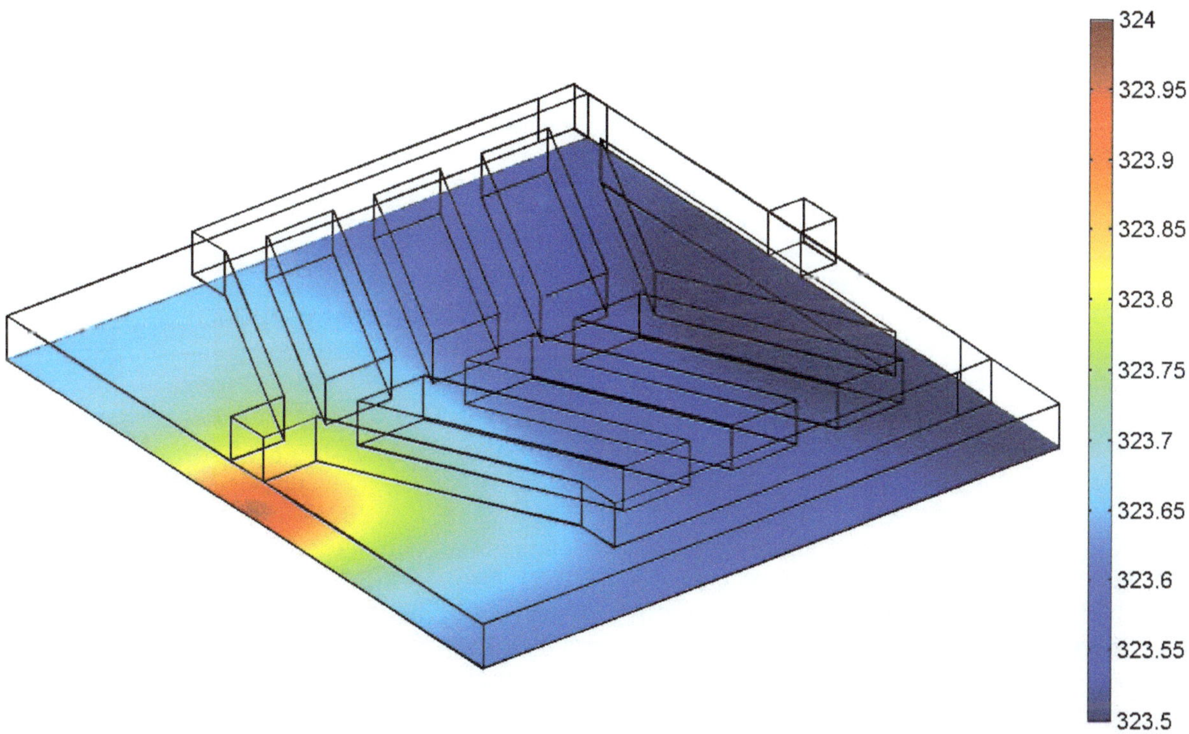

Figure 8. Temperature distribution at the cathode side catalyst layer.

4. Conclusion

Every component of the micro-fuel cells plays important role in their performance. Full three-dimensional, non-isothermal computational fluid dynamics model of all components of a micro proton exchange membrane (PEM) fuel cell with nature inspired gas flow channels has been developed and used to investigate the cell performance. This comprehensive model accounts for the major transport phenomena such as convective and diffusive heat and mass transfer, electrode kinetics, transport and phase-change mechanism of water, and potential fields in a micro PEM fuel cell with nature inspired gas flow channels. The model is shown to be able to: understand the many interacting, complex electrochemical and transport phenomena that cannot be studied experimentally; identify limiting steps and components; and provide a computer-aided tool for the design and optimization of future fuel cells to improve their lifetime with a much higher power density and lower cost.

The analysis offers valuable physical insight towards design of a cell and a cell stack, to be considered in a future study. Finally, the development of physically representative models that allow reliable simulation of the processes under realistic conditions is essential to the development and optimization of micro fuel cells, improve long-term performance and lifetime, the introduction of cheaper materials and fabrication techniques, and the design and development of novel architectures.

References

[1] Jeng, K.T., Kuo, C.P. and Lee, S.F. Modeling the catalyst layer of a PEM fuel cell cathode using a dimensionless approach. J. Power Sources, 2004, 128(2), 145–151.

[2] Jaouen, F., Lindbergh, G., and Sundholm, G. Investigation of mass-transport limitations in the solid polymer fuel cell cathode. J. Electrochem. Soc., 2002, 149(4), A437-A447.

[3] Grujicic, M., and Chittajallu, K.M. Optimization of the cathode geometry in polymer electrolyte membrane (PEM) fuel cells. Chemical Engineering Science, 2004, 59(24), 5883–5895.

[4] Grujicic, M., Zhao, C.L., Chittajallu, K.M., and Ochterbeck, J.M. Cathode and interdigitated air distributor geometry optimization in polymer electrolyte membrane (PEM) fuel cells. Materials Science and Engineering B, 2004, 108(3), 241–252.

[5] Francesco,De., Arato,E., Costa,P. Transport phenomena in membranes of PEMFC applications: an analytical approach to the calculation of membrane resistance. J.P. Sources, 2004, 132(1-2), 127-134.

[6] Maher A.R. Sadiq Al-Baghdadi. Modeling optimizes PEM fuel cell durability using three-dimensional multi-phase computational fluid dynamics model. International Journal of Energy and Environment IJEE, 2010; 1(3), 375-398.

[7] Maher A.R. Sadiq Al-Baghdadi. Mechanical behaviour of PEM fuel cell catalyst layers during regular cell operation. International Journal of Energy and Environment, 2010; 1(6), 927-936.

[8] Maher A.R. Sadiq Al-Baghdadi. A CFD analysis on the effect of ambient conditions on the hygro-thermal stresses distribution in a planar ambient air-breathing PEM fuel cell. International Journal of Energy and Environment, 2011; 2(4), 589-604.

[9] F. Brèque, J. Ramousse, Y. Dubé, K. Agbossou, P. Adzakpa. Sensibility study of flooding and drying issues to the operating conditions in PEM Fuel Cells. International Journal of Energy and Environment IJEE, 2010; 1(1), 1-20.

[10] Maher A.R. Sadiq Al-Baghdadi. Optimal design of PEM fuel cells to generate maximum power: A CFD study. International Journal of Energy and Environment, 2011; 2(6), 953-962.

[11] Maher A.R. Sadiq Al-Baghdadi. Novel design of a compacted micro-structured air-breathing PEM fuel cell as a power source for mobile phones. Int. J. of Energy and Environment IJEE, 2010; 1(4), 555-572.

[12] Siegel,N.P., Ellis,M.W., Nelson,D.J., Spakovsky,M.R. von. A two-dimensional computational model of a PEMFC with liquid water transport. J. Power Sources, 2004, 128(2), 173–184.

[13] Berning, T., Lu, D.M., and Djilali, N. Three-Dimensional Computational Analysis of Transport Phenomena in a PEM Fuel Cell. J. Power Sources, 2002, 106(1-2), 284-294.

[14] Nguyen, P.T., Berning, T., and Djilali, N. Computational Model of a PEM Fuel Cell with Serpentine Gas Flow Channels. J. Power Sources, 2004, 130(1-2), 149-157.

[15] Sukkee, Um., and Wang, C.Y. Three-dimensional analysis of transport and electrochemical reactions in polymer electrolyte fuel cells. J. Power Sources, 2004, 125(2), 40–51.

[16] Maher A.R. Sadiq Al-Baghdadi. A CFD analysis of transport phenomena and electrochemical reactions in a tubular-shaped PEM fuel cell. Int. J. of Energy and Environment, 2013; 4(1), 1-26.

[17] Fuller, E.N., Schettler, P.D., and Giddings, J.C. A new methode for prediction of binary gas-phase diffusion coefficients. Ind. Eng. Chem., 1966, 58(5), 18-27.

[18] Berning, T., and Djilali, N. A 3D, Multi-Phase, Multicomponent Model of the Cathode and Anode of a PEM Fuel Cell. J. Electrochem. Soc., 2003, 150(12), A1589-A1598.

[19] Maher A.R. Sadiq Al-Baghdadi. Novel design of a disk-shaped compacted micro-structured air-breathing PEM fuel cell. Int. Journal of Energy and Environment, 2012; 3(2), 161-180.

[20] Maher A.R. Sadiq Al-Baghdadi. Analysis of Transport Phenomena and Electrochemical Reactions in a Micro PEM Fuel Cell. International Journal of Energy and Environment, 2014; 5(1), 1-28.

7

Analytical performance investigation of parabolic trough solar collector with computed optimum air gap

Devander Kumar[1], Sudhir Kumar[2]

[1] The Technological Institute of Textile & Sciences, Bhiwani, Haryana-127021, India.
[2] National Institute of Technology, Kurukshetra, Haryana-136118, India.

Abstract
Parabolic trough collectors have a wide range of industrial as well as domestic applications. This study analytically investigates the performance of non-evacuated absorber tube with glass cover of parabolic trough collector (PTC) in terms of overall heat loss from the absorber. The impact of different parameters such as diameter of absorber tube, mean temperature of absorber tube, wind velocity, emissivity of absorber and ambient temperature have been studied and find the optimum value of air gap. The optimum value of air gap has been computed considering one-dimensional steady state model, where heat loss due to convection plus radiation is equal to heat loss due to conduction plus radiation from absorber tube to glass cover under steady state. Optimum air gap is found to be approximately 7mm and 8 mm for an absorber tube of diameter in range of 1.2-3.18cm and 4.5-7.62cm, respectively. Corresponding to optimum air gap, minimum overall heat loss has been observed. Overall heat loss increases with increase in absorber temperature, wind velocity and emissivity of absorber, whereas decreases with increase in air temperature for different absorber tube diameters. Absorber tube with diameters in the range of 3.18-4.5cm gives better performance. From the obtained data, correlations have been developed, which can be further utilized for designing the PTC system for getting desired output.

Keywords: Parabolic trough collector; Absorber tube; Glass cover; Optimum air gap; Overall heat loss.

1. Introduction

Solar energy is a permanent, environmental friendly and sustainable energy source, which can play a vital role in fulfilling the escalating energy demand and save the depletion of fossil fuel resources. Among different types of solar thermal systems [1] parabolic trough collector (PTC) is receiving much attention, despite of the requirement of solar tracking. Generally, PTC are employed for a wide range of applications from domestic hot water production [2, 3] to steam generation for power [4, 5], industrial process heat generation [6, 7] and air conditioning.

The major advantages associated with PTC are low-pressure drop and achieved temperature about 300-400°C without significant loss in the efficiency of collector. Solar Electric Generation system (SEGS) plant at Kramer Junction in California clearly illustrated that the solar thermal power plants based on PTC are currently the most successful solar technology for electricity generation [8].

Parabolic trough collector (Figure 1) consists of a parabolic reflector to reflect solar energy in to an absorber tube, which is placed at its focal point. Solar radiation is mainly absorbed at the outer surface in form of heat and further transferred partially to working fluid inside the absorber tube by conduction through tube wall and convection from inner surface of the tube to flowing fluid. Remaining transferred by combined effect of radiation, convection or conduction to the inner surface of glass cover through air and then further by conduction from inner surface to outer surface of the glass cover. The heat dissipated to surrounding by two mechanisms, convection to surrounding air and by radiation to surrounding surfaces [9]. Thus, instantaneous thermal efficiency of a PTC is given by [10]:

$$\text{Thermal efficiency} = \text{Optical efficiency} - \frac{\text{Overall heat loss}}{\text{Normal radiation incident on collector aperture}} \quad (1)$$

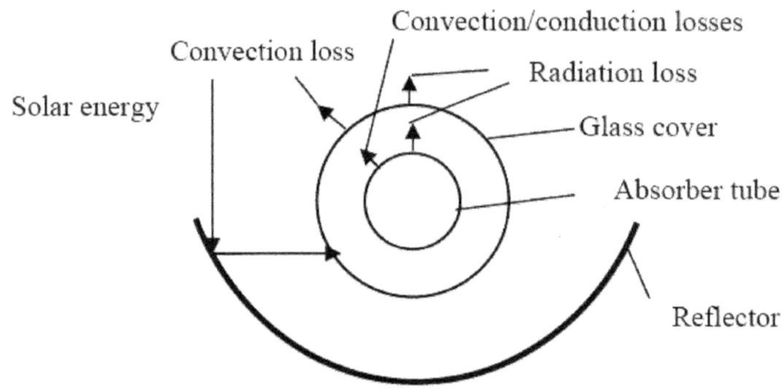

Figure 1. Configuration of Parabolic trough collector

From equation (1), it is clear that the overall heat loss through absorber tube is the prime factor to determine thermal efficiency of PTC system.

Under steady state condition, the overall heat loss from absorber tube is estimated by the correlations (2) or (3) [11-13],

$$Q_L = \frac{\pi \times D \times \sigma \times \left(T_{ab}^4 - T_c^4\right)}{\frac{1}{\in_{ab}} + \frac{D}{D_{ci}}\left(\frac{1}{\in_c} - 1\right)} + \pi \times D \times h_{cab-c} \times \left(T_{ab} - T_c\right) \quad (2)$$

$$Q_L = \frac{\pi \times D \times \sigma \times \left(T_{ab}^4 - T_c^4\right)}{\frac{1}{\in_{ab}} + \frac{D}{D_{ci}}\left(\frac{1}{\in_c} - 1\right)} + \frac{2\pi \times K_{air} \times \left(T_{ab} - T_c\right)}{\ln \frac{D_{ci}}{D}} \quad (3)$$

The overall heat loss from the glass cover is given by equation (4) as [13],

$$Q_L = \pi \times D_{co} \times \sigma \times \in_c \times (T_c^4 - T_{sky}^4) + \pi \times D_{co} \times h_{cc-a} \times (T_c - T_a) \quad (4)$$

Literature, suitably established that depending upon air gap (y) between absorber and glass cover, the overall heat loss can be computed using either equation (2) or (3) with equation (4). When convective heat transfer predominates the conduction heat transfer, then the overall heat loss (Q_L) is estimated using equations (2) and (4) otherwise equations (3) and (4) is used.

Out of different parameters, air gap is considered as the most crucial that affects the thermal performance of PTC. Treadwell [14] suggested 1cm annulus gap as the optimum gap between absorber and glass envelope, and based upon that the selection of glass tube sizes have been carried out. The sensitivity of PTC performance with change collector parameters and operating conditions has been carried out by Rabl et al. [15] and reported the optimal gap size of 0.7cm corresponding to an inner glass tube diameter of 3.9cm. Thomas and Thomas [10] presented design data for estimation of thermal loss in the receiver of parabolic trough concentrator at different parameters. The parameters considered for the analysis are:

outer diameter of absorber is 3.18 cm with emissivity 0.15, inner diameter of glass envelope is 5.5 cm having emissivity 0.90 and air gap of 1.16cm. Recently, Mohamad et al. [9] estimated theoretically the thermal performance of PTC and identified the heat losses. The above cited literature clearly enlightened the importance of air gap for optimal performance of PTC. Therefore, an attempt has been made in the present study to find the optimal air gap corresponding to different diameters of tube.

Certain assumptions (similar to Ref. [13]) have been made to carry out this investigation. These are as follows:

- Absorber tube and glass cover constitute a system of infinitely long concentric tubes.
- Flow of heat is one-dimensional.
- Negligible heat transfer via conduction in longitudinal direction.
- Temperature drop across the absorber tube and the glass cover is negligible. So, conduction through the glass cover is neglected.

In this investigation, two cases are assumed separately for finding overall heat loss from absorber to glass cover at optimum gap.

Case 1: heat loss takes place due convection and radiation, and

Case 2: heat loss takes place due to conduction and radiation.

Furthermore, the effects different design parameters and operating parameters on thermal performance in terms of overall heat losses have been discussed.

2. Estimation of optimum air gap and overall heat loss

In the present analytical investigation, a Matlab/Simulink model based on analytical expressions and their related parameters has been developed for calculation of optimum air gap and overall heat loss. The specifications of absorber tube and glass cover considered in the present investigation are reported in Table 1.

Table 1. Specifications of absorber tube and glass cover used in present study

Specifications	Data
Outside diameters of absorber tube, (D)	1.2, 2.2, 2.54, 3.18, 4.5, 5.08, 6.03, 7.62(cm) (Standard diameters of commercially available tubes)
Thickness of glass cover, (t)	0.002m
Emissivity of glass cover, ($\in_c$)	0.90
Emissivity of selective coating, ($\in_{ab}$)	0.15
Rim angle, (ϕ)	90°

The computation of optimum air gap and overall heat loss by using equations (2), (3) and (4) has been carried out as follows:

2.1 Overall heat loss between absorber tube and glass cover

Step 1: Initialization

a) Put the values of known parameters i.e. T_{ab}, T_a, $D, V_w, \in_{ab}$
b) Assumed initial guess value of air gap is 5mm.
c) Assumed value of T_c would be corrected till steady state is achieved using equations (2) and (4).
d) Further, the corrected value of T_c is kept constant and find the corrected value of air gap, where heat loss due to conduction and convection becomes equal. At this condition, the air gap is optimum and known as optimum air gap.
e) Using step (c) at optimum air gap and calculate T_c corresponding to optimum air gap.

Step 2: Calculation

a) Calculate mean temperature $\left(T_{mab-c} = \dfrac{T_{ab} + T_c}{2}\right)$.
b) At this mean temperature, calculate following properties of air between absorber and glass cover:

- Density(kg/m^3) using Ideal gas law at atmospheric pressure =101325N/m^2,
- Dynamic viscosity using the correlation given by Sutherland's [16],

$$\mu = \mu_{\circ}\left[\frac{T_{so}+C_o}{T_{mab-c}+C_o}\right]\left[\frac{T_{mab-c}}{T_{so}}\right]^{3/2} \quad (5)$$

- Thermal conductivity using the equation (6) [13],

$$K_{air} = 1.4\times10^{-11}T_{mab-c}^{3} - 5.3\times10^{-8}T_{mab-c}^{2} + 1.1\times10^{-4}T_{mab-c} - 0.0019 \quad (6)$$

- Kinematic viscosity and Prandtl number estimated from the above estimated values and the value of specific heat (C_p=1009 J/kg-K) is assumed to be constant at all T_{mab-c}.

c) Using above estimated values and data, Rayleigh number (R_a) can be calculated as [10].

$$R_a = \frac{g\times\beta\times(T_{ab}-T_c)\times Z^3\times \Pr}{\nu^2} \quad (7)$$

where, $\beta = \dfrac{1}{\dfrac{T_{ab}+T_c}{2}}$

d) Effective thermal conductivity has been estimated by Raithby and Holland correlation [17] as given below:

$$\frac{K_{eff}}{K_{air}} = 0.317(R_a^*)^{1/4} \quad (8)$$

where, $(R_a^*)^{1/4} = \dfrac{\ln\left(\dfrac{D_{ci}}{D}\right)\times R_a^{1/4}}{Z^{3/4}\left(\dfrac{1}{D^{3/5}}+\dfrac{1}{D_{ci}^{3/5}}\right)^{5/4}}$.

The limitations on using equation (8) are that R_a^* should be less than 10^7. The effective thermal conductivity K_{eff} cannot be less than the thermal conductivity K_{air}. Hence, the value of K_{eff}/K_{air} is taken as unity if equation (8) yields a value less than unity [13].

e) The value of h_{cab-c} can be computed as using equation (9) [13],

$$h_{cab-c} = \frac{2\times K_{eff}}{D\times\ln\left(\dfrac{D_{ci}}{D}\right)} \quad (9)$$

f) Substituting the calculated values of h_{cab-c} and other parameters in equations (2) and (3), to get the overall heat loss (Q_L) between absorber tube and glass cover at optimum air gap.

2.2 Overall heat loss between glass cover and surrounding air

Following steps have been followed:

Step 1: Same as above step 1.

Step 2: Calculation

a) Calculate mean temperature, $T_{mc-a} = \left(\frac{T_c + T_a}{2}\right)$

b) Use the correlations used in step 2 (b) from to estimate above properties of air on outside surface of glass cover at mean temperature, T_{mc-a}.

c) For the assumed value of wind velocity V_w, the Reynold number (Re) was computed using equation (10),

$$\mathrm{Re} = \frac{\rho_{air-c} \times D_{co} \times V_w}{\mu_{air-c}} \tag{10}$$

d) To compute Nusselt number, Hilpert's correlation has been used [13, 18]

$$Nu = C_1 \,\mathrm{Re}^n \tag{11}$$

where C_1 and n are constant having following values:

- for 40 < Re < 4000, C_1 = 0.615, n = 0.466;
- for 4000 < Re < 40000, C_1 = 0.174, n = 0.618;
- for 40000 < Re < 400000, C_1 = 0.0239, n = 0.805.

e) h_{cc-a} can be calculated using equation (12).

$$h_{cc-a} = \frac{Nu \times K_{air-c}}{D_{co}} \tag{12}$$

f) Substituting the calculated values of h_{cc-a} and other parameters in equation (4), to get the overall heat loss (Q_L) between glass cover and surrounding air at optimum air gap.

3. Results and discussions

The value of computed optimum air gap for different diameter of absorber tube and parameters are presented in Table 2.

Table 2. Optimum air gap for different diameter of absorber tube at T_{ab} = 250°C, T_a = 10°C, V_w = 1 m/s

D (cm)	y_{og} (mm)	D_{ci} (cm)	D_{co} (cm)	Q_{L-cd} (W/m)	Q_{L-cv} (W/m)	Q_L (W/m)
1.20	5.83	2.366	2.766	66.87	68	88.29
2.20	6.43	3.486	3.886	95.39	94.77	131.2
2.54	6.64	3.868	4.268	103.6	102.9	144.8
3.18	7.00	4.58	4.98	117.9	117.2	169.2
4.50	7.65	6.03	6.43	144.3	143.3	216.8
5.08	7.85	6.65	7.05	155.8	154.8	236.7
6.03	8.10	7.67	8.07	173.2	172.5	269.3
7.62	8.68	9.356	9.756	201.3	200.3	322

As the diameter of absorber tube increases, the optimum air gap increases slightly as observed from Table 2 and Figure 2. Optimum air gap is found to be approximately 7mm and 8mm for an absorber tube of diameter in range of 1.2-3.18cm and 4.5-7.62cm, respectively for optimal performance. Heat loss due to conduction is nearly equal to heat loss due to convection at optimum air gap for given diameter of absorber tube is also clearly observed from Table 2. Further inspection of Figure 2 depicts the overall heat loss increases directly with increase in absorber tube diameter. Too larger diameter gives a higher intercept factor but simultaneously enhance the overall heat losses [19]. Therefore, the final selection of absorber tube diameter has been made on the basis of thermal analysis and a trade-off between increased interception of reflected solar energy and acceptably less heat losses.

Further, to reduce the overall heat loss, it is necessary to find the optimal air gap corresponding to different absorber tube diameters. That's the reason, which attributed the development of correlations presented in Table 3 for optimum air gap and overall heat loss with the variation of diameter of absorber tube. This procedure can also be utilized to compute design data and correlations at other parameters.

For estimating the accuracy of the developed correlations, error analysis has been carried out for the data obtained from the analytical expressions and the proposed correlations. It is varying in between 0.28% to 1.08% for optimum air gap and 0.17% to 1.31% for overall heat loss corresponding to the absorber tube diameter ranging from 1.2cm to 7.62cm.

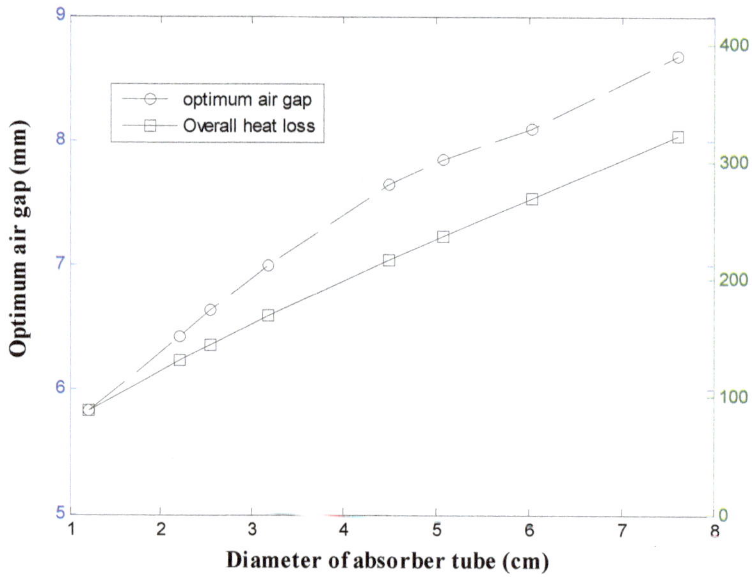

Figure 2. Variation of optimum air gap and overall heat loss with diameter of absorber tube at T_{ab}=250°C, T_a=10°C, V_w = 1 m/s

Table 3. Correlations for optimum air gap and overall heat loss with diameter of absorber tube

$y_{og} = -0.031D^2 + 0.71D + 5$
$Q_L = -0.82D^2 + 43D + 39$

Figure 3 shows the variation of overall heat with air temperature for different diameters of absorber tube. From the analysis of Figure 3, it is clear that with increase in temperature of air, overall heat loss decreases for a fixed absorber tube diameter at optimum air gap. As the diameter of absorber tube increases from 1.2cm to 7.62cm, the overall heat loss increases at a given air temperature.

From Figure 4, it has been clearly observed that the overall heat loss increases with increase in wind velocity for the given diameter of absorber tube with their computed optimum air gap. The overall heat loss also increases when diameter of absorber increases for same wind velocity.

The effect of emissivity of selective coating of absorber tube on overall heat loss has been shown in Figure 5. It reveals that with increase in emissivity of absorber tube, overall heat loss also increases from absorber to air for different diameters of absorber tube. Overall heat loss also increases when diameter of absorber increases for same emissivity of absorber surface. The range of emissivity considered in the present investigation is nearer to the range of emissivity of some selective surfaces which varies from 0.09 to 0.17 [20].

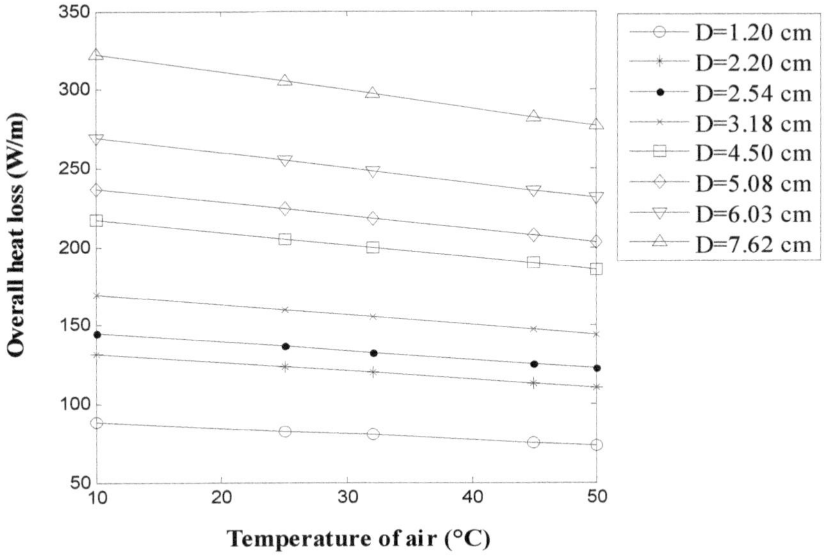

Figure 3. Variation of overall heat loss with temperature of air at T_{ab} =250°C, V_w =1 m/s

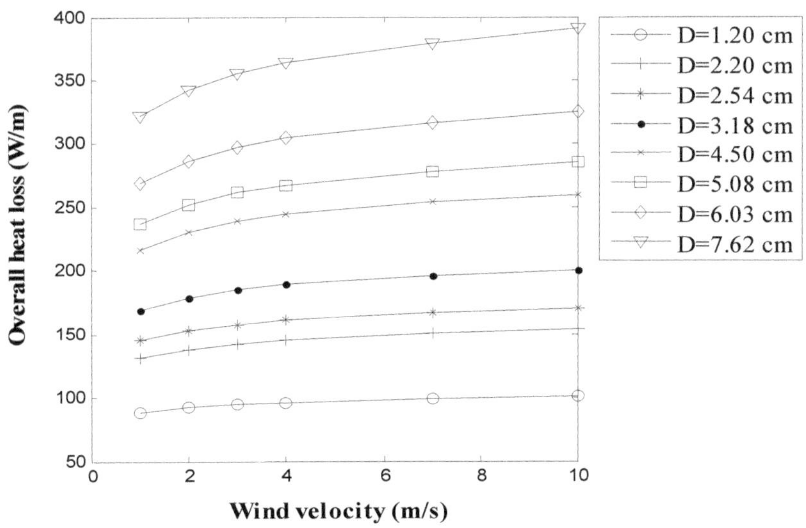

Figure 4. Variation of overall heat loss with wind velocity at T_{ab} =250°C, T_a=10°C

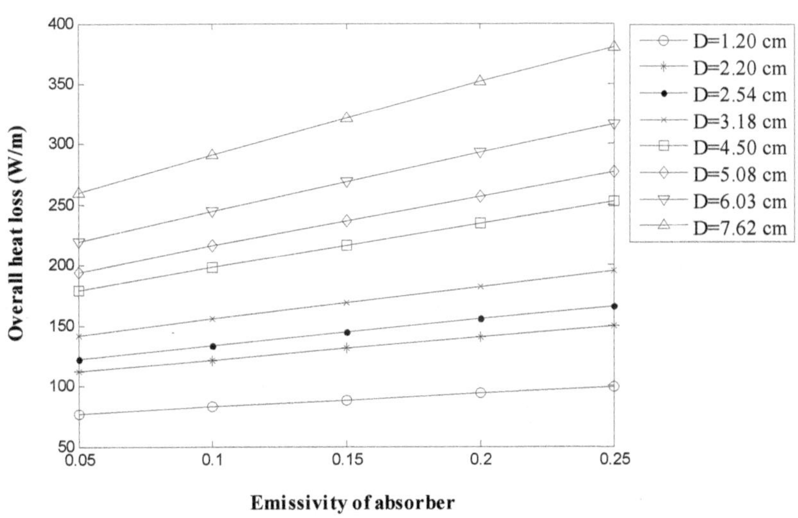

Figure 5. Variation of overall heat loss with emissivity of absorber at T_{ab} =250 °C, T_a=10°C, V_w =1 m/s

The temperature of absorber tube has significant effect on overall heat loss (Figure 6). Overall heat loss increases with increase in temperature of absorber tube. Overall heat loss also increases when diameter of absorber increases for same temperature of absorber.

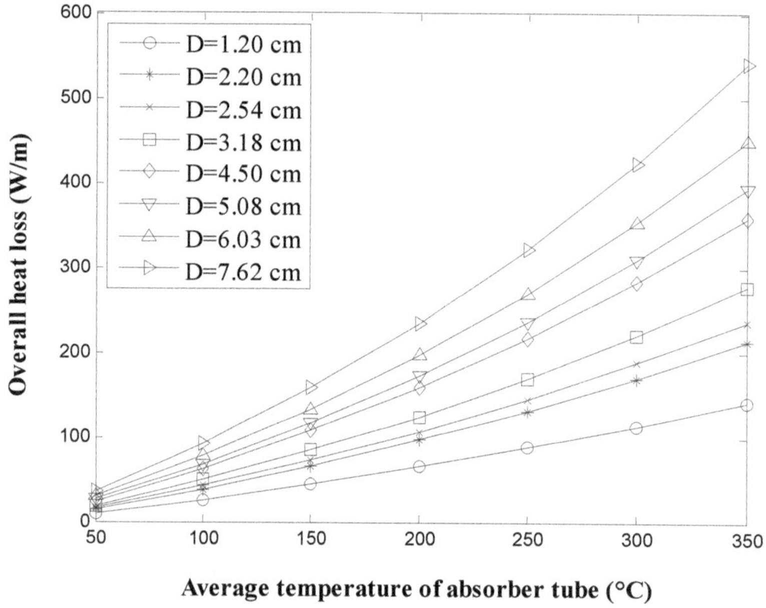

Figure 6. Variation of overall heat loss with average temperature of absorber tube at T_a=10 °C, V_w =1m/s

4. Validation of results

The correctness of the proposed model has been evaluated by comparing the results in terms of heat loss with the results of Thomas and Thomas [10] at D=3.18, D_{ci} =5.5 and D_{co} =6, Air gap (y)=11.6mm (Figure 7) The values of other parameters have been kept constant as per Ref. [10] to validate the results (T_{ab} =250°C, V_w =1m/s, $\in_{ab}$ =0.15). From the inspection of Figure 7, it is clear that the proposed model gives close agreement (variation within ≤ 1.2 %) with Ref. [10]. The optimum air gap is found to be approximately 7mm for absorber tube diameter of 2.54cm, which is similar to that Ref. [15] and [16].

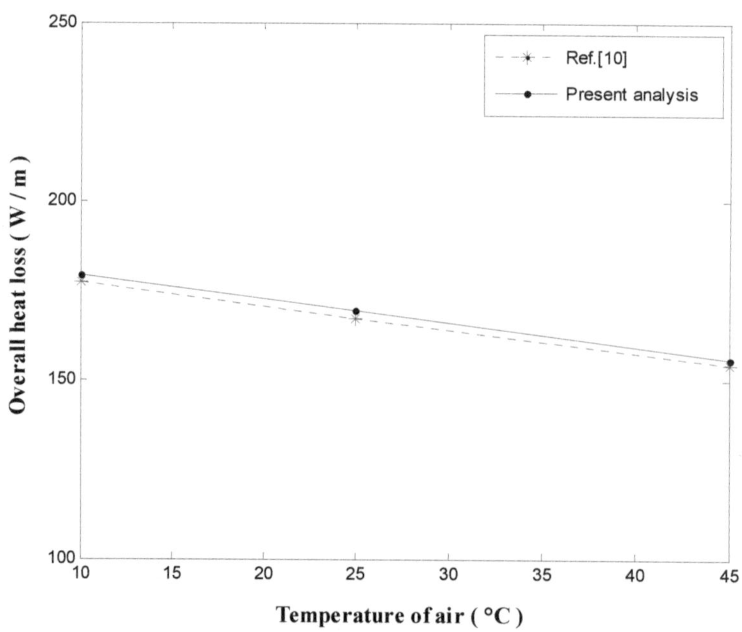

Figure 7. Variation of overall heat loss with temperature of air at D=3.18cm, D_{ci} =5.5cm and D_{co} =6cm, Air gap (y)=11.6mm

5. Conclusion

The present analysis and proposed correlations will be useful for computing optimum air gap in order to minimize overall heat loss and finally enhancing the performance of PTC system. Here, the procedure for finding optimum air gap and overall heat loss has been developed for non- evacuated absorber tube. At optimum air gap, non-evacuated configuration of PTC system performs better than at any other value of air gap, which results in improved efficiency as well as reasonable economic benefits. Overall heat loss increases with increase in absorber temperature, wind velocity, emissivity of absorber and decreases with increase in air temperature. From the overall analysis, it has been observed that the absorber tube diameter ranging from 3.18 to 4.5cm provides considerable overall heat losses keeping in view of interception of reflected solar energy at that absorber tube diameter. The benefits of such non-evacuated configuration at optimum air gap are simple design for installation and flexible for the developing countries.

Nomenclature

C_o	Sutherland's constant for air = 120 K
D	outer diameter of absorber tube (m)
D_{ci}	inner diameter of glass cover (m)
D_{co}	outer diameter of glass cover (m)
g	acceleration due to gravity = 9.81 (m/s^2)
h_{cab-c}	convective heat transfer coefficient between absorber tube and glass cover (Wm^{-2}K^{-1})
h_{cc-a}	convective heat transfer coefficient of outside surface of glass cover (Wm^{-2}K^{-1})
K_{air}	thermal conductivity of air between absorber tube and glass cover (Wm^{-1}K^{-1})
K_{eff}	effective thermal conductivity of air between absorber tube and glass cover (Wm^{-1}K^{-1})
K_{air-c}	thermal conductivity of air on outside surface of glass cover (Wm^{-1}K^{-1})
Pr	Prandtl number
Q_L	overall heat loss rate per unit length from absorber tube (W/m)
Q_{L-cd}	heat loss due to conduction from the absorber tube (W/m)
Q_{L-cv}	heat loss due to convection from the absorber tube (W/m)
Re	Reynolds number
R	gas constant for air = 287 J/kg-K
R_a^*	modified Rayleigh number
t	thickness of glass cover(m)
T_{ab}	average temperature of the absorber tube (K)
T_c	average temperature of glass cover (K)
T_a	ambient/air temperature (K)
T_{mab-c}	mean temperature between absorber and glass cover (K)
T_{mc-a}	mean temperature between glass cover and ambient air (K)
T_{sky}	sky temperature (K) $= T_a - 5$ [11]
T_{so}	reference temperature in Sutherland's formula = 291.15 K
V_w	wind velocity (m/s)
y	Air gap between absorber tube and glass cover(m)
y_{og}	optimum air gap by analytical expressions(m)
Z	Characteristic dimension (m) = $(D_{ci} - D)/2$

Greek symbols

$\in_{ab}$	emissivity of selective coating of absorber tube
$\in_c$	emissivity of glass cover
σ	Stefan-Boltzmann constant = 5.67×10^{-8} (Wm^{-2}K^{-4})
μ	dynamic viscosity at input temperature T_{mab-c} (N-s/m^2)
μ_o	reference viscosity at T_{so}=18.27×10^{-6} (N-s/m^2)
ρ_{air-c}	density of air on outside surface of glass cover (kg/m^3)
μ_{air-c}	dynamic viscosity of air on outside surface of glass cover (N-s/m^2)
ν	kinematic viscosity of air between absorber tube and glass cover (m^2/s)

References

[1] Sharma N., Varun, Siddhartha. Stochastic technique used for optimization in solar systems: A review. Renewable and Sustainable Energy Reviews 2012, 16, 1399-1411.

[2] Kalogirou S., Lloyd S. Use of solar parabolic trough collectors for hot water production in Cyprus – A feasibility study. Renewable Energy 1992, 2, 117–124.

[3] Kruger D., Heller A., Hennecke K., Duer K. Parabolic trough collectors for district heating systems at high latitudes – a case study. In: Proceedings of Eurosun'2000 on CD ROM, 2000.

[4] Lippke F. Direct steam generation in parabolic trough solar power plants: numerical investigation of the transients and the control of a once-through system. Journal of Solar Energy Engineering 1996, 118, 9–14.

[5] Thomas A. Solar steam generating systems using parabolic trough concentrators. Energy Conversion and Management 1996, 37, 215–245.

[6] May E.K., Murphy L.M. Performance benefits of the direct generation of steam in line-focus solar collectors. Journal of Solar Energy Engineering 1983, 105, 126–133.

[7] Kalogirou S. Parabolic trough collectors for industrial process heat in Cyprus. Energy 2002, 27, 813–830.

[8] Arasu A. Valan, Sornakumar T. Performance characteristics of parabolic trough solar collector system for hot water generation. International Energy Journal 2006, Vol. 7, No. 2, 137-146.

[9] Mohamad A., Orfi J., Alansary H. Heat losses from parabolic trough solar collectors. Int. J. Energy Res. 2014, 38, 20-28.

[10] Thomas A., Thomas S.A. Design data for the computation of thermal loss in the receiver of a parabolic trough concentrator. Energy convers. Mgmt. 1994, 35(7), 555-568.

[11] Daniel P., Joshi Y., Das A.K. Numerical investigation of parabolic trough receiver performance with outer vacuum shell. Solar Energy 2011, 85, 1910-1914.

[12] Ratzel A.C. Evaluation of the evacuated solar annular receivers used at mid temperature solar systems test facility (MSSTF). SAND 78-0983, Sandia Laboratories, Albuquerque, N.M., USA, 1979.

[13] Sukhatme S.P. Principal of thermal collection and storage. Second ed., New Delhi, TMGH Pub. Com. Ltd., 1996.

[14] Treadwell G.W. Design consideration for parabolic cylindrical solar collectors. SAND 76-0082, Sandia Laboratories, Albuquerque, N.M., USA, 1976.

[15] Rabl A, Bendt P, Gaul H.W. Optimization of parabolic trough solar collectors. Solar Energy 1982, 29(5), 407-417.

[16] Yadav A. Experimental and numerical investigation of solar powered solid desiccant dehumidifier. Ph.D. Thesis, Department of Mech. Engg., NIT, Kurukshetra, India, 2012.

[17] Raithby G.D., Holland K.G.T. A general method of obtaining approximate solution to laminar and turbulent free convection problems. Adv. Heat Transf. 1933, 11, 265.

[18] Hilpert R. Rohren Wameabgabe Vongeheizen drahten und Rohren, Forsch, Geb. Ingenieur. 1933; 4:220.

[19] Kalogirou S.A., Lloyd S., Ward J., Eleftheriou P. Design and performance characteristics of a parabolic-trough solar-collector system. Applied energy 1994, 47, 341-354.

[20] Duffie J.A., Beckman W.A. Solar Engineering of Thermal Processes. First ed., John Wiley, New York, 1980.

8

Effect of heat transfer law on the finite-time exergoeconomic performance of a generalized irreversible carnot heat engine

Yi Zhang[1], Lingeng Chen[2,3,4], Guozhong Chai[1]

[1] College of Mechanical Engineering, Zhejiang University of Technology, Hangzhou, 310014, China.
[2] Institute of Thermal Science and Power Engineering, Naval University of Engineering, Wuhan 430033, China.
[3] Military Key Laboratory for Naval Ship Power Engineering, Naval University of Engineering, Wuhan 430033, China.
[4] College of Power Engineering, Naval University of Engineering, Wuhan 430033, China.

Abstract
The analytical expression for profit rate of a generalized irreversible Carnot heat engine cycle based on a generalized radiative heat transfer law $q \propto \Delta(T^n)$ is derived by applying the finite time exergoeconomic method, taking into account several additional irreversibilities, such as heat resistance, heat leakage and other undesirable irreversible factors. The compromise optimization between economics (profit rate) and the efficiency was obtained by searching the efficiency at maximum profit rate, which is termed as the finite time exergoeconomic performance bound.

Keywords: Finite-time thermodynamics; Generalized irreversible Carnot heat engine; Exergoeconomic performance; Generalized thermodynamic optimization; Heat transfer law.

1. Introduction

Recently, the intensive consumption of energy and the exhaustion of resources lead to the rising costs for energy. Hence, from the economic perspective, improvement of engine performance is urgently required. Finite-time thermodynamics [1-8] is a powerful tool often used to optimize thermodynamic parameters including power, efficiency, entropy generation, effectiveness, cooling load, heating load, loss of exergy, etc.

Nowadays, systems like heat engines are analyzed and designed based on the consideration of both thermodynamic parameters and cost accounting requirements after the research of Salamon and Nitzan [9, 10], which was to maximize the profit of an endoreversible heat engine by a combination of a thermodynamic analysis with an engineer economic analysis. In order to distinguish this method from the endoreversible analysis optimizing pure thermodynamic objectives, Chen *et al.* [11-17] analyzed the profit rate of thermal systems by attributing costs to input and output exergy and termed this method as finite-time exergoeconomic analysis and its performance bound at maximum profit as finite-time exergoeconomic performance bound. Other researches seeking for best economic performance of thermal systems were carried out on endoreversible engines, refrigerators and heat pumps by Ibrahim *et al.* [18], De Vos [19, 20] and Bejan [21], with the only irreversibility restricted to the heat transfer between the working fluid and the heat reservoirs. De Vos [19, 20] applied the Newton (linear) heat transfer law to derive the relation between the optimal efficiency and economic returns when carrying out thermoeconomics analysis for heat engine. Chen *et al.* [22] investigated the endoreversible thermoeconomic performance of heat engine with the heat transfer between the working fluid and the heat reservoirs obeying linear phenomenological law. Sahin *et al.* [23-26] proposed an optimization criterion considering thermodynamic parameters per unit total cost.

In many pioneer works concerning finite-time exergoeconomic optimization for heat engines, the basic thermodynamic model is endoreversible. However, a real heat engine will operate in an irreversible power cycle which incorporates several internal and external irreversibilities, such as heat resistance, bypass heat leakage, friction, turbulence and other undesirable irreversibility factors. Considering external and internal irreversibilities, Chen *et al.* [16-27] established a generalized irreversible Carnot heat engine model. As heat transfer is not necessarily Newtonian or linear phenomenological, a further step made in this paper is to establish a fundamental optimal relationship between profit and efficiency of the generalized irreversible Carnot heat engine based on generalized radiative heat transfer law $q \propto \Delta(T^n)$. The result obtained by searching the optimum efficiency at maximum profit involved three common heat transfer laws: Newton's law ($n = 1$), the linear phenomenological law ($n = -1$), and the radiative heat transfer law ($n = 4$). The relative studies can be seen in Refs. [28-35].

2. Cycle model and performance analysis

In order to conduct the simulation closer to the performance of an actual heat engine, Chen, *et al.* [16, 27] established a generalized irreversible steady flow Carnot heat engine cycle model as shown in Figure 1, considering heat resistance, heat leakage, and internal irreversibilities. The working fluid in this generalized irreversible engine with constant-temperature heat-reservoirs flows steadily. The system undergoes a cycle which consists of four irreversible processes, two isothermal and two adiabatic. External irreversibilities are caused by the heat resistance existed in the high- and low-temperature heat-exchangers. Heat-transfer between the heat engine and its surrounding heat reservoirs leads to the difference between the working fluid temperature (T_{HC} and T_{LC}) and the heat-reservoir temperature (T_H and T_L). These temperatures are related to one another in the following order:

$$T_H > T_{HC} > T_{LC} > T_L \tag{1}$$

A constant rate of heat leakage (q) from the heat source at the temperature T_H to the heat sink at T_L is assumed for this system, which yields,

$$Q_H = Q_{HC} + q \tag{2}$$

$$Q_L = Q_{LC} + q \tag{3}$$

where Q_{HC} and Q_{LC} are the rates of heat-transfer supplied by the heat source and released to the heat sink by the working fluid, respectively; Q_H and Q_L are the real rates of heat-supply and heat-release, respectively. Assuming the heat-transfer law obeys $q \propto \Delta(T^n)$, the rate of heat leakage can be expressed as

$$q = C_i(T_H^n - T_L^n) \tag{4}$$

where C_i is the heat leakage coefficient.

When analyzing actual heat engines, heat resistance and heat leakage discussed above are not the only irreversibilities. Irreversibilities caused by friction, turbulence, and non-equilibrium activities inside the working fluid are also required to be considered. Thus when compared with an endoreversible Carnot heat engine of the same heat input, the generalized irreversible Carnot engine can deliver less power and release more heat to the heat sink. Hence, the rate of heat flow (Q_{LC}) to the heat sink for the generalized irreversible Carnot engine is larger than that (Q'_{LC}) for the endoreversible Carnot engine with the same input. A constant coefficient (φ) is introduced in the following expression to generally characterize the additional miscellaneous irreversible effects

$$\varphi = Q_{LC}/Q'_{LC} \geq 1 \tag{5}$$

Application of the second law of thermodynamics yields,

$$Q'_{LC}/T_{LC} = Q_{HC}/T_{HC} \tag{6}$$

Combining Eqs. (5) and (6) gives

$$Q_{LC} = \varphi Q_{HC}/x \tag{7}$$

where $x = T_{HC}/T_{LC}$ ($1 \leq x \leq T_H/T_L$) is the temperature ratio of the working fluid.
Application of the first law of thermodynamics gives the expressions of power output and thermal efficiency, respectively

$$P = Q_H - Q_L = Q_{HC} - Q_{LC} \tag{8}$$

$$\eta = P/Q_H = (Q_{HC} - Q_{LC})/(Q_{HC} + q) \tag{9}$$

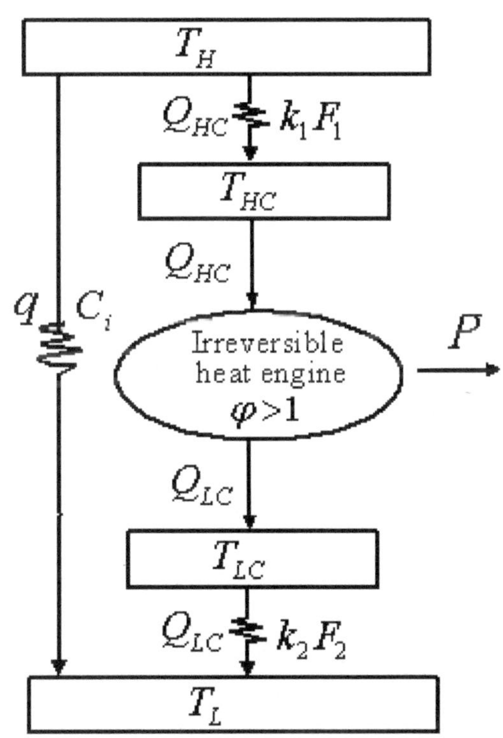

Figure 1. The generalized irreversible Carnot heat engine cycle model

Assuming the rates of the heat flow in the heat-exchangers follow the generalized radiative heat transfer law, $q \propto \Delta(T^n)$, where n is a heat transfer exponent, with $n = 1$ representing the Newton's law, $n = -1$ representing the linear phenomenological law and $n = 4$ representing the radiative heat transfer law. Then

$$Q_{HC} = k_1 F_1 (T_H^n - T_{HC}^n) \tag{10}$$

$$Q_{LC} = k_2 F_2 (T_{LC}^n - T_L^n) \tag{11}$$

where k_1 and k_2 are the overall heat-transfer coefficients of high- and low-temperature side heat-exchangers, F_1 and F_2 are the surface areas of high- and low-temperature side heat-exchangers. The total heat transfer surface area of the two heat exchangers is taken as a constant , that is

$$F_1 + F_2 = F_T \tag{12}$$

And a ratio (f) of heat exchanger area is defined as

$$f = F_1 / F_2 \tag{13}$$

Assuming that the prices of the work output and exergy input are ψ_1 and ψ_2 respectively, the profit rate (profit per unit time) of the generalized irreversible Carnot heat engine is [11]

$$\pi = \psi_1 P - \psi_2 A \tag{14}$$

where A is the rate of exergy input of the heat engine which can be expressed as

$$A = Q_H (1 - T_0/T_H) - Q_L (1 - T_0/T_L) = Q_H \varepsilon_1 - Q_L \varepsilon_2 \tag{15}$$

where ε_i is the Carnot coefficient of the reservoir and T_0 is the environmental temperature.
Combining Eqs. (2)-(3) and (7)-(15)gives

$$\eta = B'(x - \varphi)[T_H^n - (xT_L)^n] / \{B' x [T_H^n - (xT_L)^n] + q\} \tag{16}$$

$$\pi = B' \psi_1 [T_H^n - (xT_L)^n][x(1 - \varepsilon_1 \psi_2/\psi_1) - \varphi(1 - \varepsilon_2 \psi_2/\psi_1)] + q\psi_2(\varepsilon_2 - \varepsilon_1) \tag{17}$$

Where $B' = k_1 f F_T / [(1 + f)(x + x^n \varphi f k_1/k_2)]$.
Maximizing η and π with respect to f by setting $d\eta/df = 0$ and $d\pi/df = 0$ using Eqs. (16) and (17) yields the same optimal ratio of heat-exchanger area (f_{opt})

$$f = f_{opt} = [k_2 / (x^{n-1} \varphi k_1)]^{0.5} \tag{18}$$

Substituting Eq. (18) into Eqs. (16) and (17), respectively, yields the optimal efficiency and profit rate in the following forms:

$$\eta = B(x - \varphi)[T_H^n - (xT_L)^n] / \{Bx[T_H^n - (xT_L)^n] + q\} \tag{19}$$

$$\pi = B\psi_1 [T_H^n - (xT_L)^n][x(1 - \varepsilon_1 \psi_2/\psi_1) - \varphi(1 - \varepsilon_2 \psi_2/\psi_1)] + q\psi_2(\varepsilon_2 - \varepsilon_1) \tag{20}$$

where $B = k_1 F_T / [x^{0.5} + (x^n \varphi k_1/k_2)^{0.5}]^2$.
Maximum profit rate and maximum efficiency with respect to temperature ratio can be derived by taking derivatives of Eq.(19) and Eq.(20) with respect to x . However, in general, the optimal temperature ratio

x_π at maximum profit rate π_{max} does not equal to the optimal temperature ratio x_η at maximum efficiency η_{max}. The optimal temperature ratio x_π at maximum profit rate π_{max} can be derived by taking the derivative of the profit rate with temperature ratio and setting it equal to zero ($d\pi/dx = 0$). By substituting the optimal temperature ratio x_π into Eq. (20), the maximum profit rate can be achieved. Furthermore, the finite-time exergoeconomic bound of the generalized irreversible Carnot heat engine will be obtained by substituting the optimal temperature ratio x_π with respect to maximum profit rate into Eq. (19).

3. Discussions

3.1 Effects of various losses on the performance

If $\varphi = 1$ and $q > 0$, Eq.(19) and Eq.(20) become

$$\eta = B_{en}(x-1)[T_H^n - (xT_L)^n]/\{B_{en}x[T_H^n - (xT_L)^n] + q\} \tag{21}$$

$$\pi = B_{en}\psi_1[T_H^n - (xT_L)^n][x(1-\varepsilon_1\psi_2/\psi_1) - (1-\varepsilon_2\psi_2/\psi_1)] + q\psi_2(\varepsilon_2 - \varepsilon_1) \tag{22}$$

where $B_{en} = k_1F_T/[x^{0.5} + (x^n k_1/k_2)^{0.5}]^2$. Eqs. (21) and (22) are the relations between profit rate and efficiency of the irreversible Carnot heat engine with heat resistance and heat leakage losses.

If $\varphi > 1$ and $q = 0$, Combining Eq.(19) and Eq.(20) gives

$$\pi = B_{en}\phi\psi_1\{T_H^n - [\phi T_L/(1-\eta)]^n\}[(1-\varepsilon_1\psi_2/\psi_1)/(1-\eta) - (1-\varepsilon_2\psi_2/\psi_1)] \tag{23}$$

Eq. (23) is the relation between profit rate and efficiency of the irreversible Carnot heat engine with heat resistance and internal irreversibility losses.

If $\varphi = 1$ and $q = 0$, Eq. (23) is reduced to

$$\pi = B_{en}\psi_1\{T_H^n - [T_L/(1-\eta)]^n\}[(1-\varepsilon_1\psi_2/\psi_1)/(1-\eta) - (1-\varepsilon_2\psi_2/\psi_1)] \tag{24}$$

Eq. (24) is the relation between profit rate and efficiency of the endoreversible Carnot heat engine [11].

3.2 Special cases

(1) Case of $n = 1$

In the case of $n = 1$, Eq.(19) and Eq.(20) become:

$$\eta = (x-\varphi)(T_H x^{-1} - T_L)/(T_H - xT_L + qB^{-1}) \tag{25}$$

$$\pi = B\psi_1(T_H x^{-1} - T_L)[x(1-\varepsilon_1\psi_2/\psi_1) - \varphi(1-\varepsilon_2\psi_2/\psi_1)] + q\psi_2(\varepsilon_2 - \varepsilon_1) \tag{26}$$

where $B = k_1F_T/[1 + (\varphi k_1/k_2)^{0.5}]^2$. Maximizing π with respect to x by setting $d\pi/dx = 0$ in Eq. (26) yields the optimal temperature ratio and the maximum profit rate of the heat engine:

$$x_{opt} = (\varphi \frac{T_H}{T_L} \frac{\psi_1 - \varepsilon_2\psi_2}{\psi_1 - \varepsilon_1\psi_2})^{0.5} \tag{27}$$

$$\pi_{max} = B\{[T_H(\psi_1 - \varepsilon_1\psi_2)]^{0.5} - [\varphi T_L(\psi_1 - \varepsilon_2\psi_2)]^{0.5}\}^2 + \psi_2 q(\varepsilon_2 - \varepsilon_1) \tag{28}$$

Substituting Eq. (27) into Eq. (25) gives η_π, which is the finite-time exergoeconomic bound of generalized irreversible Carnot heat engine with Newton's heat transfer law

$$\eta_\pi = \frac{T_H + \varphi T_L - (\varphi T_H T_L)^{0.5}\{[(\psi_1 - \varepsilon_1\psi_2)/(\psi_1 - \varepsilon_2\psi_2)]^{0.5} + [(\psi_1 - \varepsilon_2\psi_2)/(\psi_1 - \varepsilon_1\psi_2)]^{0.5}\}}{T_H - (\varphi T_H T_L)^{0.5}[(\psi_1 - \varepsilon_2\psi_2)/(\psi_1 - \varepsilon_1\psi_2)]^{0.5} + qB^{-1}} \tag{29}$$

(2) Case of $n = -1$

In the case of $n = -1$, Eq.(19) and Eq.(20) become

$$\eta = B_1(x - \phi)(T_L^{-1} - xT_H^{-1})/[B_1 x(T_L^{-1} - xT_H^{-1}) + q] \tag{30}$$

$$\pi = B_1\psi_1[T_L^{-1} - xT_H^{-1}][x(1 - \varepsilon_1\psi_2/\psi_1) - \phi(1 - \varepsilon_2\psi_2/\psi_1)] + q\psi_2(\varepsilon_2 - \varepsilon_1) \tag{31}$$

where $B_1 = k_1 F_T/[x + (\phi k_1/k_2)^{0.5}]^2$. Maximizing π with respect to x by setting $d\pi/dx = 0$ in Eq. (29) yields the optimal temperature ratio and the maximum profit rate of the heat engine

$$x_{opt} = \frac{2T_H\phi(1 - \varepsilon_2\psi_2/\psi_1) + (\varphi k_1/k_2)^{0.5}[T_L\phi(1 - \varepsilon_2\psi_2/\psi_1) + T_H(1 - \varepsilon_1\psi_2/\psi_1)]}{2T_L(\varphi k_1/k_2)^{0.5}(1 - \varepsilon_1\psi_2/\psi_1) + T_L\phi(1 - \varepsilon_2\psi_2/\psi_1) + T_H(1 - \varepsilon_1\psi_2/\psi_1)} \tag{32}$$

$$\pi_{\max} = B_\pi\psi_1[T_L^{-1} - x_{opt}T_H^{-1}][x_{opt}(1 - \varepsilon_1\psi_2/\psi_1) - \varphi(1 - \varepsilon_2\psi_2/\psi_1)] + q\psi_2(\varepsilon_2 - \varepsilon_1) \tag{33}$$

Substituting Eq. (32) into Eq. (30) gives η_π, which is the finite-time exergoeconomic bound of generalized irreversible Carnot heat engine based on linear phenomenological heat transfer law

$$\eta_\pi = (x_{opt} - \varphi)/\{x_{opt} + qT_H T_L/[B_\pi(T_H - x_{opt}T_L)]\} \tag{34}$$

where $B_\pi = k_1 F_T/[x_{opt} + (\varphi k_1/k_2)^{0.5}]^2$.

(3) Case of $n = 4$

In the case of $n = 4$, Eq.(19) and Eq.(20) become

$$\eta = B_4(x - \varphi)[T_H^4 - (xT_L)^4]/\{B_4 x[(T_H)^4 - (xT_L)^4] + q\} \tag{35}$$

$$\pi = B_4\psi_1[T_H^4 - (xT_L)^4][x(1 - \varepsilon_1\psi_2/\psi_1) - \varphi(1 - \varepsilon_2\psi_2/\psi_1)] + q\psi_2(\varepsilon_2 - \varepsilon_1) \tag{36}$$

where $B_4 = k_1 F_T/[x^{0.5} + x^2(\varphi k_1/k_2)^{0.5}]^2$. Eqs (35) and (36) are the relations between profit rate and efficiency of the irreversible Carnot heat engine based on the radiative heat transfer law.

The relationships between the profit rate and efficiency of the irreversible Carnot heat engine for all discussed cases are shown in Figure 2 and Figure 3. It can be concluded from the figure that the profit rate versus efficiency is a loop-shaped curve for all cases with heat leakage. For the cases without heat leakage, the profit rate decreases when the irreversibility factor φ increases with the shape of the curve remaining parabolic. For the case of $n < 0$, the optimal efficiency at the maximum profit rate increases with the increase of n as shown in Figure 2. While, for the case of $n > 0$, the optimal efficiency at the maximum profit rate decreases with the increase of n as shown in Figure 3. When n increases, the influence of temperature on power becomes more remarkable. Hence, when n is relatively large, by slightly sacrificing the efficiency, a significant increase of power can be achieved.

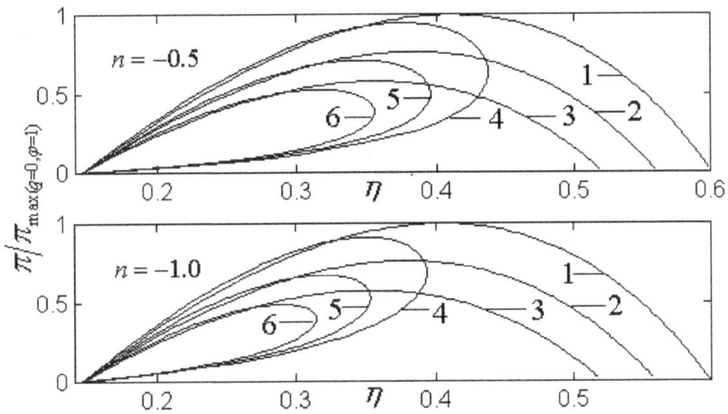

1、$C_i=0$，$\varphi=1.0$； 2、$C_i=0$，$\varphi=1.1$； 3、$C_i=0$，$\varphi=1.2$；
4、$C_i>0$，$\varphi=1.0$； 5、$C_i>0$，$\varphi=1.1$； 6、$C_i>0$，$\varphi=1.2$。

Figure 2. The influences of heat leak, internal irreversibility and heat transfer law on $\pi-\eta$ characteristic for $n<0$

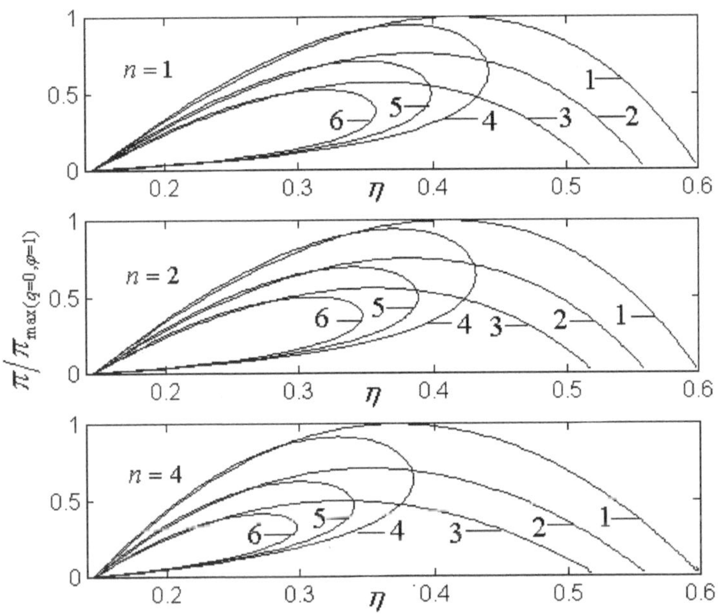

1、$C_i=0$，$\varphi=1.0$； 2、$C_i=0$，$\varphi=1.1$； 3、$C_i=0$，$\varphi=1.2$；
4、$C_i>0$，$\varphi=1.0$； 5、$C_i>0$，$\varphi=1.1$； 6、$C_i>0$，$\varphi=1.2$。

Figure 3. The influences of heat leak, internal irreversibility and heat transfer law on $\pi-\eta$ characteristic for $n>0$

3.3 The effect of price ratio ψ_2/ψ_1

The finite-time exergoeconomic performance bound at the maximum profit rate is different from the classical reversible bound and the finite-time thermodynamic bound. It is dependent on T_H, T_L, T_0 and ψ_2/ψ_1. In order to ensure the process being potential profitable, $0<\psi_2/\psi_1<1$ is required.

As the price of work output becomes very large compared with that of exergy input, i.e., $\psi_2/\psi_1 \to 0$, the function of the profit rate becomes

$$\pi = B\psi_1(x-\varphi)[T_H^n - (xT_L)^n] = \psi_1(Q_{Hc} - Q_{Lc}) = \psi_1 P \tag{37}$$

The optimization of the profit rate also leads to the maximization of the power output P of the generalized irreversible heat engine cycle.

On the other hand, with the price of work output approaching the price of the exergy input, i.e. $\psi_2/\psi_1 \to 1$, the function of the profit rate becomes

$$\pi = -\psi_1 T_0[(Q_{LC}+q)/T_L - (Q_{HC}+q)/T_H] = -\psi_1 T_0 \sigma \tag{38}$$

where σ is the rate of entropy production of the generalized irreversible heat engine cycle. When maximizing the profit under this condition, minimization of the losses of exergy can be achieved. Eq. (38) indicates that the heat engine is always operating at a loss, unless it operates reversibly to reach the break-even point.

Therefore, for any intermediate values of ψ_2/ψ_1, the finite-time exergoeconomic performance bound (η_π) lies between the finite-time thermodynamic performance bound and the reversible performance bound.

4. Conclusion

The relationship between the optimal profit rate and efficiency of a generalized irreversible Carnot heat engine is derived based on generalized radiative heat transfer law. The influence of different heat transfer laws and irreversibilities on this relationship has been discussed. The results are helpful for establishing a link among finite-time exergoeconomic performance bound, finite-time thermodynamic performance bound and the reversible performance bound.

Acknowledgments

This paper is supported by National Natural Science Foundation of China (Project No. 10905093).

References

[1] Andresen, B., Finite-Time Thermodynamics. 1983: University of Copenhagen Copenhagen.

[2] Bejan, A., Entropy generation minimization: The new thermodynamics of finite-size devices and finite?time processes. Journal of Applied Physics, 1996, 79(3): p. 1191-1218.

[3] Chen, L., C. Wu, and F. Sun, Finite time thermodynamic optimization or entropy generation minimization of energy systems. Journal of Non-Equilibrium Thermodynamics, 1999, 24(4): p. 327-359.

[4] Berry, R.S., et al., Thermodynamic Optimization of Finite-Time Processes. 2000: Wiley Chichester.

[5] Chen, L. and F. Sun, Advances in Finite Time Thermodynamics: Analysis and Optimization. 2004: Nova Publishers.

[6] Durmayaz, A., et al., Optimization of thermal systems based on finite-time thermodynamics and thermoeconomics. Progress in Energy and Combustion Science, 2004, 30(2): p. 175-217.

[7] Andresen, B., Current Trends in Finite-Time Thermodynamics. Angewandte Chemie International Edition, 2011, 50(12): p. 2690-2704.

[8] Le Roux, W., T. Bello-Ochende, and J. Meyer, A review on the thermodynamic optimisation and modelling of the solar thermal Brayton cycle. Renewable and Sustainable Energy Reviews, 2013, 28: p. 677-690.

[9] Berry, R.S., P. Salamon, and G. Heal, On a relation between economic and thermodynamic optima. Resources and Energy, 1978, 1(2): p. 125-137.

[10] Salamon, P. and A. Nitzan, Finite time optimizations of a Newton's law Carnot cycle. The Journal of Chemical Physics, 1981, 74: p. 3546.

[11] Wu, C., L. Chen, and F. Sun, Effect of the heat transfer law on the finite-time, exergoeconomic performance of heat engines. Energy, 1996, 21(12): p. 1127-1134.

[12] Chen, L., F. Sun, and C. Wu, Effect of heat transfer law on the performance of a generalized irreversible Carnot engine. Journal of Physics D: Applied Physics, 1999, 32(2): p. 99.

[13] Chen, L., F. Sun, and C. Wu,, Maximum profit performance of a three heat-reservoir heat pump. International Journal of Energy Research, 1999, 23(9): p. 773-777.

[14] Wu, F., et al., Finite-time exergoeconomic performance bound for a quantum Stirling engine. International Journal of Engineering Science, 2000, 38(2): p. 239-247.

[15] Chen, L., C. Wu, and F. Sun, Effect of heat transfer law on the finite-time exergoeconomic performance of a Carnot refrigerator. Exergy, An International Journal, 2001, 1(4): p. 295-302.

[16] Chen, L., F. Sun, and C. Wu, Maximum-profit performance for generalized irreversible Carnot-engines. Applied energy, 2004, 79(1): p. 15-25.

[17] Qin, X., et al., Thermo-economic optimization of an endoreversible four-heat-reservoir absorption-refrigerator. Applied Energy, 2005, 81(4): p. 420-433.

[18] Ibrahim, O., S. Klein, and J. Mitchell, Effects of irreversibility and economics on the performance of a heat engine. Trans. ASME Journal of Solar Energy Engineering, 1992, 114(4) 786-790.

[19] De Vos, A., Endoreversible thermoeconomics. Energy Conversion and Management, 1995, 36(1): p. 1-5.

[20] De Vos, A., Endoreversible economics. Energy Conversion and Management, 1997, 38(4): p. 311-317.

[21] Bejan, A., Power and refrigeration plants for minimum heat exchanger inventory. Trans. ASME Journal of Energy Resources Technology, 1993, 115(2): p. 148-150.

[22] Chen, L., F. Sun, and C. Wu, Endoreversible thermoeconomics for heat engines. Applied Energy, 2005, 81(4): p. 388-396.

[23] Sahin, B. and A. Kodal, Finite time thermoeconomic optimization for endoreversible refrigerators and heat pumps. Energy Conversion and Management, 1999, 40(9): p. 951-960.

[24] Sahin, B. and A. Kodal, Performance analysis of an endoreversible heat engine based on a new thermoeconomic optimization criterion. Energy Conversion and Management, 2001, 42(9): p. 1085-1093.

[25] Sahin, B. and A. Kodal, Thermoeconomic optimization of a two stage combined refrigeration system: a finite-time approach. International Journal of Refrigeration, 2002, 25(7): p. 872-877.

[26] Sahin, B., A. Kodal, and A. Koyun, Optimal performance characteristics of a two-stage irreversible combined refrigeration system under maximum cooling load per unit total cost conditions. Energy Conversion and Management, 2001, 42(4): p. 451-465.

[27] Chen, L., F. Sun, and C. Wu, A generalised model of a real heat engine and its performance. Journal of the Institute of Energy, 1996, 69(481): p. 214-222.

[28] Feng H, Chen L, Sun F. Finite time exergoeconomic performance optimization for an irreversible universal steady flow variable-temperature heat reservoir heat pump cycle model. International Journal of Energy and Environment, 2010, 1(6): 969-986.

[29] Kan X, Chen L, Sun F, Wu F. Finite time exergoeconomic performance optimization of a thermoacoustic heat engine. International Journal of Energy and Environment, 2011, 2(1): 85-98.

[30] Li J, Chen L, Sun F. Finite-time exergoeconomic performance of an endoreversible Carnot heat engine with complex heat transfer law. International Journal of Energy and Environment, 2011, 2(1): 171-178.

[31] Chen L, Yang B, Sun F. Exergoeconomic performance optimization of an endoreversible intercooled regenerated Brayton cogeneration plant. Part 1: thermodynamic model and parameter analyses. International Journal of Energy and Environment, 2011, 2(2): 199-210

[32] Yang B, Chen L, Sun F. Exergoeconomic performance optimization of an endoreversible intercooled regenerated Brayton cogeneration plant. Part 2: heat conductance allocation and pressure ratio optimization. International Journal of Energy and Environment, 2011, 2(2): 211-218.

[33] Chen L, Kan X, Wu F, Sun F. Finite time exergoeconomic performance optimization of a thermoacoustic cooler with a complex heat transfer exponent. International Journal of Energy and Environment, 2012, 3(1): 19-32.

[34] Yang B, Chen L, Sun F. Exergoeconomic performance optimization of an endoreversible intercooled regenerative Brayton combined heat and power plant coupled to variable- temperature heat reservoirs. International Journal of Energy and Environment, 2012, 3(4): 505-520.

[35] Chen L, Kan X, Sun F, Wu F. Exergoeconomic performance optimization for a steady-flow endoreversible refrigerator model including six typical cycles. International Journal of Energy and Environment, 2013, 4(1): 93-102.

Process simulation and maximization of energy output in chemical-looping combustion using ASPEN plus

Xiao Zhang, Subhodeep Banerjee, Ling Zhou, Ramesh Agarwal

Department of Mechanical Engineering & Materials Science, Washington University in St. Louis, 1 Brookings Drive, St. Louis, MO 63130, USA.

Abstract
Chemical-looping combustion (CLC) is currently considered as a leading technology for reducing the economic cost of CO_2 capture. In this paper, several process simulations of chemical-looping combustion are conducted using the ASPEN Plus software. The entire CLC process from the beginning of coal gasification to the reduction and oxidation of the oxygen carrier is modeled and validated against experimental data. The energy balance of each major component of the CLC process, e.g., the fuel and air reactors and air/flue gas heat exchangers is examined. Different air flow rates and oxygen carrier feeding rates are used in the simulations to obtain the optimum ratio of coal, air, and oxygen carrier that produces the maximum power. Two scaled-up simulations are also conducted to investigate the influence of increase in coal feeding on power generation. It is demonstrated that the optimum ratio of coal, air supply, and oxygen carrier for maximum power generation remains valid for scaled-up cases with substantially larger coal feeding rates; the maximum power generation scales up linearly by using the process simulation models in ASPEN Plus. The energy output from four different types of coals is compared, and the optimum ratio of coal, air supply and oxygen carrier for maximum power generation for each type of coal is determined.

Keywords: Carbon capture; Process simulation; Chemical-looping combustion; Maximum energy output; Optimization; Scaled-up simulation.

1. Introduction

Coal-fired power plants contribute to significant CO_2 emissions; this reality has driven research in recent years on investigation of combustion processes that can capture CO_2 with reduced energy penalty. One technology that is showing great promise for high-efficiency low-cost carbon capture is the Chemical-Looping Combustion (CLC) process. In contrast to other methods for CO_2 separation from flue gas such as oxy-combustion, chemical absorption, and physical adsorption, the CLC is an advanced technology that creates and captures an almost pure and concentrated CO_2 stream with relatively less energy requirement [1, 2]. Several theoretical and experimental studies have demonstrated the potential of CLC to capture almost pure CO_2 very efficiently [3-6]. CLC employs a dual fluidized bed system with circulating fluidized bed process where an oxygen carrier (OC) is used as a bed material providing the oxygen for combustion in the fuel reactor. The reduced OC is then transferred to a second bed and re-oxidized by the atmospheric air [7-9] in an air reactor before it is returned to the fuel reactor to complete the loop. Because of the absence of air in the fuel reactor, the combustion products are not diluted by

other gases (e.g., N_2), resulting in high purity CO_2 available at the outlet of the fuel reactor. Thus, the CLC process for power generation provides a sequestration-ready CO_2 stream directly after combustion, without the need for using costly gas separation techniques to purify CO_2 from the flue stream. CLC therefore holds significant promise as a next generation combustion technology due to its potential for pre-capturing almost pure CO_2 emission with very limited effect on the efficiency of the power plant.
ASPEN Plus is a process simulation software that simulates chemical processes at system level using basic engineering relationships such as mass and energy balance, and multi-phase and chemical reaction models. It consists of flow sheet simulations to calculate stream properties such as flow rate and mass composition given various chemical processes and operating conditions. In this paper, a system level model of CLC process is developed to conduct parametric studies for optimal energy output.. These studies provide valuable insight into the design and operating conditions required in an industrial-scale CLC plant and to assess the feasibility of deploying CLC as an economically viable solution for electricity generation and carbon capture.

2. Validation of the CLC process simulation with experiment

The CLC process simulation in ASPEN Plus was validated against the experimental work of Sahir et al [10]. The physical and chemical properties of the Colombian coal used as the solid fuel in the experiment are summarized in Table 1.

Table 1. Physical and chemical properties of Colombian coal

Proximate Analysis (wt. %)				Ultimate Analysis (wt. %)					Energy
Moisture	Volatile matter	Fixed carbon	Ash	C	H	N	S	O	LHV (MJ/kg)
3.3	37.0	54.5	5.2	80.7	5.5	1.7	0.6	11.5	29.1

The schematic of the flow sheet for this simulation is shown in Figure 1. The coal is first pulverized and dried before it is pressurized and introduced into a shell gasifier to be partially oxidized to form syngas. The molar ratio of steam and carbon is maintained at unity for the process model. The syngas composition at the gasifier outlet is 34.5% CO, 50.3% H_2, 12.3% H_2O, and 2.4% CO_2. The syngas is converted completely to CO_2 and H_2O in the fuel reactor while the Fe_2O_3 in the oxygen carrier is reduced to Fe_3O_4. The oxygen carrier material used is a mixture of 60 wt. % Fe_2O_3 and 40 wt. % inert Al_2O_3 as support. The outflow from the fuel reactor is a concentrated stream of H_2O and CO_2. After condensing the stream, high purity CO_2 is obtained. The reduced oxygen carrier is fed into the air reactor where the oxidation reaction takes place with an 80% conversion of Fe_3O_4 to Fe_2O_3.

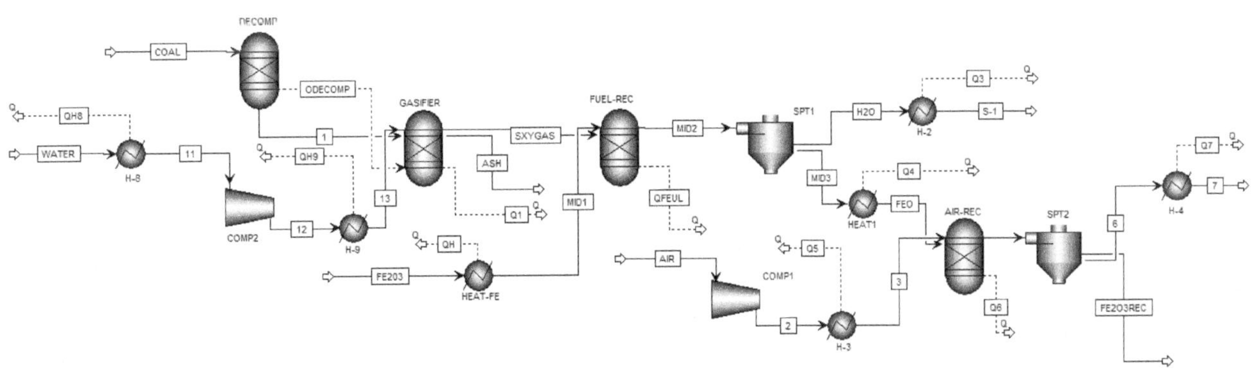

Figure 1. Flow sheet of the CLC model in ASPEN Plus

The various process models used in the ASPEN Plus shown in flow sheet in Figure 1 are summarized in Table 2. The coal devolatilization is defined by the RYIELD reactor, followed by the gasification of coal represented by the RGIBBS reactor. Another RGIBBS reactor defines the actual syngas combustion and the corresponding reduction of the oxygen carrier. These blocks together represent the fuel reactor. The flow sheet within the ASPEN Plus simulation package cannot model this entire reaction with one reactor. As a result, the fuel reactor simulation is broken down into several different reactor simulations. The air reactor is also modeled as an RGIBBS reactor.

Table 2. Process models used in different parts of the CLC process in ASPEN Plus

Name	Model	Function	Reaction formula
DECOMP	RYIELD	Coal devolatilization	Coal $\rightarrow$ volatile matter + char
BURN	RGIBBS	Gasification	Char + volatile matter $\rightarrow CO_2 + H_2O$
FUEL-R	RSTOIC	Carrier reduction reaction	$3Fe_2O_3 + CO/H_2 \rightarrow 2Fe_3O_4 + CO_2/H_2O$
AIR-R	RSTOIC	Carrier oxidation reaction	$4Fe_3O_4 + O_2 \rightarrow 6Fe_2O_3$

For the purpose of validation, the energy balance of the CLC process model was analyzed using the input values from the experiment of Sahir et al [10]. The input values and the energy requirements for the various units and streams in Figure 1 are presented in Table 3; this will be referred to as the baseline case in rest of the paper. Energy is consumed mainly in the compressor processes. Compressed air is required in the air reactor to regenerate Fe_2O_3 from Fe_3O_4. The air compressor for the combustor compresses air to 18atm. Another compressor is used to compress the steam for the gasifier. There is a large amount of energy produced in the air reactor, but the fuel reactor needs to be supplied with energy. This is because the net heat work in the fuel reactor is the summation of the heat work from the DECOMP, GASIFER, and FUEL-R blocks. Although FUEL-R produces energy because of the combustion of syngas, the combined energy requirement of DECOMP and GASIFIER are more than the energy produced in FUEL-R. Summing the energy requirements of each individual stream, the total energy obtained from the CLC process is 554.2 kW.

Table 3. Input values and energy balance for the baseline case corresponding to the experiment of Sahir et al. [10]

Input values	Coal (kg/h)	100
	Steam (kg/h)	140
	Air Flow Rate (kg/h)	71
	Temperature of Fuel Reactor (°C)	950
	Temperature of Air Reactor (°C)	935
	Fe_2O_3 flow in the Fuel Reactor (kg/h)	5921
	Al_2O_3 in the System (kg/h)	3951
	Particle Density (kg/m³)	3200
Energy Balance (kW)	Fuel reactor	-161.8
	Air reactor	688.0
	Cool air reactor exhaust	135.4
	Cool flue gas	148.3
	Cool OC for air reactor	40.9
	Reheat OC for fuel reactor	-42.7
	Heat steam	-69.8
	Heat air	-184.1
	Net	554.2

The results shown in Table 3 for the baseline case with a coal feed rate of 100 kg/h are in excellent agreement with those reported by Sahir et al [10]. These calculations validate our CLC model developed in ASPEN Plus.

3. Investigation of the effect of various parameters on the energy output of the CLC process simulation

With the successful validation of the process simulation of the CLC experiment of Sahir et al in the previous section, the ASPEN Plus simulation is expanded to consider the effect of varying the air flow rate and the oxygen carrier feeding rate. Additional scaled-up simulations are also conducted to determine these effects on an industrial scale power plant.

3.1 Effect of varying the air flow rate on energy output of baseline case with 100 kg/h coal feeding rate

The recent paper of Mukherjee et al. [11] suggests that it is favorable to operate the air reactor of the CLC process at higher temperatures with excess air supply in order to achieve higher power efficiency.

In order to evaluate the effect of air supply on energy output, we consider the baseline case of Table 3 and vary the air flow rate. The results are presented in Figure 2 and Table A1. From Figure 2, it can be seen that with an increase in the air flow rate, the net energy output increases and achieves a maximum for a certain air flow rate. If the air flow rate is further increased from its maximum value (i.e., value corresponding to maximum energy output), the energy output starts decreasing albeit very slowly. This result implies that there exists a certain rate of air supply around 900 kg/h to obtain the maximum energy output for 100 kg/h of coal supply. At this flow rate in the air reactor, 131.06 kW of additional energy is generated, which is 23.6% more than the baseline case given in Table 2 indicating that the reaction in the air reactor is not complete for the baseline case. Excess air supply ensures the 80% conversion of Fe_3O_4 to Fe_2O_3.

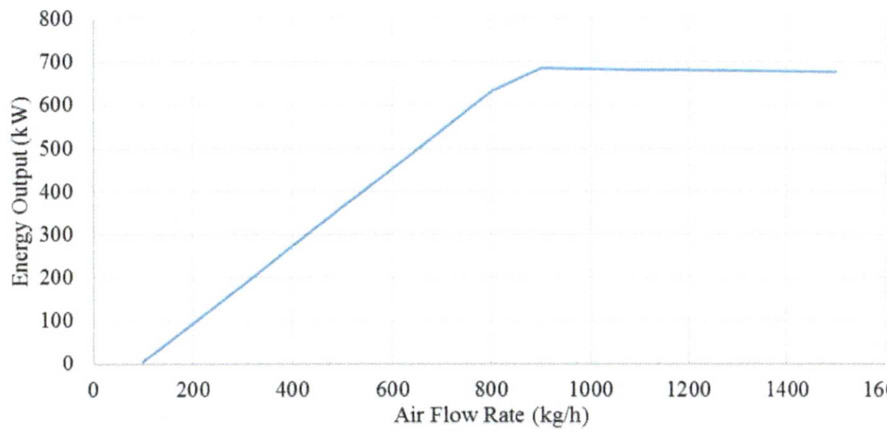

Figure 2. Energy output for various air flow rates for 100 kg/h of coal supply

3.2 Effect of varying the oxygen carrier feeding rate on energy output of baseline case with 100 kg/h coal feed rate

The oxygen carrier (OC) plays a vital role in the CLC process; it reacts with the syngas in the fuel reactor and reacts with the air in the air reactor. Both of these reactions contribute a large amount to the net energy output. Figure 3 and Tables A1–A6 present the energy output for different OC feeding rates in the system with varying air flow rates. As expected, Figure 3 shows that for a given air flow rate, a higher OC feeding rate yields more energy output. However, when the OC feeding rate increases above a certain threshold value, the marginal increase in energy output by increasing the OC rate becomes extremely small. The red line in Figure 3 represents the baseline case (Fe_2O_3 at 5921 kg/h), for which the maximum energy output is 685.26 kW with 900 kg/h air flow rate. For the threshold Fe_2O_3 rate of 7000 kg/h, the maximum energy output of 824.33 kW occurs at the 1000 kg/h air flow rate. 138.97 kW of additional energy output is obtained by increasing the OC rate from 5921 kg/h to 7000 kg/h. Therefore, for maximum energy output with a coal feeding rate of 100 kg/h, the optimum rates of air flow and OC feeding are 1000 kg/h and 7000 kg/h respectively. In other words, the optimum ratio of Coal: Air: OC is 1: 10: 70.

3.3 Scaled-up simulation

Scaling up is an essential step for the realization and optimization of industrial-scale power plants. Two different scaled-up simulations were conducted to investigate the relationship between the coal feeding rate and energy output. The first scaled-up simulation employed the initial values of the baseline case and the other was based on the optimum values of coal: air supply: oxygen carrier rate. The details of the scaled-up simulations are given in Table A7 and Table A8 respectively. In both cases, the coal feeding rate is scaled up by a factor of up to 12. The OC feeding rate and air supply rate are also scaled-up accordingly to meet the demand of the increased coal feeding. Other modeling parameters such as the reactor efficiency and coal decomposition rate are considered unchanged for both the scaled up simulations. The total thermal power output for the scaled-up simulations is summarized in Figure 4 and Table 4 below. From Figure 4, it can be seen that the total power output increases linearly with increase in coal feeding rate. Considering the principles of energy and mass balance on which the ASPEN Plus modeling is based, linearity in the scaled-up results is expected since the non-linear effects (e.g., the energy loss at multiple locations in the flow sheet) are omitted in the modeling process.

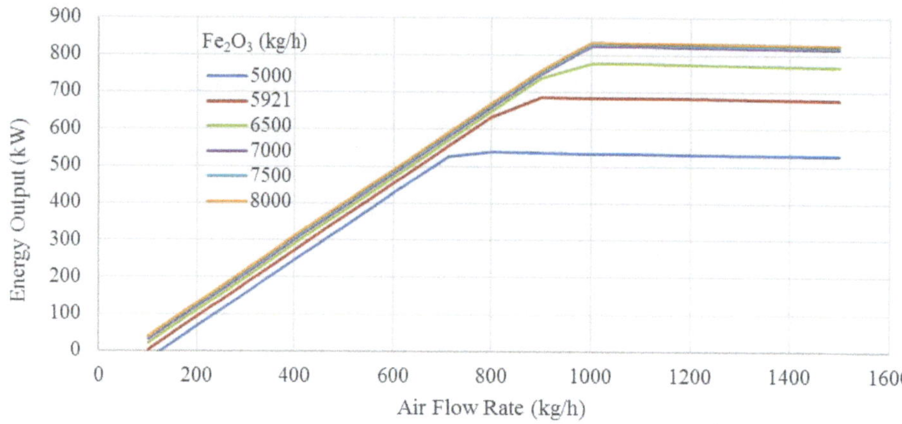

Figure 3. Energy output for various oxygen carrier feeding rates and air flow rates for 100 kg/h of coal supply

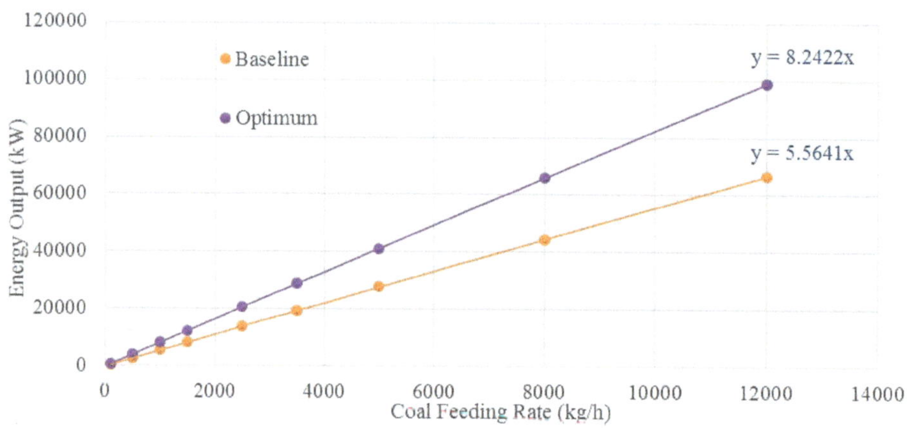

Figure 4. Energy output of two scaled-up simulations for various coal feeding rates

Table 4. Results of two scaled-up simulations for different ratios of Coal: Air: OC

Coal (kg/h)		100	500	1000	1500	2500	3500	5000	8000	12000
Energy output (kW)	Baseline	554.2	2782	5564	8346	13910	19474	27820	44513	66769
	Optimum	824.2	4121	8242	12363	20606	28847	41211	65936	98907

Based on these scaled-up simulations, the energy output for the baseline case is given by the equation

$$\text{Energy output} = 5.5641 \times \text{Coal feeding rate} \quad (1)$$

and the energy output for the optimum case is given by the equation

$$\text{Energy output} = 8.2422 \times \text{Coal feeding rate} \quad (2)$$

3.4 Validation of optimum values of air flow rate and oxygen carrier feeding rate for scaled-up simulation

To demonstrate that the optimum values of air flow rate and OC feeding rate for maximum energy output are valid for the scaled-up simulations, four more cases with 12,000 kg/h coal feeding rate and varying rates of air flow and OC were studied, which are presented in Figure 5 and Tables A9-A11. Figure 5 shows that the maximum energy output occurs at 120,000 kg/h of air flow rate, and 840,000 kg/h of

Fe_2O_3 feeding rate. This suggests that the optimum ratio of Coal: Air: OC in the system still holds for the scaled-up simulations; it is given by

Coal feeding rate: Air flow rate: OC feeding rate=1:10:70 (3)

Equation (3) is an important relationship among these three input parameters for obtaining the maximum energy output from a CLC-based power plant. This relationship can be used for the initial estimates in designing a CLC-based industrial-scale power plant.

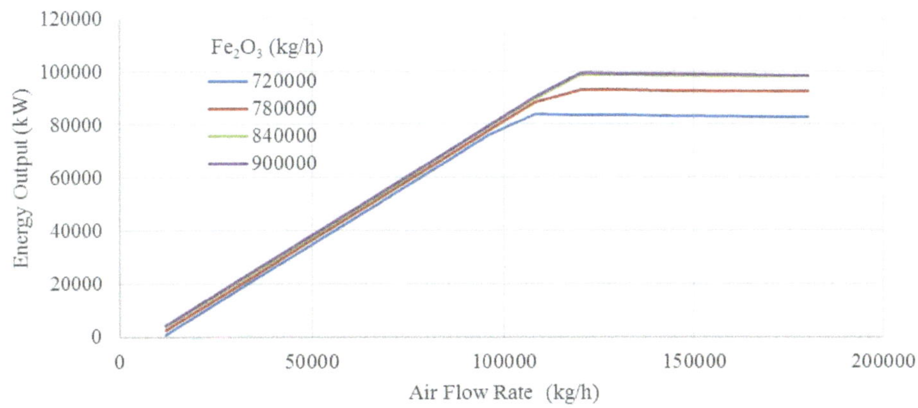

Figure 5. Energy output for different airflow rates and OC rates for a 12000 kg/h coal feeding rate

4. Energy output of different types of coals

All the results above are dependent on the certain type of coal, the Colombian coal of which the physical and chemical properties are listed in Table 1. Now it is interesting to investigate the performance of different types of coal. Four types of coals are used in this paper which are Colombian, Bituminous, Anthracite and Lignite. The proximate analysis and ultimate analysis of Colombian coal is given in Table 1 and that of other three coals are summarized in Table 5.

Table 5. Physical and chemical properties of bituminous, anthracite and lignite coals

Coal name	Proximate Analysis (wt. %)				Ultimate Analysis (wt. %)						Energy
	Moisture	Volatile matter	Fixed carbon	Ash	C	H	N	S	O	Ash	LHV (kJ/kg)
Bituminous	2.3	33.0	55.9	8.8	65.8	3.3	1.6	0.6	17.6	11.1	21899
Anthracite	1.0	7.5	59.9	31.6	60.7	2.1	0.9	1.3	2.4	32.6	21900
Lignite	12.6	28.6	33.6	25.2	45.4	2.5	0.6	5.2	8.5	37.8	16250

4.1 Effect of varying the air flow rate on energy output of four types of coals with 100 kg/h coal feeding rate

Again in order to evaluate the effect of air supply on energy output, we conduct the same process modeling as described in section 3.1 by varying the air flow rate with coal feeding rate of 100 kg/h for four different types of coals. The results are presented in Figure 6 and Table A1 for Colombian coal and in Tables A12-A14 for Bituminous, Anthracite and Lignite coal. From Figure 6, it can be seen that with an increase in air flow rate, all four types of coal show the same trend that net energy output increases and achieves a maximum for a certain air flow rate. Every coal type has a different inflection point which corresponds to the maximum energy output on the y-axis for a certain air flow rate shown on x-axis. It can be seen that the inflection point is different depending upon the type of coal which is expected because of different properties of the coals as given in Table 5. By qualitative analysis, one can infer that higher the concentration of fixed carbon in a coal gives more fuel to burn, and the higher concentration of volatile matter and ash cost less energy to decompose the coal. Next, we determine the optimal ratio of Coal: Air: OC for the other three types of coal.

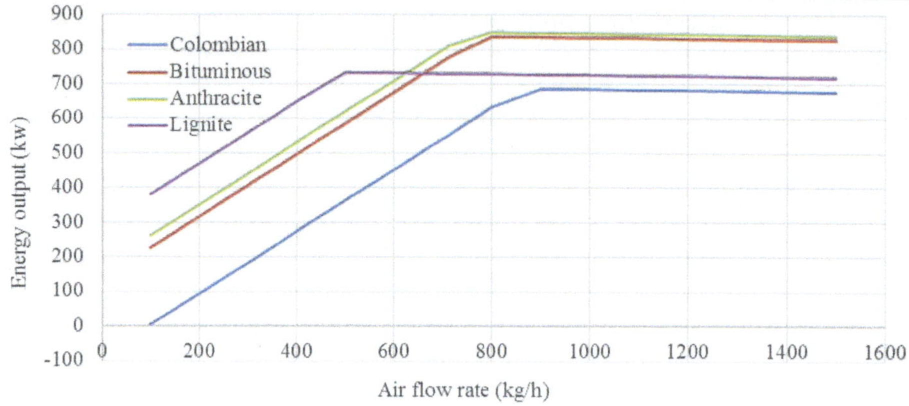

Figure 6. Energy output for various air flow rates for 100 kg/h of four types of coals

4.2 Effect of varying the oxygen carrier feeding rate on energy output of four types of coals with 100 kg/h coal feeding rate

The effect of varying the oxygen carrier feeding rate on energy output of Colombian coal was shown in Figure 3 and Tables A1-A6. The results of other three types of coal, Bituminous, Anthracite and Lignite are presented in Figures 7-9 and Tables A12-A24. As expected, as with the Colombian coal, there is the maximum energy output based on optimal coal feeding rate: air flow rate: OC feeding rate for Bituminous, Anthracite and Lignite coal as well. Table 6 summarizes the maximum energy output and optimal ratio of Coal: Air: OC for four types of coal with 100 kg/h coal feed rate.

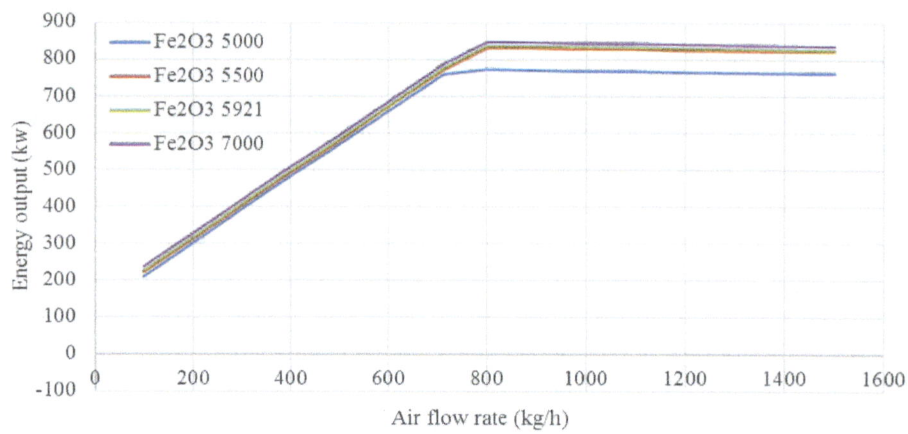

Figure 7. Energy output for various oxygen carrier feeding rates and air flow rates for 100 kg/h of Bituminous coal

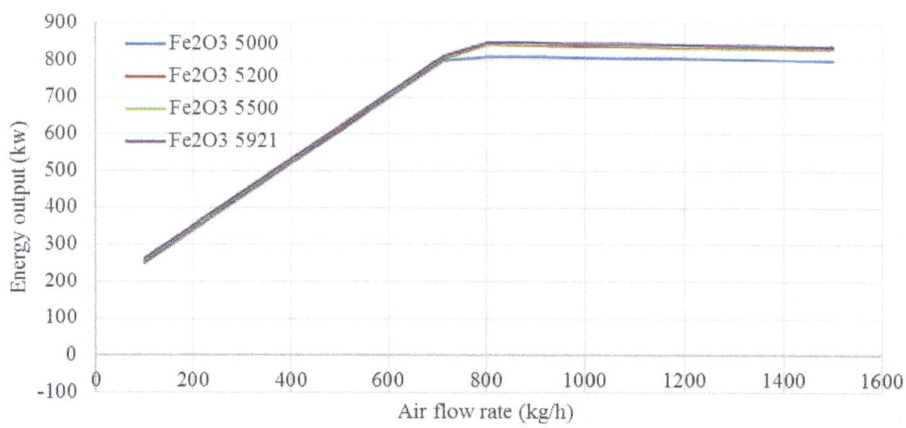

Figure 8. Energy output for various oxygen carrier feeding rates and air flow rates for 100 kg/h of Anthracites coal

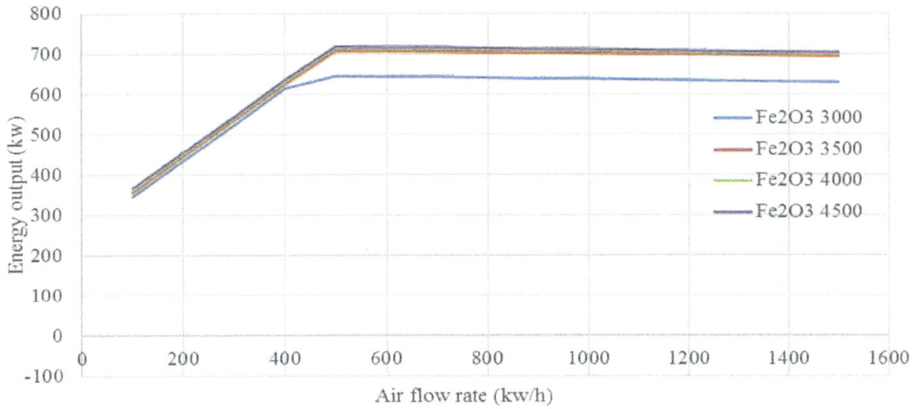

Figure 9. Energy output for various oxygen carrier feeding rates and air flow rates for 100 kg/h of Lignite coal

Table 6. Maximum energy output and optimal ratio of Coal: Air: OC for four types of coal with 100 kg/h coal feed rate

Coal Name	Maximum Energy (kw)	Optimal Ratio of Coal: Air: OC
Colombian	824.229	1: 70: 10
Bituminous	832.373	1: 55: 8
Anthracite	841.258	1: 52: 8
Lignite	707.905	1: 35: 5

5. Conclusions

In this paper, ASPEN Plus is employed to model and study the complete CLC process from the coal gasification to the reduction and eventual re-oxidation of the oxygen carrier (OC). The CLC process model is validated against the experiment of Sahir et al [10] and shows good agreement between the experimental data and the simulation results. Based on further studies on the effect of varying air flow rates and OC feeding rates, it is found that the maximization of energy output from CLC can be accomplished by using the optimum ratio of Coal: Air: OC in the system equal to 1: 10: 70 for Colombian coal. Compared to the baseline case based on the experiment of Sahir et al [10], a net increase in power of 48% can be obtained by increasing the air flow rate by 40.25% and the OC feeding rate by 18.22% to attain this optimum ratio for the Colombian coal for the given coal feeding rate of 100kg/h. Scaled-up simulations are also conducted using different coal feeding rates. The results show that the total power output is linear with increase in coal feeding rate. In general, such linearity is not expected for actual industrial-level scale-up since the ASPEN Plus system modeling software neglects miscellaneous energy losses in the system due to changes in the hydrodynamic characteristics of the two fluidized bed reactors. To account for the changes in the hydrodynamics characteristics, detailed hydrodynamic simulations are needed using the computational fluid dynamics software. Three other types of coal (Bituminous, Anthracite, and Lignite) are also investigated, and the optimal ratio of coal: airflow: OC is determined for each of these coal types. There are other parameters that may also influence the energy output such as the temperature and pressure of the reactors, particle size, etc., which are not investigated in this paper.

Reference

[1] Cuadrat A., Abad A., de Diego L.F. (2012) Prompt considerations on the design of chemical-looping combustion of coal from experimental tests, Fuel 97, pp. 219-232.

[2] Wang J., Anthony E.J. (2008) Clean combustion of solid fuels, Applied Energy 85 (2), pp. 73-79.

[3] Gnanapragasam N.V., Reddy B.V., Rosen M.A. (2009) Hydrogen production from coal using coal direct chemical looping and syngas chemical looping combustion systems: assessment of system operation and resource requirements, Int. J. Hydrogen Energy 34 (6), pp. 2606-2615.

[4] Adánez J., Gayán P., Celaya J. (2006) Chemical looping combustion in a 10 kWth prototype using a CuO/Al2O3 oxygen carrier: Effect of operating conditions on methane combustion, Industrial & Engineering Chemistry Research 45 (17), pp. 6075-6080.

[5] Arjmand M., Azad A.M., Leion H., Lyngfelt A., Mattisson T. (2011) Prospects of Al2O3 and MgAl2O4-supported CuO oxygen carriers in chemical-looping combustion (CLC) and chemical-looping with oxygen uncoupling (CLOU), Energy & Fuels 25 (11), pp. 5493-5502.

[6] Leion H., Lyngfelt A., Johansson M., Jerndal E., Mattisson T. (2008) The use of ilmenite as an oxygen carrier in chemical-looping combustion, Chemical Engineering Research and Design 86 (9), pp. 1017-1026.

[7] Leion H., Mattisson T., Lyngfelt A. (2009) Using chemical-looping with oxygen uncoupling (CLOU) for combustion of six different solid fuels, Energy Procedia 1 (1), pp. 447-453.

[8] Mattisson T., Lyngfelt A., Leion H. (2009) Chemical-looping with oxygen uncoupling for combustion of solid fuels, Int. J. Greenhouse Gas Control 3 (1), pp. 11-19.

[9] Cuadrat A., Abad A., Adánez J., de Diego L.F. (2012) Behavior of ilmenite as oxygen carrier in chemical-looping combustion, Fuel Processing Technology 94 (1), pp. 101-112.

[10] Sahir A.H., Cadore A.L., Dansie J.K. (2012) Process analysis of chemical looping with oxygen uncoupling (CLOU) and chemical looping combustion (CLC) for solid fuels, 2nd International Conference on Chemical looping, Darmstadt, Germany.

[11] Mukherjee S., Kumar P., Hosseini A. (2014) Comparative assessment of gasification based coal power plants with various CO_2 capture technologies producing electricity and hydrogen Energy & Fuels 28 (2), pp. 1028-1040.

Appendix

Table A1. CLC process simulation results for different air flow rates with Colombian coal at 100 kg/h and Fe_2O_3/Al_2O_3 at 5921/3951 kg/h

Initial values	Coal (kg/h)	100	100	100	100	100	100
	Water (kg/h)	140	140	140	140	140	140
	Air Flow Rate (kg/h)	100	300	400	500	600	713
	Temperature of Fuel Reactor (°C)	950	950	950	950	950	950
	Temperature of Air Reactor (°C)	935	935	935	935	935	935
	Fe_2O_3 flow in the Fuel Reactor (kg/h)	5921	5921	5921	5921	5921	5921
	Al_2O_3 in the System (kg/h)	3951	3951	3951	3951	3951	3951
	Particle Density (kg/m^3)	3200	3200	3200	3200	3200	3200
Energy balance (kW)	Fuel Reactor	-161.8	-161.8	-161.8	-161.8	-161.8	-161.8
	Air Reactor	96.498	289.5	386	482.49	578.99	688
	Cool air reactor exhaust	18.996	56.988	75.985	94.981	113.98	135.4
	Cool flue gas	148.3	148.3	148.3	148.3	148.3	148.3
	Cool OC for air reactor	40.9	40.9	40.9	40.9	40.9	40.9
	Reheat OC for fuel reactor	-42.7	-42.7	-42.7	-42.7	-42.7	-42.7
	Heat steam	-69.8	-69.8	-69.8	-69.8	-69.8	-69.8
	Heat air	-25.82	-77.47	-103.3	-129.1	-154.9	-184.1
	Net	4.57	183.92	273.59	363.26	452.93	554.2
Initial values (continued)	Coal (kg/h)	100	100	100	100	100	100
	Water (kg/h)	140	140	140	140	140	140
	Air Flow Rate (kg/h)	800	900	1000	1100	1200	1500
	Temperature of Fuel Reactor (°C)	950	950	950	950	950	950
	Temperature of Air Reactor (°C)	935	935	935	935	935	935
	Fe_2O_3 flow in the Fuel Reactor (kg/h)	5921	5921	5921	5921	5921	5921
	Al_2O_3 in the System (kg/h)	3951	3951	3951	3951	3951	3951
	Particle Density (kg/m^3)	3200	3200	3200	3200	3200	3200
Energy balance (kW) (continued)	Fuel Reactor	-161.8	-161.8	-161.8	-161.8	-161.8	-161.8
	Air Reactor	771.99	829.67	829.79	829.93	830.07	830.49
	Cool air reactor exhaust	151.97	173.08	197.33	221.58	245.83	318.57
	Cool flue gas	148.33	148.33	148.33	148.33	148.33	148.33
	Cool OC for air reactor	40.9	40.9	40.9	40.9	40.9	40.9
	Reheat OC for fuel reactor	-42.7	-42.7	-42.7	-42.7	-42.7	-42.7
	Heat steam	-69.8	-69.8	-69.8	-69.8	-68.8	-69.8
	Heat air	-206.6	-232.4	-258.2	-284.1	-309.9	-387.3
	Net	632.3	685.26	683.81	682.37	681.94	676.63

Table A2. CLC process simulation results for different air flow rates with Colombian coal at 100 kg/h and Fe_2O_3/Al_2O_3 at 5000/3000 kg/h

Initial values	Coal (kg/h)	100	100	100	100	100	100
	Water (kg/h)	140	140	140	140	140	140
	Air Flow Rate (kg/h)	100	300	400	500	600	713
	Temperature of Fuel Reactor (°C)	950	950	950	950	950	950
	Temperature of Air Reactor (°C)	935	935	935	935	935	935
	Fe_2O_3 flow in the Fuel Reactor (kg/h)	5000	5000	5000	5000	5000	5000
	Al_2O_3 in the System (kg/h)	3000	3000	3000	3000	3000	3000
	Particle Density (kg/m^3)	3200	3200	3200	3200	3200	3200
Energy balance (kW)	Fuel Reactor	-183.2	-183.2	-183.2	-183.2	-183.2	-183.2
	Air Reactor	96.498	289.5	386	482.49	578.99	688
	Cool air reactor exhaust	18.996	56.999	75.985	94.981	113.98	135.4
	Cool flue gas	142.63	142.63	142.63	142.63	142.63	142.63
	Cool OC for air reactor	32.792	32.792	32.792	32.792	32.792	32.792
	Reheat OC for fuel reactor	-34.27	-34.27	-34.27	-34.27	-34.27	-34.27
	Heat steam	-69.8	-69.8	-69.8	-69.8	-69.8	-69.8
	Heat air	-25.82	-77.47	-103.3	-129.1	-154.9	-184.1
	Net	-22.21	157.14	246.8	336.47	426.14	527.41
Initial values (continued)	Coal (kg/h)	100	100	100	100	100	100
	Water (kg/h)	140	140	140	140	140	140
	Air Flow Rate (kg/h)	800	900	1000	1100	1200	1500
	Temperature of Fuel Reactor (°C)	950	950	950	950	950	950
	Temperature of Air Reactor (°C)	935	935	935	935	935	935
	Fe_2O_3 flow in the Fuel Reactor (kg/h)	5000	5000	5000	5000	5000	5000
	Al_2O_3 in the System (kg/h)	3000	3000	3000	3000	3000	3000
	Particle Density (kg/m^3)	3200	3200	3200	3200	3200	3200
Energy balance (kW) (continued)	Fuel Reactor	-183.2	-183.2	-183.2	-183.2	-183.2	-183.2
	Air Reactor	700.66	700.79	700.93	701.11	701.21	701.64
	Cool air reactor exhaust	155.86	180.1	204.35	228.6	252.85	325.59
	Cool flue gas	142.63	142.63	142.63	142.63	142.63	142.63
	Cool OC for air reactor	32.792	32.792	32.792	32.792	32.792	32.792
	Reheat OC for fuel reactor	-34.27	-34.27	-34.27	-34.27	-34.27	-34.27
	Heat steam	-69.8	-69.8	-69.8	-69.8	-69.8	-69.8
	Heat air	-206.6	-232.4	-258.2	-284.1	-309.9	-387.3
	Net	538.05	536.6	535.16	533.77	532.29	527.99

Table A3. CLC process simulation results for different air flow rates with Colombian coal at 100 kg/h and Fe_2O_3/Al_2O_3 at 6500/4500 kg/h

Initial values	Coal (kg/h)	100	100	100	100	100	100
	Water (kg/h)	140	140	140	140	140	140
	Air Flow Rate (kg/h)	100	300	400	500	600	713
	Temperature of Fuel Reactor (°C)	950	950	950	950	950	950
	Temperature of Air Reactor (°C)	935	935	935	935	935	935
	Fe_2O_3 flow in the Fuel Reactor (kg/h)	6500	6500	6500	6500	6500	6500
	Al_2O_3 in the System (kg/h)	4500	4500	4500	4500	4500	4500
	Particle Density (kg/m^3)	3200	3200	3200	3200	3200	3200
Energy balance (kW)	Fuel Reactor	-148.4	-148.4	-148.4	-148.4	-148.4	-148.4
	Air Reactor	96.498	289.5	386	482.49	578.99	688.04
	Cool air reactor exhaust	18.996	56.999	75.985	94.981	113.98	135.4
	Cool flue gas	151.96	151.96	151.96	151.96	151.96	151.96
	Cool OC for air reactor	45.781	45.781	45.781	45.781	45.781	45.781
	Reheat OC for fuel reactor	-47.71	-47.71	-47.71	-47.71	-47.71	-47.71
	Heat steam	-69.8	-69.8	-69.8	-69.8	-69.8	-69.8
	Heat air	-25.82	-77.47	-103.3	-129.1	-154.9	-184.1
	Net	21.473	200.83	290.49	380.16	469.83	571.14

Initial values (continued)	Coal (kg/h)	100	100	100	100	100	100
	Water (kg/h)	140	140	140	140	140	140
	Air Flow Rate (kg/h)	800	900	1000	1100	1200	1500
	Temperature of Fuel Reactor (°C)	950	950	950	950	950	950
	Temperature of Air Reactor (°C)	935	935	935	935	935	935
	Fe_2O_3 flow in the Fuel Reactor (kg/h)	6500	6500	6500	6500	6500	6500
	Al_2O_3 in the System (kg/h)	4500	4500	4500	4500	4500	4500
	Particle Density (kg/m³)	3200	3200	3200	3200	3200	3200
Energy balance (kW) (continued)	Fuel Reactor	-148.4	-148.4	-148.4	-148.4	-148.4	-148.4
	Air Reactor	771.99	868.49	910.81	910.94	911.07	911.49
	Cool air reactor exhaust	151.97	170.97	192.91	217.16	241.41	314.15
	Cool flue gas	151.96	151.96	151.96	151.96	151.96	151.96
	Cool OC for air reactor	45.781	45.781	45.781	45.781	45.781	45.781
	Reheat OC for fuel reactor	-47.71	-47.71	-47.71	-47.71	-47.71	-47.71
	Heat steam	-69.8	-69.8	-69.8	-69.8	-69.8	-69.8
	Heat air	-206.6	-232.4	-258.2	-284.1	-309.9	-387.3
	Net	649.18	738.85	777.29	775.85	774.41	770.1

Table A4. CLC process simulation results for different air flow rates with Colombian coal at 100 kg/h and Fe_2O_3/Al_2O_3 at 7000/5000 kg/h

Initial values	Coal (kg/h)	100	100	100	100	100	100
	Water (kg/h)	140	140	140	140	140	140
	Air Flow Rate (kg/h)	100	300	400	500	600	713
	Temperature of Fuel Reactor (°C)	950	950	950	950	950	950
	Temperature of Air Reactor (°C)	935	935	935	935	935	935
	Fe_2O_3 flow in the Fuel Reactor (kg/h)	7000	7000	7000	7000	7000	7000
	Al_2O_3 in the System (kg/h)	5000	5000	5000	5000	5000	5000
	Particle Density (kg/m³)	3200	3200	3200	3200	3200	3200
Energy balance (kW)	Fuel Reactor	-139.6	-139.6	-139.6	-139.6	-139.6	-139.6
	Air Reactor	96.498	289.5	386	482.49	578.99	688.04
	Cool air reactor exhaust	18.996	56.999	75.985	94.981	113.98	135.4
	Cool flue gas	153.71	153.71	153.71	153.71	153.71	153.71
	Cool OC for air reactor	50.175	50.175	50.175	50.175	50.175	50.175
	Reheat OC for fuel reactor	-52.19	-52.19	-52.19	-52.19	-52.19	-52.19
	Heat steam	-69.8	-69.8	-69.8	-69.8	-69.8	-69.8
	Heat air	-25.82	-77.47	-103.3	-129.1	-154.9	-184.1
	Net	31.917	211.27	300.93	390.6	480.28	581.58
Initial values (continued)	Coal (kg/h)	100	100	100	100	100	100
	Water (kg/h)	140	140	140	140	140	140
	Air Flow Rate (kg/h)	800	900	1000	1100	1200	1500
	Temperature of Fuel Reactor (°C)	950	950	950	950	950	950
	Temperature of Air Reactor (°C)	935	935	935	935	935	935
	Fe_2O_3 flow in the Fuel Reactor (kg/h)	7000	7000	7000	7000	7000	7000
	Al_2O_3 in the System (kg/h)	5000	5000	5000	5000	5000	5000
	Particle Density (kg/m³)	3200	3200	3200	3200	3200	3200
Energy balance (kW) (continued)	Fuel Reactor	-139.6	-139.6	-139.6	-139.6	-139.6	-139.6
	Air Reactor	771.99	868.49	949.41	949.52	949.66	950.07
	Cool air reactor exhaust	151.97	170.97	190.81	215.06	239.31	312.05
	Cool flue gas	153.71	153.71	153.71	153.71	153.71	153.71
	Cool OC for air reactor	50.175	50.175	50.175	50.175	50.175	50.175
	Reheat OC for fuel reactor	-52.19	-52.19	-52.19	-52.19	-52.19	-52.19
	Heat steam	-69.8	-69.8	-69.8	-69.8	-68.8	-69.8
	Heat air	-206.6	-232.4	-258.2	-284.1	-309.9	-387.3
	Net	659.62	749.29	824.23	822.77	822.33	817.02

Table A5. CLC process simulation results for different air flow rates with Colombian coal at 100 kg/h and Fe_2O_3/Al_2O_3 at 7500/5500 kg/h

Initial values	Coal (kg/h)	100	100	100	100	100	100
	Water (kg/h)	140	140	140	140	140	140
	Air Flow Rate (kg/h)	100	300	400	500	600	713
	Temperature of Fuel Reactor (°C)	950	950	950	950	950	950
	Temperature of Air Reactor (°C)	935	935	935	935	935	935
	Fe_2O_3 flow in the Fuel Reactor (kg/h)	7500	7500	7500	7500	7500	7500
	Al_2O_3 in the System (kg/h)	5500	5500	5500	5500	5500	5500
	Particle Density (kg/m³)	3200	3200	3200	3200	3200	3200
Energy balance (kW)	Fuel Reactor	-134.6	-134.6	-134.6	-134.6	-134.6	-134.6
	Air Reactor	96.498	289.5	386	482.49	578.99	688.04
	Cool air reactor exhaust	18.996	56.999	75.985	94.981	113.98	135.4
	Cool flue gas	153.71	153.71	153.71	153.71	153.71	153.71
	Cool OC for air reactor	54.647	54.647	54.647	54.647	54.647	54.647
	Reheat OC for fuel reactor	-56.67	-56.67	-56.67	-56.67	-56.67	-56.67
	Heat steam	-69.8	-69.8	-69.8	-69.8	-69.8	-69.8
	Heat air	-25.82	-77.47	-103.3	-129.1	-154.9	-184.1
	Net	36.942	216.3	305.96	395.63	485.3	586.61
Initial values (continued)	Coal (kg/h)	100	100	100	100	100	100
	Water (kg/h)	140	140	140	140	140	140
	Air Flow Rate (kg/h)	800	900	1000	1100	1200	1500
	Temperature of Fuel Reactor (°C)	950	950	950	950	950	950
	Temperature of Air Reactor (°C)	935	935	935	935	935	935
	Fe_2O_3 flow in the Fuel Reactor (kg/h)	7500	7500	7500	7500	7500	7500
	Al_2O_3 in the System (kg/h)	5500	5500	5500	5500	5500	5500
	Particle Density (kg/m³)	3200	3200	3200	3200	3200	3200
Energy balance (kW) (continued)	Fuel Reactor	-134.6	-134.6	-134.6	-134.6	-134.6	-134.6
	Air Reactor	771.99	868.49	949.41	949.52	949.66	950.07
	Cool air reactor exhaust	151.97	170.97	190.81	215.06	239.31	312.05
	Cool flue gas	153.71	153.71	153.71	153.71	153.71	153.71
	Cool OC for air reactor	54.647	54.647	54.647	54.647	54.647	54.647
	Reheat OC for fuel reactor	-56.67	-56.67	-56.67	-56.67	-56.67	-56.67
	Heat steam	-69.8	-69.8	-69.8	-69.8	-69.8	-69.8
	Heat air	-206.6	-232.4	-258.2	-284.1	-309.9	-387.3
	Net	664.65	754.32	829.25	827.8	826.35	822.04

Table A6. CLC process simulation results for different air flow rates with Colombian coal at 100 kg/h and Fe_2O_3/Al_2O_3 at 8000/6000 kg/h

Initial values	Coal (kg/h)	100	100	100	100	100	100
	Water (kg/h)	140	140	140	140	140	140
	Air Flow Rate (kg/h)	100	300	400	500	600	713
	Temperature of Fuel Reactor (°C)	950	950	950	950	950	950
	Temperature of Air Reactor (°C)	935	935	935	935	935	935
	Fe_2O_3 flow in the Fuel Reactor (kg/h)	8000	8000	8000	8000	8000	8000
	Al_2O_3 in the System (kg/h)	6000	6000	6000	6000	6000	6000
	Particle Density (kg/m³)	3200	3200	3200	3200	3200	3200
Energy balance (kW)	Fuel Reactor	-129.6	-129.6	-129.6	-129.6	-129.6	-129.6
	Air Reactor	96.498	289.5	386	482.49	578.99	688.04
	Cool air reactor exhaust	18.996	56.999	75.985	94.981	113.98	135.4
	Cool flue gas	153.71	153.71	153.71	153.71	153.71	153.71
	Cool OC for air reactor	59.121	59.121	59.121	59.121	59.121	59.121
	Reheat OC for fuel reactor	-61.15	-61.15	-61.15	-61.15	-61.15	-61.15
	Heat steam	-69.8	-69.8	-69.8	-69.8	-69.8	-69.8
	Heat air	-25.82	-77.47	-103.3	-129.1	-154.9	-184.1
	Net	41.968	221.32	310.98	400.66	490.33	591.63

Initial values (continued)	Coal (kg/h)	100	100	100	100	100	100
	Water (kg/h)	140	140	140	140	140	140
	Air Flow Rate (kg/h)	800	900	1000	1100	1200	1500
	Temperature of Fuel Reactor (°C)	950	950	950	950	950	950
	Temperature of Air Reactor (°C)	935	935	935	935	935	935
	Fe_2O_3 flow in the Fuel Reactor (kg/h)	8000	8000	8000	8000	8000	8000
	Al_2O_3 in the System (kg/h)	6000	6000	6000	6000	6000	6000
	Particle Density (kg/m³)	3200	3200	3200	3200	3200	3200
Energy balance (kW) (continued)	Fuel Reactor	-129.6	-129.6	-129.6	-129.6	-129.6	-129.6
	Air Reactor	771.99	868.49	949.41	949.52	949.66	950.07
	Cool air reactor exhaust	151.97	170.97	190.81	215.06	239.31	312.05
	Cool flue gas	153.71	153.71	153.71	153.71	153.71	153.71
	Cool OC for air reactor	59.121	59.121	59.121	59.121	59.121	59.121
	Reheat OC for fuel reactor	-61.15	-61.15	-61.15	-61.15	-61.15	-61.15
	Heat steam	-69.8	-69.8	-69.8	-69.8	-68.8	-69.8
	Heat air	-206.6	-232.4	-258.2	-284.1	-309.9	-387.3
	Net	669.67	759.34	834.28	832.82	832.38	827.07

Table A7. Scaled-up simulation results for different coal feeding rates using the baseline ratios of air flow rate and oxygen carrier flow rate from the experiment of Sahir et al [10]

Initial values	Coal (kg/h)	100	500	1000	1500	2500
	Water (kg/h)	140	700	1400	2100	3500
	Air Flow Rate (kg/h)	713	3565	7130	10695	17825
	Temperature of Fuel Reactor (°C)	950	950	950	950	950
	Temperature of Air Reactor (°C)	935	935	935	935	935
	Fe_2O_3 flow in the Fuel Reactor (kg/h)	5921	30000	60000	90000	150000
	Al_2O_3 in the System (kg/h)	3951	20000	40000	60000	100000
	Particle Density (kg/m³)	3200	3200	3200	3200	3200
Energy balance (KW)	Fuel Reactor	-161.8	-800.8	-1602	-2402	-4.004
	Air Reactor	688	3440.2	6880.4	10321	17.201
	Cool air reactor exhaust	135.4	677.22	1354.4	2031.7	3.3861
	Cool flue gas	148.3	744.09	1488.2	2232.3	3.7205
	Cool OC for air reactor	40.9	207.26	414.52	621.77	1.0363
	Reheat OC for fuel reactor	-42.7	-216.2	-432.3	-648.5	-1.081
	Heat steam	-69.8	-349.1	-698.3	-1047	-1.746
	Heat air	-184.1	-920.6	-1841	-2762	-4.603
	Net	554.2	2782	5564.1	8346.1	13910
Initial values (continued)	Coal (kg/h)	3500	5000	8000	12000	
	Water (kg/h)	4900	7000	11200	16800	
	Air Flow Rate (kg/h)	24955	35650	57040	85560	
	Temperature of Fuel Reactor (°C)	950	950	950	950	
	Temperature of Air Reactor (°C)	935	935	935	935	
	Fe_2O_3 flow in the Fuel Reactor (kg/h)	210000	300000	480000	720000	
	Al_2O_3 in the System (kg/h)	140000	200000	320000	480000	
	Particle Density (kg/m³)	3200	3200	3200	3200	
Energy balance (KW) (continued)	Fuel Reactor	-5.606	-8.008	-12.81	-19.22	
	Air Reactor	24.081	34.402	55.043	82.564	
	Cool air reactor exhaust	4.7405	6.7722	10.836	16.253	
	Cool flue gas	5.2087	7.4409	11.906	17.858	
	Cool OC for air reactor	1.4508	2.0726	3.3161	4.9742	
	Reheat OC for fuel reactor	-1.513	-2.162	-3.459	-5.188	
	Heat steam	-2.444	-3.491	-5.586	-8.379	
	Heat air	-6.444	-9.206	-14.73	-22.09	
	Net	19474	27820	44513	66769	

Table A8. Scaled-up simulation results for different coal feeding rates using the optimum ratios of air flow rate and oxygen carrier flow rate

Initial values	Coal (kg/h)	100	500	1000	1500	2500
	Water (kg/h)	140	700	1400	2100	3500
	Air Flow Rate (kg/h)	1000	5000	10000	15000	25000
	Temperature of Fuel Reactor (°C)	950	950	950	950	950
	Temperature of Air Reactor (°C)	935	935	935	935	935
	Fe_2O_3 flow in the Fuel Reactor (kg/h)	7000	35000	70000	105000	175000
	Al_2O_3 in the System (kg/h)	5000	25000	50000	75000	125000
	Particle Density (kg/m³)	3200	3200	3200	3200	3200
Energy balance (KW)	Fuel Reactor	-139.6	-698.2	-1484	-2226	-3711
	Air Reactor	949.41	4747	9108.1	13662	22770
	Cool air reactor exhaust	190.81	954.06	1929.1	2893.7	4822.9
	Cool flue gas	153.71	768.54	1519.6	2279.4	3799
	Cool OC for air reactor	50.175	250.87	457.81	686.72	1144.5
	Reheat OC for fuel reactor	-52.19	-261	-477.1	-715.7	-1193
	Heat steam	-69.8	-349.1	-698.3	-1047	-1746
	Heat air	-258.2	-1291	-2582	-3873	-6456
	Net	824.23	4121	8242.3	12363	20606
Initial values (continued)	Coal (kg/h)	3500	5000	8000	12000	
	Water (kg/h)	4900	7000	11200	16800	
	Air Flow Rate (kg/h)	35000	50000	80000	120000	
	Temperature of Fuel Reactor (°C)	950	950	950	950	
	Temperature of Air Reactor (°C)	935	935	935	935	
	Fe_2O_3 flow in the Fuel Reactor (kg/h)	245000	350000	560000	840000	
	Al_2O_3 in the System (kg/h)	175000	250000	400000	600000	
	Particle Density (kg/m³)	3200	3200	3200	3200	
Energy balance (KW) (continued)	Fuel Reactor	-5195	-7421	-11874	-17811	
	Air Reactor	31878	45541	72865	109297	
	Cool air reactor exhaust	6752	9645.7	15433	23150	
	Cool flue gas	5318.6	7598.1	12157	18235	
	Cool OC for air reactor	1602.3	2289.1	3662.5	5493.7	
	Reheat OC for fuel reactor	-1670	-2386	-3817	-5725	
	Heat steam	-2444	-3491	-5586	-8379	
	Heat air	-9038	-12912	-20659	-30988	
	Net	28847	41211	65936	98907	

Table A9. CLC process simulation results for different air flow rates with Colombian coal at 12000 kg/h and Fe_2O_3/Al_2O_3 at 780000/540000 kg/h

Initial values	Coal (kg/h)	12000	12000	12000	12000	12000	12000
	Water (kg/h)	16800	16800	16800	16800	16800	16800
	Air Flow Rate (kg/h)	12000	36000	48000	60000	72000	84000
	Temperature of Fuel Reactor (°C)	950	950	950	950	950	950
	Temperature of Air Reactor (°C)	935	935	935	935	935	935
	Fe_2O_3 flow in the Fuel Reactor (kg/h)	780000	780000	780000	780000	780000	780000
	Al_2O_3 in the System (kg/h)	540000	540000	540000	540000	540000	540000
	Particle Density (kg/m³)	3200	3200	3200	3200	3200	3200
Energy balance (KW)	Fuel Reactor	-17811	-17811	-17811	-17811	-17811	-17811
	Air Reactor	11580	34740	46320	57899	69479	81059
	Cool air reactor exhaust	2279.6	6838.7	9118.2	11398	13677	15957
	Cool flue gas	18235	18235	18235	18235	18235	18235
	Cool OC for air reactor	5493.7	5493.7	5493.7	5493.7	5493.7	5493.7
	Reheat OC for fuel reactor	-5725	-5725	-5725	-5725	-5725	-5725
	Heat steam	-8379	-8379	-8379	-8379	-8379	-8379
	Heat air	-3099	-9296	-12395	-15494	-18593	-21692
	Net	2573.3	24095	34855	45616	56376	67137

Initial values (continued)	Coal (kg/h)	12000	12000	12000	12000	12000	12000
	Water (kg/h)	16800	16800	16800	16800	16800	16800
	Air Flow Rate (kg/h)	96000	108000	120000	132000	144000	180000
	Temperature of Fuel Reactor (°C)	950	950	950	950	950	950
	Temperature of Air Reactor (°C)	935	935	935	935	935	935
	Fe_2O_3 flow in the Fuel Reactor (kg/h)	780000	780000	780000	780000	780000	780000
	Al_2O_3 in the System (kg/h)	540000	540000	540000	540000	540000	540000
	Particle Density (kg/m³)	3200	3200	3200	3200	3200	3200
Energy balance (KW) (continued)	Fuel Reactor	-17811	-17811	-17811	-17811	-17811	-17811
	Air Reactor	92639	104219	109297	109313	109329	109379
	Cool air reactor exhaust	18236	20516	23150	26059	28969	37698
	Cool flue gas	18235	18235	18235	18235	18235	18235
	Cool OC for air reactor	5493.7	5493.7	5493.7	5493.7	5493.7	5493.7
	Reheat OC for fuel reactor	-5725	-5725	-5725	-5725	-5725	-5725
	Heat steam	-8379	-8379	-8379	-8379	-8379	-8379
	Heat air	-24790	-27889	-30988	-34087	-37186	-46482
	Net	77898	88658	93271	93098	92925	92408

Table A10. CLC process simulation results for different air flow rates with Colombian coal at 12000 kg/h and Fe_2O_3/Al_2O_3 at 840000/600000 kg/h

Initial values	Coal (kg/h)	12000	12000	12000	12000	12000	12000
	Water (kg/h)	16800	16800	16800	16800	16800	16800
	Air Flow Rate (kg/h)	12000	36000	48000	60000	72000	84000
	Temperature of Fuel Reactor (°C)	950	950	950	950	950	950
	Temperature of Air Reactor (°C)	935	935	935	935	935	935
	Fe_2O_3 flow in the Fuel Reactor (kg/h)	840000	840000	840000	840000	840000	840000
	Al_2O_3 in the System (kg/h)	600000	600000	600000	600000	600000	600000
	Particle Density (kg/m³)	3200	3200	3200	3200	3200	3200
Energy balance (KW)	Fuel Reactor	-16757	-16757	-16757	-16757	-16757	-16757
	Air Reactor	11580	34740	46320	57899	69479	81059
	Cool air reactor exhaust	2279.6	6838.7	9118.2	11398	13677	15957
	Cool flue gas	18445	18445	18445	18445	18445	18445
	Cool OC for air reactor	6021	6021	6021	6021	6021	6021
	Reheat OC for fuel reactor	-6263	-6263	-6263	-6263	-6263	-6263
	Heat steam	-8379	-8379	-8379	-8379	-8379	-8379
	Heat air	-3099	-9296	-12395	-15494	-18593	-21692
	Net	3826.8	25348	36109	46869	57630	68390
Initial values (continued)	Coal (kg/h)	12000	12000	12000	12000	12000	12000
	Water (kg/h)	16800	16800	16800	16800	16800	16800
	Air Flow Rate (kg/h)	96000	108000	120000	132000	144000	180000
	Temperature of Fuel Reactor (°C)	950	950	950	950	950	950
	Temperature of Air Reactor (°C)	935	935	935	935	935	935
	Fe_2O_3 flow in the Fuel Reactor (kg/h)	840000	840000	840000	840000	840000	840000
	Al_2O_3 in the System (kg/h)	600000	600000	600000	600000	600000	600000
	Particle Density (kg/m³)	3200	3200	3200	3200	3200	3200
Energy balance (KW) (continued)	Fuel Reactor	-16757	-16757	-16757	-16757	-16757	-16757
	Air Reactor	92639	104219	113929	113943	113959	114009
	Cool air reactor exhaust	18236	20516	22897	25807	28717	37446
	Cool flue gas	18445	18445	18445	18445	18445	18445
	Cool OC for air reactor	6021	6021	6021	6021	6021	6021
	Reheat OC for fuel reactor	-6263	-6263	-6263	-6263	-6263	-6263
	Heat steam	-8379	-8379	-8379	-8379	-8379	-8379
	Heat air	-24790	-27889	-30988	-34087	-37186	-46482
	Net	79151	89912	98904	98730	98556	98039

Table A11. CLC process simulation results for different air flow rates with Colombian coal at 12000 kg/h and Fe_2O_3/Al_2O_3 at 900000/660000 kg/h

Initial values	Coal (kg/h)	12000	12000	12000	12000	12000	12000
	Water (kg/h)	16800	16800	16800	16800	16800	16800
	Air Flow Rate (kg/h)	12000	36000	48000	60000	72000	84000
	Temperature of Fuel Reactor (°C)	950	950	950	950	950	950
	Temperature of Air Reactor (°C)	935	935	935	935	935	935
	Fe_2O_3 flow in the Fuel Reactor (kg/h)	900000	900000	900000	900000	900000	900000
	Al_2O_3 in the System (kg/h)	660000	660000	660000	660000	660000	660000
	Particle Density (kg/m³)	3200	3200	3200	3200	3200	3200
Energy balance (KW)	Fuel Reactor	-16154	-16154	-16154	-16154	-16154	-16154
	Air Reactor	11580	34740	46320	57899	69479	81059
	Cool air reactor exhaust	2279.6	6838.7	9118.2	11398	13677	15957
	Cool flue gas	18445	18445	18445	18445	18445	18445
	Cool OC for air reactor	6557.7	6557.7	6557.7	6557.7	6557.7	6557.7
	Reheat OC for fuel reactor	-6801	-6801	-6801	-6801	-6801	-6801
	Heat steam	-8379	-8379	-8379	-8379	-8379	-8379
	Heat air	-3099	-9296	-12395	-15494	-18593	-21692
	Net	4429.8	25951	36712	47472	58233	68993
Initial values (continued)	Coal (kg/h)	12000	12000	12000	12000	12000	12000
	Water (kg/h)	16800	16800	16800	16800	16800	16800
	Air Flow Rate (kg/h)	96000	108000	120000	132000	144000	180000
	Temperature of Fuel Reactor (°C)	950	950	950	950	950	950
	Temperature of Air Reactor (°C)	935	935	935	935	935	935
	Fe_2O_3 flow in the Fuel Reactor (kg/h)	900000	900000	900000	900000	900000	900000
	Al_2O_3 in the System (kg/h)	660000	660000	660000	660000	660000	660000
	Particle Density (kg/m³)	3200	3200	3200	3200	3200	3200
Energy balance (KW) (continued)	Fuel Reactor	-16154	-16154	-16154	-16154	-16154	-16154
	Air Reactor	92639	104219	113929	113943	113959	114009
	Cool air reactor exhaust	18236	20516	22897	25807	28717	37446
	Cool flue gas	18445	18445	18445	18445	18445	18445
	Cool OC for air reactor	6557.7	6557.7	6557.7	6557.7	6557.7	6557.7
	Reheat OC for fuel reactor	-6801	-6801	-6801	-6801	-6801	-6801
	Heat steam	-8379	-8379	-8379	-8379	-8379	-8379
	Heat air	-24790	-27889	-30988	-34087	-37186	-46482
	Net	79754	90515	99507	99333	99159	98642

Table A12. CLC process simulation results for different air flow rates with Bituminous coal at 100 kg/h and Fe_2O_3/Al_2O_3 at 5921/3951 kg/h

Initial values	Coal (kg/h)	100	100	100	100	100	100
	Water (kg/h)	140	140	140	140	140	140
	Air Flow Rate (kg/h)	100	300	400	500	600	713
	Temperature of Fuel Reactor (°C)	950	950	950	950	950	950
	Temperature of Air Reactor (°C)	935	935	935	935	935	935
	Fe_2O_3 flow in the Fuel Reactor (kg/h)	5921	5921	5921	5921	5921	5921
	Al_2O_3 in the System (kg/h)	3951	3951	3951	3951	3951	3951
	Particle Density (kg/m³)	3200	3200	3200	3200	3200	3200
Energy balance (kW)	Fuel Reactor	72.119	72.119	72.119	72.119	72.119	72.119
	Air Reactor	96.498	289.497	385.995	482.494	578.993	688
	Cool air reactor exhaust	18.996	56.988	75.985	94.981	113.978	135.4
	Cool flue gas	137.064	137.064	137.064	137.064	137.064	137.064
	Cool OC for air reactor	41.082	41.082	41.082	41.082	41.082	41.082
	Reheat OC for fuel reactor	-42.7	-42.7	-42.7	-42.7	-42.7	-42.7

	Heat steam	-69.8	-69.8	-69.8	-69.8	-69.8	-69.8
	Heat air	-25.823	-77.469	-103.29	-129.12	-154.94	-184.1
	Net	227.436	406.781	496.452	586.124	675.796	777.065
Initial values (continued)	Coal (kg/h)	100	100	100	100	100	100
	Water (kg/h)	140	140	140	140	140	140
	Air Flow Rate (kg/h)	800	900	1000	1100	1200	1500
	Temperature of Fuel Reactor (°C)	950	950	950	950	950	950
	Temperature of Air Reactor (°C)	935	935	935	935	935	935
	Fe_2O_3 flow in the Fuel Reactor (kg/h)	5921	5921	5921	5921	5921	5921
	Al_2O_3 in the System (kg/h)	3951	3951	3951	3951	3951	3951
	Particle Density (kg/m³)	3200	3200	3200	3200	3200	3200
Energy balance (kW) (continued)	Fuel Reactor	72.119	72.119	72.119	72.119	72.119	72.119
	Air Reactor	752.554	752.667	752.812	752.951	753.091	753.515
	Cool air reactor exhaust	153.029	177.277	201.524	225.772	250.02	322.764
	Cool flue gas	137.064	137.064	137.064	137.064	137.064	137.064
	Cool OC for air reactor	41.082	41.082	41.082	41.082	41.082	41.082
	Reheat OC for fuel reactor	-42.7	-42.7	-42.7	-42.7	-42.7	-42.7
	Heat steam	-69.8	-69.8	-69.8	-69.8	-69.8	-69.8
	Heat air	-206.59	-232.41	-258.23	-284.06	-309.88	-387.35
	Net	836.762	835.299	833.868	832.432	830.996	826.695

Table A13. CLC process simulation results for different air flow rates with Anthracite coal at 100 kg/h and Fe_2O_3/Al_2O_3 at 5921/3951 kg/h

Initial values	Coal (kg/h)	100	100	100	100	100	100
	Water (kg/h)	140	140	140	140	140	140
	Air Flow Rate (kg/h)	100	300	400	500	600	713
	Temperature of Fuel Reactor (°C)	950	950	950	950	950	950
	Temperature of Air Reactor (°C)	935	935	935	935	935	935
	Fe_2O_3 flow in the Fuel Reactor (kg/h)	5921	5921	5921	5921	5921	5921
	Al_2O_3 in the System (kg/h)	3951	3951	3951	3951	3951	3951
	Particle Density (kg/m³)	3200	3200	3200	3200	3200	3200
Energy balance (kW)	Fuel Reactor	116.145	116.145	116.145	116.145	116.145	116.145
	Air Reactor	96.498	289.497	385.995	482.494	578.993	688
	Cool air reactor exhaust	18.996	56.988	75.985	94.981	113.978	135.4
	Cool flue gas	127.834	127.834	127.834	127.834	127.834	127.834
	Cool OC for air reactor	41.132	41.132	41.132	41.132	41.132	41.132
	Reheat OC for fuel reactor	-42.7	-42.7	-42.7	-42.7	-42.7	-42.7
	Heat steam	-69.8	-69.8	-69.8	-69.8	-69.8	-69.8
	Heat air	-25.823	-77.469	-103.29	-129.11	-154.94	-184.1
	Net	262.282	441.627	531.298	620.97	710.642	811.911
Initial values (continued)	Coal (kg/h)	100	100	100	100	100	100
	Water (kg/h)	140	140	140	140	140	140
	Air Flow Rate (kg/h)	800	900	1000	1100	1200	1500
	Temperature of Fuel Reactor (°C)	950	950	950	950	950	950
	Temperature of Air Reactor (°C)	935	935	935	935	935	935
	Fe_2O_3 flow in the Fuel Reactor (kg/h)	5921	5921	5921	5921	5921	5921
	Al_2O_3 in the System (kg/h)	3951	3951	3951	3951	3951	3951
	Particle Density (kg/m³)	3200	3200	3200	3200	3200	3200
Energy balance (kW)	Fuel Reactor	116.145	116.145	116.145	116.145	116.145	116.145
	Air Reactor	728.287	728.416	728.553	728.692	728.832	729.257
	Cool air reactor exhaust	154.351	178.599	202.847	227.094	251.342	324.086

(continued)	Cool flue gas	127.834	127.834	127.834	127.834	127.834	127.834
	Cool OC for air reactor	41.132	41.132	41.132	41.132	41.132	41.132
	Reheat OC for fuel reactor	-42.7	-42.7	-42.7	-42.7	-42.7	-42.7
	Heat steam	-69.8	-69.8	-69.8	-69.8	-69.8	-69.8
	Heat air	-206.58	-232.41	-258.23	-284.05	-309.88	-387.34
	Net	848.663	847.216	845.778	844.341	842.905	838.605

Table A14. CLC process simulation results for different air flow rates with Lignite coal at 100 kg/h and Fe_2O_3/Al_2O_3 at 5921/3951 kg/h

Initial values	Coal (kg/h)	100	100	100	100	100	100
	Water (kg/h)	140	140	140	140	140	140
	Air Flow Rate (kg/h)	100	300	400	500	600	713
	Temperature of Fuel Reactor (°C)	950	950	950	950	950	950
	Temperature of Air Reactor (°C)	935	935	935	935	935	935
	Fe_2O_3 flow in the Fuel Reactor (kg/h)	5921	5921	5921	5921	5921	5921
	Al_2O_3 in the System (kg/h)	3951	3951	3951	3951	3951	3951
	Particle Density (kg/m³)	3200	3200	3200	3200	3200	3200
Energy balance (kW)	Fuel Reactor	246.014	246.014	246.014	246.014	246.014	246.014
	Air Reactor	96.498	289.497	385.995	475.591	475.716	475.872
	Cool air reactor exhaust	18.996	56.988	75.985	95.357	119.605	147.005
	Cool flue gas	115.4	115.4	115.4	115.4	115.4	115.4
	Cool OC for air reactor	41.649	41.649	41.649	41.649	41.649	41.649
	Reheat OC for fuel reactor	-42.7	-42.7	-42.7	-42.7	-42.7	-42.7
	Heat steam	-69.8	-69.8	-69.8	-69.8	-69.8	-69.8
	Heat air	-25.823	-77.46	-103.29	-129.11	-154.94	-184.1
	Net	380.234	559.579	649.25	732.395	730.944	729.34
Initial values (continued)	Coal (kg/h)	100	100	100	100	100	100
	Water (kg/h)	140	140	140	140	140	140
	Air Flow Rate (kg/h)	800	900	1000	1100	1200	1500
	Temperature of Fuel Reactor (°C)	950	950	950	950	950	950
	Temperature of Air Reactor (°C)	935	935	935	935	935	935
	Fe_2O_3 flow in the Fuel Reactor (kg/h)	5921	5921	5921	5921	5921	5921
	Al_2O_3 in the System (kg/h)	3951	3951	3951	3951	3951	3951
	Particle Density (kg/m³)	3200	3200	3200	3200	3200	3200
Energy balance (kW) (continued)	Fuel Reactor	246.014	246.014	246.014	246.014	246.014	246.014
	Air Reactor	475.994	476.136	476.278	476.42	476.563	476.992
	Cool air reactor exhaust	168.101	192.349	216.596	240.884	265.092	337.836
	Cool flue gas	115.4	115.4	115.4	115.4	115.4	115.4
	Cool OC for air reactor	41.649	41.649	41.649	41.649	41.649	41.649
	Reheat OC for fuel reactor	-42.7	-42.7	-42.7	-42.7	-42.7	-42.7
	Heat steam	-69.8	-69.8	-69.8	-69.8	-69.8	-69.8
	Heat air	-206.58	-232.41	-258.23	-284.05	-309.88	-387.34
	Net	728.072	726.638	725.204	723.811	722.338	718.042

Table A15. CLC process simulation results for different air flow rates with Bituminous coal at 100 kg/h and Fe_2O_3/Al_2O_3 at 5000/3000 kg/h

Initial values	Coal (kg/h)	100	100	100	100	100	100
	Water (kg/h)	140	140	140	140	140	140

	Air Flow Rate (kg/h)	100	300	400	500	600	713
	Temperature of Fuel Reactor (°C)	950	950	950	950	950	950
	Temperature of Air Reactor (°C)	935	935	935	935	935	935
	Fe_2O_3 flow in the Fuel Reactor (kg/h)	5000	5000	5000	5000	5000	5000
	Al_2O_3 in the System (kg/h)	3000	3000	3000	3000	3000	3000
	Particle Density (kg/m^3)	3200	3200	3200	3200	3200	3200
Energy balance (kW)	Fuel Reactor	57.719	57.719	57.719	57.719	57.719	57.719
	Air Reactor	96.498	289.497	385.995	482.494	578.993	688
	Cool air reactor exhaust	18.996	56.988	75.985	94.981	113.978	135.4
	Cool flue gas	134.734	134.734	134.734	134.734	134.734	134.734
	Cool OC for air reactor	32.792	32.792	32.792	32.792	32.792	32.792
	Reheat OC for fuel reactor	-34.273	-34.273	-34.273	-34.273	-34.273	-34.273
	Heat steam	-69.8	-69.8	-69.8	-69.8	-69.8	-69.8
	Heat air	-25.82	-77.46	-103.29	-129.11	-154.94	-184.1
	Net	210.843	390.188	479.859	569.531	659.203	760.472
Initial values (continued)	Coal (kg/h)	100	100	100	100	100	100
	Water (kg/h)	140	140	140	140	140	140
	Air Flow Rate (kg/h)	800	900	1000	1100	1200	1500
	Temperature of Fuel Reactor (°C)	950	950	950	950	950	950
	Temperature of Air Reactor (°C)	935	935	935	935	935	935
	Fe_2O_3 flow in the Fuel Reactor (kg/h)	5000	5000	5000	5000	5000	5000
	Al_2O_3 in the System (kg/h)	3000	3000	3000	3000	3000	3000
	Particle Density (kg/m^3)	3200	3200	3200	3200	3200	3200
Energy balance (kW) (continued)	Fuel Reactor	57.719	57.719	57.719	57.719	57.719	57.719
	Air Reactor	700.662	700.794	700.932	701.072	701.213	701.639
	Cool air reactor exhaust	155.856	180.104	204.352	228.6	252.848	325.591
	Cool flue gas	134.734	134.734	134.734	134.734	134.734	134.734
	Cool OC for air reactor	32.792	32.792	32.792	32.792	32.792	32.792
	Reheat OC for fuel reactor	-34.273	-34.273	-34.273	-34.273	-34.273	-34.273
	Heat steam	-69.8	-69.8	-69.8	-69.8	-69.8	-69.8
	Heat air	-206.58	-232.41	-258.23	-284.05	-309.88	-387.34
	Net	771.104	769.66	768.223	766.788	765.353	761.053

Table A16. CLC process simulation results for different air flow rates with Bituminous coal at 100 kg/h and Fe_2O_3/Al_2O_3 at 5500/3500 kg/h

Initial values	Coal (kg/h)	100	100	100	100	100	100
	Water (kg/h)	140	140	140	140	140	140
	Air Flow Rate (kg/h)	100	300	400	500	600	713
	Temperature of Fuel Reactor (°C)	950	950	950	950	950	950
	Temperature of Air Reactor (°C)	935	935	935	935	935	935
	Fe_2O_3 flow in the Fuel Reactor (kg/h)	5500	5500	5500	5500	5500	5500
	Al_2O_3 in the System (kg/h)	3500	3500	3500	3500	3500	3500
	Particle Density (kg/m^3)	3200	3200	3200	3200	3200	3200
Energy balance (kW)	Fuel Reactor	67.705	67.705	67.705	67.705	67.705	67.705
	Air Reactor	96.498	289.497	385.995	482.494	578.993	688.037
	Cool air reactor exhaust	18.9963	56.9989	75.985	94.981	113.978	135.4
	Cool flue gas	137.064	137.064	137.064	137.064	137.064	137.064
	Cool OC for air reactor	37.159	37.159	37.159	37.159	37.159	37.159
	Reheat OC for fuel reactor	-38.752	-38.752	-38.752	-38.752	-38.752	-38.752
	Heat steam	-69.8	-69.8	-69.8	-69.8	-69.8	-69.8

	Heat air	-25.82	-77.46	-103.29	-129.11	-154.94	-184.12
	Net	223.047	402.4029	492.063	581.735	671.407	772.693
Initial values (continued)	Coal (kg/h)	100	100	100	100	100	100
	Water (kg/h)	140	140	140	140	140	140
	Air Flow Rate (kg/h)	800	900	1000	1100	1200	1500
	Temperature of Fuel Reactor (°C)	950	950	950	950	950	950
	Temperature of Air Reactor (°C)	935	935	935	935	935	935
	Fe_2O_3 flow in the Fuel Reactor (kg/h)	5500	5500	5500	5500	5500	5500
	Al_2O_3 in the System (kg/h)	3500	3500	3500	3500	3500	3500
	Particle Density (kg/m³)	3200	3200	3200	3200	3200	3200
Energy balance (kW) (continued)	Fuel Reactor	67.705	67.705	67.705	67.705	67.705	67.705
	Air Reactor	752.554	752.677	752.812	752.951	753.091	753.515
	Cool air reactor exhaust	153.029	177.277	201.524	225.773	250.02	322.764
	Cool flue gas	137.064	137.064	137.064	137.064	137.064	137.064
	Cool OC for air reactor	37.159	37.159	37.159	37.159	37.159	37.159
	Reheat OC for fuel reactor	-38.752	-38.752	-38.752	-38.752	-38.752	-38.752
	Heat steam	-69.8	-69.8	-69.8	-69.8	-69.8	-69.8
	Heat air	-206.58	-232.41	-258.23	-284.05	-309.88	-387.34
	Net	832.373	830.92	829.479	828.044	826.607	822.306

Table A17. CLC process simulation results for different air flow rates with Bituminous coal at 100 kg/h and Fe_2O_3/Al_2O_3 at 7000/5000 kg/h

Initial values	Coal (kg/h)	100	100	100	100	100	100
	Water (kg/h)	140	140	140	140	140	140
	Air Flow Rate (kg/h)	100	300	400	500	600	713
	Temperature of Fuel Reactor (°C)	950	950	950	950	950	950
	Temperature of Air Reactor (°C)	935	935	935	935	935	935
	Fe_2O_3 flow in the Fuel Reactor (kg/h)	7000	7000	7000	7000	7000	7000
	Al_2O_3 in the System (kg/h)	5000	5000	5000	5000	5000	5000
	Particle Density (kg/m³)	3200	3200	3200	3200	3200	3200
Energy balance (kW)	Fuel Reactor	82.8	82.8	82.8	82.8	82.8	82.8
	Air Reactor	96.498	289.497	385.995	482.494	578.993	688.037
	Cool air reactor exhaust	18.9963	56.9989	75.985	94.981	113.978	135.4
	Cool flue gas	137.064	137.064	137.064	137.064	137.064	137.064
	Cool OC for air reactor	50.577	50.577	50.577	50.577	50.577	50.577
	Reheat OC for fuel reactor	-52.192	-52.192	-52.192	-52.192	-52.192	-52.192
	Heat steam	-69.8	-69.8	-69.8	-69.8	-69.8	-69.8
	Heat air	-25.82	-77.46	-103.29	-129.11	-154.94	-184.12
	Net	238.12	417.4759	507.136	596.808	686.48	787.766
Initial values (continued)	Coal (kg/h)	100	100	100	100	100	100
	Water (kg/h)	140	140	140	140	140	140
	Air Flow Rate (kg/h)	800	900	1000	1100	1200	1500
	Temperature of Fuel Reactor (°C)	950	950	950	950	950	950
	Temperature of Air Reactor (°C)	935	935	935	935	935	935
	Fe_2O_3 flow in the Fuel Reactor (kg/h)	7000	7000	7000	7000	7000	7000
	Al_2O_3 in the System (kg/h)	5000	5000	5000	5000	5000	5000
	Particle Density (kg/m³)	3200	3200	3200	3200	3200	3200
Energy balance (kW)	Fuel Reactor	82.8	82.8	82.8	82.8	82.8	82.8
	Air Reactor	752.554	752.677	752.812	752.951	753.091	753.515
	Cool air reactor exhaust	153.029	177.277	201.524	225.772	250.02	322.764

(continued)	Cool flue gas	137.064	137.064	137.064	137.064	137.064	137.064
	Cool OC for air reactor	50.577	50.577	50.577	50.577	50.577	50.577
	Reheat OC for fuel reactor	-52.192	-52.192	-52.192	-52.192	-52.192	-52.192
	Heat steam	-69.8	-69.8	-69.8	-69.8	-69.8	-69.8
	Heat air	-206.58	-232.41	-258.23	-284.05	-309.88	-387.34
	Net	847.446	845.993	844.552	843.116	841.68	837.379

Table A18. CLC process simulation results for different air flow rates with Anthracite coal at 100 kg/h and Fe_2O_3/Al_2O_3 at 5000/3000 kg/h

Initial values	Coal (kg/h)	100	100	100	100	100	100
	Water (kg/h)	140	140	140	140	140	140
	Air Flow Rate (kg/h)	100	300	400	500	600	713
	Temperature of Fuel Reactor (°C)	950	950	950	950	950	950
	Temperature of Air Reactor (°C)	935	935	935	935	935	935
	Fe_2O_3 flow in the Fuel Reactor (kg/h)	5000	5000	5000	5000	5000	5000
	Al_2O_3 in the System (kg/h)	3000	3000	3000	3000	3000	3000
	Particle Density (kg/m³)	3200	3200	3200	3200	3200	3200
Energy balance (kW)	Fuel Reactor	104.05	104.05	104.05	104.05	104.05	104.05
	Air Reactor	96.50	289.50	386.00	482.49	578.99	688.04
	Cool air reactor exhaust	19.00	56.99	75.99	94.98	113.98	135.44
	Cool flue gas	126.59	126.59	126.59	126.59	126.59	126.59
	Cool OC for air reactor	32.79	32.79	32.79	32.79	32.79	32.79
	Reheat OC for fuel reactor	-34.27	-34.27	-34.27	-34.27	-34.27	-34.27
	Heat steam	-69.80	-69.80	-69.80	-69.80	-69.80	-69.80
	Heat air	-25.82	-77.47	-103.29	-129.12	-154.94	-184.12
	Net	249.03	428.37	518.05	607.72	697.39	798.72
Initial values (continued)	Coal (kg/h)	100	100	100	100	100	100
	Water (kg/h)	140	140	140	140	140	140
	Air Flow Rate (kg/h)	800	900	1000	1100	1200	1500
	Temperature of Fuel Reactor (°C)	950	950	950	950	950	950
	Temperature of Air Reactor (°C)	935	935	935	935	935	935
	Fe_2O_3 flow in the Fuel Reactor (kg/h)	5000	5000	5000	5000	5000	5000
	Al_2O_3 in the System (kg/h)	3000	3000	3000	3000	3000	3000
	Particle Density (kg/m³)	3200	3200	3200	3200	3200	3200
Energy balance (kW) (continued)	Fuel Reactor	104.05	104.05	104.05	104.05	104.05	104.05
	Air Reactor	700.66	700.79	700.93	701.07	701.21	701.64
	Cool air reactor exhaust	155.86	180.10	204.35	228.60	252.85	325.59
	Cool flue gas	126.59	126.59	126.59	126.59	126.59	126.59
	Cool OC for air reactor	32.79	32.79	32.79	32.79	32.79	32.79
	Reheat OC for fuel reactor	-34.27	-34.27	-34.27	-34.27	-34.27	-34.27
	Heat steam	-69.80	-69.80	-69.80	-69.80	-69.80	-69.80
	Heat air	-206.59	-232.41	-258.23	-284.06	-309.88	-387.35
	Net	809.29	807.85	806.41	804.97	803.54	799.24

Table A19. CLC process simulation results for different air flow rates with Anthracite coal at 100 kg/h and Fe_2O_3/Al_2O_3 at 5200/3200 kg/h

Initial values	Coal (kg/h)	100	100	100	100	100	100
	Water (kg/h)	140	140	140	140	140	140

	Air Flow Rate (kg/h)	100	300	400	500	600	713
	Temperature of Fuel Reactor (°C)	950	950	950	950	950	950
	Temperature of Air Reactor (°C)	935	935	935	935	935	935
	Fe_2O_3 flow in the Fuel Reactor (kg/h)	5200	5200	5200	5200	5200	5200
	Al_2O_3 in the System (kg/h)	3200	3200	3200	3200	3200	3200
	Particle Density (kg/m³)	3200	3200	3200	3200	3200	3200
Energy balance (kW)	Fuel Reactor	108.71	108.71	108.71	108.71	108.71	108.71
	Air Reactor	96.50	289.50	386.00	482.49	578.99	688.04
	Cool air reactor exhaust	19.00	56.99	75.99	94.98	113.98	135.44
	Cool flue gas	127.83	127.83	127.83	127.83	127.83	127.83
	Cool OC for air reactor	34.53	34.53	34.53	34.53	34.53	34.53
	Reheat OC for fuel reactor	-36.07	-36.07	-36.07	-36.07	-36.07	-36.07
	Heat steam	-69.80	-69.80	-69.80	-69.80	-69.80	-69.80
	Heat air	-25.82	-77.47	-103.29	-129.12	-154.94	-184.12
	Net	254.88	434.22	523.89	613.57	703.24	804.57
Initial values (continued)	Coal (kg/h)	100	100	100	100	100	100
	Water (kg/h)	140	140	140	140	140	140
	Air Flow Rate (kg/h)	800	900	1000	1100	1200	1500
	Temperature of Fuel Reactor (°C)	950	950	950	950	950	950
	Temperature of Air Reactor (°C)	935	935	935	935	935	935
	Fe_2O_3 flow in the Fuel Reactor (kg/h)	5200	5200	5200	5200	5200	5200
	Al_2O_3 in the System (kg/h)	3200	3200	3200	3200	3200	3200
	Particle Density (kg/m³)	3200	3200	3200	3200	3200	3200
Energy balance (kW) (continued)	Fuel Reactor	108.71	108.71	108.71	108.71	108.71	108.71
	Air Reactor	728.29	728.42	728.55	728.69	728.83	729.26
	Cool air reactor exhaust	154.35	178.60	202.85	227.09	251.34	324.09
	Cool flue gas	127.83	127.83	127.83	127.83	127.83	127.83
	Cool OC for air reactor	34.53	34.53	34.53	34.53	34.53	34.53
	Reheat OC for fuel reactor	-36.07	-36.07	-36.07	-36.07	-36.07	-36.07
	Heat steam	-69.80	-69.80	-69.80	-69.80	-69.80	-69.80
	Heat air	-206.59	-232.41	-258.23	-284.06	-309.88	-387.35
	Net	841.26	839.81	838.37	836.94	835.50	831.20

Table A20. CLC process simulation results for different air flow rates with Anthracite coal at 100 kg/h and Fe_2O_3/Al_2O_3 at 5500/3500 kg/h

Initial values	Coal (kg/h)	100	100	100	100	100	100
	Water (kg/h)	140	140	140	140	140	140
	Air Flow Rate (kg/h)	100	300	400	500	600	713
	Temperature of Fuel Reactor (°C)	950	950	950	950	950	950
	Temperature of Air Reactor (°C)	935	935	935	935	935	935
	Fe_2O_3 flow in the Fuel Reactor (kg/h)	5500	5500	5500	5500	5500	5500
	Al_2O_3 in the System (kg/h)	3500	3500	3500	3500	3500	3500
	Particle Density (kg/m³)	3200	3200	3200	3200	3200	3200
Energy balance (kW)	Fuel Reactor	111.73	111.73	111.73	111.73	111.73	111.73
	Air Reactor	96.50	289.50	386.00	482.49	578.99	688.04
	Cool air reactor exhaust	19.00	56.99	75.99	94.98	113.98	135.40
	Cool flue gas	127.83	127.83	127.83	127.83	127.83	127.83
	Cool OC for air reactor	37.03	37.03	37.03	37.03	37.03	37.03
	Reheat OC for fuel reactor	-38.75	-38.75	-38.75	-38.75	-38.75	-38.75
	Heat steam	-69.80	-69.80	-69.80	-69.80	-69.80	-69.80

	Heat air	-25.82	-77.47	-103.29	-129.12	-154.94	-184.12
	Net	257.71	437.06	526.73	616.40	706.07	807.36
Initial values (continued)	Coal (kg/h)	100	100	100	100	100	100
	Water (kg/h)	140	140	140	140	140	140
	Air Flow Rate (kg/h)	800	900	1000	1100	1200	1500
	Temperature of Fuel Reactor (°C)	950	950	950	950	950	950
	Temperature of Air Reactor (°C)	935	935	935	935	935	935
	Fe_2O_3 flow in the Fuel Reactor (kg/h)	5500	5500	5500	5500	5500	5500
	Al_2O_3 in the System (kg/h)	3500	3500	3500	3500	3500	3500
	Particle Density (kg/m³)	3200	3200	3200	3200	3200	3200
Energy balance (kW) (continued)	Fuel Reactor	111.73	111.73	111.73	111.73	111.73	111.73
	Air Reactor	728.29	728.42	728.55	728.69	728.83	729.26
	Cool air reactor exhaust	154.35	178.60	202.85	227.09	251.34	324.09
	Cool flue gas	127.83	127.83	127.83	127.83	127.83	127.83
	Cool OC for air reactor	37.03	37.03	37.03	37.03	37.03	37.03
	Reheat OC for fuel reactor	-38.75	-38.75	-38.75	-38.75	-38.75	-38.75
	Heat steam	-69.80	-69.80	-69.80	-69.80	-69.80	-69.80
	Heat air	-206.59	-232.41	-258.23	-284.06	-309.88	-387.35
	Net	844.09	842.65	841.21	839.77	838.34	834.04

Table A21. CLC process simulation results for different air flow rates with Lignite coal at 100 kg/h and Fe_2O_3/Al_2O_3 at 3000/1000 kg/h

Initial values	Coal (kg/h)	100	100	100	100	100	100
	Water (kg/h)	140	140	140	140	140	140
	Air Flow Rate (kg/h)	100	300	400	500	600	713
	Temperature of Fuel Reactor (°C)	950	950	950	950	950	950
	Temperature of Air Reactor (°C)	935	935	935	935	935	935
	Fe_2O_3 flow in the Fuel Reactor (kg/h)	3000	3000	3000	3000	3000	3000
	Al_2O_3 in the System (kg/h)	1000	1000	1000	1000	1000	1000
	Particle Density (kg/m³)	3200	3200	3200	3200	3200	3200
Energy balance (kW)	Fuel Reactor	212.04	212.04	212.04	212.04	212.04	212.04
	Air Reactor	96.50	289.50	386.00	420.42	420.60	420.72
	Cool air reactor exhaust	19.00	56.99	75.99	98.36	122.61	150.01
	Cool flue gas	113.10	113.10	113.10	113.10	113.10	113.10
	Cool OC for air reactor	15.47	15.47	15.47	15.47	15.47	15.47
	Reheat OC for fuel reactor	-16.35	-16.35	-16.35	-16.35	-16.35	-16.35
	Heat steam	-69.80	-69.80	-69.80	-69.80	-69.80	-69.80
	Heat air	-25.82	-77.47	-103.29	-129.12	-154.94	-184.10
	Net	344.13	523.48	613.15	644.13	642.73	641.09
Initial values (continued)	Coal (kg/h)	100	100	100	100	100	100
	Water (kg/h)	140	140	140	140	140	140
	Air Flow Rate (kg/h)	800	900	1000	1100	1200	1500
	Temperature of Fuel Reactor (°C)	950	950	950	950	950	950
	Temperature of Air Reactor (°C)	935	935	935	935	935	935
	Fe_2O_3 flow in the Fuel Reactor (kg/h)	3000	3000	3000	3000	3000	3000
	Al_2O_3 in the System (kg/h)	1000	1000	1000	1000	1000	1000
	Particle Density (kg/m³)	3200	3200	3200	3200	3200	3200
Energy balance (kW)	Fuel Reactor	212.04	212.04	212.04	212.04	212.04	212.04
	Air Reactor	420.84	420.98	421.13	421.27	421.41	421.84
	Cool air reactor exhaust	171.11	195.36	219.60	243.85	268.10	340.84

(continued)	Cool flue gas	113.10	113.10	113.10	113.10	113.10	113.10
	Cool OC for air reactor	15.47	15.47	15.47	15.47	15.47	15.47
	Reheat OC for fuel reactor	-16.35	-16.35	-16.35	-16.35	-16.35	-16.35
	Heat steam	-69.80	-69.80	-69.80	-69.80	-69.80	-69.80
	Heat air	-206.59	-232.41	-258.23	-284.06	-309.88	-387.35
	Net	639.82	638.39	636.96	635.52	634.09	629.79

Table A22. CLC process simulation results for different air flow rates with Lignite coal at 100 kg/h and Fe_2O_3/Al_2O_3 at 3500/1500 kg/h

Initial values	Coal (kg/h)	100	100	100	100	100	100
	Water (kg/h)	140	140	140	140	140	140
	Air Flow Rate (kg/h)	100	300	400	500	600	713
	Temperature of Fuel Reactor (°C)	950	950	950	950	950	950
	Temperature of Air Reactor (°C)	935	935	935	935	935	935
	Fe_2O_3 flow in the Fuel Reactor (kg/h)	3500	3500	3500	3500	3500	3500
	Al_2O_3 in the System (kg/h)	1500	1500	1500	1500	1500	1500
	Particle Density (kg/m³)	3200	3200	3200	3200	3200	3200
Energy balance (kW)	Fuel Reactor	221.47	221.47	221.47	221.47	221.47	221.47
	Air Reactor	96.50	289.50	386.00	475.59	475.72	475.87
	Cool air reactor exhaust	19.00	56.99	75.99	95.36	119.61	147.01
	Cool flue gas	115.40	115.40	115.40	115.40	115.40	115.40
	Cool OC for air reactor	19.83	19.83	19.83	19.83	19.83	19.83
	Reheat OC for fuel reactor	-20.83	-20.83	-20.83	-20.83	-20.83	-20.83
	Heat steam	-69.80	-69.80	-69.80	-69.80	-69.80	-69.80
	Heat air	-25.82	-77.47	-103.29	-129.12	-154.94	-184.10
	Net	355.74	535.09	624.76	707.91	706.45	704.85
Initial values (continued)	Coal (kg/h)	100	100	100	100	100	100
	Water (kg/h)	140	140	140	140	140	140
	Air Flow Rate (kg/h)	800	900	1000	1100	1200	1500
	Temperature of Fuel Reactor (°C)	950	950	950	950	950	950
	Temperature of Air Reactor (°C)	935	935	935	935	935	935
	Fe_2O_3 flow in the Fuel Reactor (kg/h)	3500	3500	3500	3500	3500	3500
	Al_2O_3 in the System (kg/h)	1500	1500	1500	1500	1500	1500
	Particle Density (kg/m³)	3200	3200	3200	3200	3200	3200
Energy balance (kW) (continued)	Fuel Reactor	221.47	221.47	221.47	221.47	221.47	221.47
	Air Reactor	475.99	476.14	476.28	476.42	476.56	476.99
	Cool air reactor exhaust	168.10	192.35	216.60	240.88	265.09	337.84
	Cool flue gas	115.40	115.40	115.40	115.40	115.40	115.40
	Cool OC for air reactor	19.83	19.83	19.83	19.83	19.83	19.83
	Reheat OC for fuel reactor	-20.83	-20.83	-20.83	-20.83	-20.83	-20.83
	Heat steam	-69.80	-69.80	-69.80	-69.80	-69.80	-69.80
	Heat air	-206.59	-232.41	-258.23	-284.06	-309.88	-387.35
	Net	703.58	702.15	700.71	699.32	697.85	693.55

Table A23. CLC process simulation results for different air flow rates with Lignite coal at 100 kg/h and Fe_2O_3/Al_2O_3 at 4000/2000 kg/h

Initial values	Coal (kg/h)	100	100	100	100	100	100
	Water (kg/h)	140	140	140	140	140	140
	Air Flow Rate (kg/h)	100	300	400	500	600	713

	Temperature of Fuel Reactor (°C)	950	950	950	950	950	950
	Temperature of Air Reactor (°C)	935	935	935	935	935	935
	Fe_2O_3 flow in the Fuel Reactor (kg/h)	4000	4000	4000	4000	4000	4000
	Al_2O_3 in the System (kg/h)	2000	2000	2000	2000	2000	2000
	Particle Density (kg/m³)	3200	3200	3200	3200	3200	3200
Energy balance (kW)	Fuel Reactor	226.51	226.51	226.51	226.51	226.51	226.51
	Air Reactor	96.50	289.50	386.00	475.59	475.72	475.87
	Cool air reactor exhaust	19.00	56.99	75.99	95.36	119.61	147.01
	Cool flue gas	115.40	115.40	115.40	115.40	115.40	115.40
	Cool OC for air reactor	24.31	24.31	24.31	24.31	24.31	24.31
	Reheat OC for fuel reactor	-25.31	-25.31	-25.31	-25.31	-25.31	-25.31
	Heat steam	-69.80	-69.80	-69.80	-69.80	-69.80	-69.80
	Heat air	-25.82	-77.47	-103.29	-129.12	-154.94	-184.10
	Net	360.77	540.11	629.79	712.93	711.48	709.88
Initial values (continued)	Coal (kg/h)	100	100	100	100	100	100
	Water (kg/h)	140	140	140	140	140	140
	Air Flow Rate (kg/h)	800	900	1000	1100	1200	1500
	Temperature of Fuel Reactor (°C)	950	950	950	950	950	950
	Temperature of Air Reactor (°C)	935	935	935	935	935	935
	Fe_2O_3 flow in the Fuel Reactor (kg/h)	4000	4000	4000	4000	4000	4000
	Al_2O_3 in the System (kg/h)	2000	2000	2000	2000	2000	2000
	Particle Density (kg/m³)	3200	3200	3200	3200	3200	3200
Energy balance (kW) (continued)	Fuel Reactor	226.51	226.51	226.51	226.51	226.51	226.51
	Air Reactor	475.99	476.14	476.28	476.42	476.56	476.99
	Cool air reactor exhaust	168.10	192.35	216.60	240.88	265.09	337.84
	Cool flue gas	115.40	115.40	115.40	115.40	115.40	115.40
	Cool OC for air reactor	24.31	24.31	24.31	24.31	24.31	24.31
	Reheat OC for fuel reactor	-25.31	-25.31	-25.31	-25.31	-25.31	-25.31
	Heat steam	-69.80	-69.80	-69.80	-69.80	-69.80	-69.80
	Heat air	-206.59	-232.41	-258.23	-284.06	-309.88	-387.35
	Net	708.61	707.17	705.74	704.35	702.87	698.58

Table A24. CLC process simulation results for different air flow rates with Lignite coal at 100 kg/h and Fe_2O_3/Al_2O_3 at 4500/2500 kg/h

Initial values	Coal (kg/h)	100	100	100	100	100	100
	Water (kg/h)	140	140	140	140	140	140
	Air Flow Rate (kg/h)	100	300	400	500	600	713
	Temperature of Fuel Reactor (°C)	950	950	950	950	950	950
	Temperature of Air Reactor (°C)	935	935	935	935	935	935
	Fe_2O_3 flow in the Fuel Reactor (kg/h)	4500	4500	4500	4500	4500	4500
	Al_2O_3 in the System (kg/h)	2500	2500	2500	2500	2500	2500
	Particle Density (kg/m³)	3200	3200	3200	3200	3200	3200
Energy balance (kW)	Fuel Reactor	231.54	231.54	231.54	231.54	231.54	231.54
	Air Reactor	96.50	289.50	386.00	475.59	475.72	475.87
	Cool air reactor exhaust	19.00	56.99	75.99	95.36	119.61	147.01
	Cool flue gas	115.40	115.40	115.40	115.40	115.40	115.40
	Cool OC for air reactor	28.78	28.78	28.78	28.78	28.78	28.78
	Reheat OC for fuel reactor	-29.79	-29.79	-29.79	-29.79	-29.79	-29.79
	Heat steam	-69.80	-69.80	-69.80	-69.80	-69.80	-69.80
	Heat air	-25.82	-77.47	-103.29	-129.12	-154.94	-184.10
	Net	365.79	545.14	634.81	717.96	716.50	714.90
Initial values (continued)	Coal (kg/h)	100	100	100	100	100	100
	Water (kg/h)	140	140	140	140	140	140
	Air Flow Rate (kg/h)	800	900	1000	1100	1200	1500
	Temperature of Fuel Reactor (°C)	950	950	950	950	950	950
	Temperature of Air Reactor (°C)	935	935	935	935	935	935

	Fe_2O_3 flow in the Fuel Reactor (kg/h)	4500	4500	4500	4500	4500	4500
	Al_2O_3 in the System (kg/h)	2500	2500	2500	2500	2500	2500
	Particle Density (kg/m³)	3200	3200	3200	3200	3200	3200
Energy balance (kW) (continued)	Fuel Reactor	231.54	231.54	231.54	231.54	231.54	231.54
	Air Reactor	475.99	476.14	476.28	476.42	476.56	476.99
	Cool air reactor exhaust	168.10	192.35	216.60	240.88	265.09	337.84
	Cool flue gas	115.40	115.40	115.40	115.40	115.40	115.40
	Cool OC for air reactor	28.78	28.78	28.78	28.78	28.78	28.78
	Reheat OC for fuel reactor	-29.79	-29.79	-29.79	-29.79	-29.79	-29.79
	Heat steam	-69.80	-69.80	-69.80	-69.80	-69.80	-69.80
	Heat air	-206.59	-232.41	-258.23	-284.06	-309.88	-387.35
	Net	713.63	712.20	710.76	709.37	707.90	703.60

Process simulation of oxy-combustion for maximization of energy output using ASPEN plus

Subhodeep Banerjee, Xiao Zhang, Suraj K. Puvvada, Ramesh K. Agarwal

Department of Mechanical Engineering & Materials Science, Washington University in St. Louis, 1 Brookings Drive, St. Louis, MO 63130, USA.

Abstract

Oxy-fuel combustion is a next-generation combustion technology that shows promise to address the need of low-cost carbon capture from fossil fueled power plants. Oxy-fuel combustion requires expensive pre-processing in an air separation unit to separate pure oxygen from air for the combustion process, which reduces the overall efficiency of the process. This paper employs ASPEN Plus process simulation software to model a simple oxy-fuel combustor and investigates the effect of various parameters on the energy output. The composition of the flue gas is carefully examined. The results of this study provide a starting point for optimized oxy-fuel combustion operation for maximum energy output, which will be crucial for future deployment of oxy-fuel combustion technology.

Keywords: Carbon capture; Process simulation; Oxy-fuel combustion; Optimization.

1. Introduction

Recent years have seen considerable effort devoted towards the development of carbon capture technology to prevent or significantly reduce carbon dioxide emissions resulting from the combustion of fossil fuels in electricity-generation power plants. One such technology is oxy-fuel combustion, which involves burning coal using pure oxygen instead of air. The theoretical benefits of this approach are twofold. The significant nitrogen component (around 80%) of the air does not need to be heated to the combustion temperature, which reduces the energy input into the system. Additionally, the flue gases from the combustor consist primarily of carbon dioxide and steam and are not contaminated by the presence of nitrogen and other gases; a stream of almost pure carbon dioxide can be captured without the need for expensive gas separation techniques. However, there is significant expense associated with pre-processing of the air in the air separation unit (ASU) to separate pure oxygen from air for the combustion process, as a result of which the overall efficiency of the oxy-fuel combustion process becomes relatively lower than the standard combustion and therefore the technology has not been widely accepted for deployment on a commercial scale; however the situation may change in the future.

ASPEN Plus is a process simulation software that simulates chemical processes at system level using basic engineering relationships such as mass and energy balance, and multi-phase and chemical reaction models. It consists of flow sheet simulations to calculate stream properties such as flow rate and mass composition given various chemical processes and operating conditions. In this paper, a simple model of the oxy-fuel combustor is developed to conduct parametric studies for optimal energy output; the air separation unit and the steam cycle to extract heat from the flue gases are not considered. This study provides valuable insight into the design and operating conditions required in an industrial-scale oxy-fuel plant to increase combustor efficiency and to assess the feasibility of deploying oxy-fuel combustion as an economically viable solution for electricity generation and carbon capture.

2. Materials and methods

The solid fuel used in the ASPEN Plus model is a bituminous Colombian coal "El Cerrejon", subjected to thermal pre-treatment (i.e., heated at 180°C in atmospheric air for 28 hours) to avoid coal swelling and bed agglomeration. The coal parameters in the model are set up based on the proximate and ultimate analysis of the pre-treated "El Cerrejon" coal; these values are obtained from the work of Zhou et al. [1] and are presented in Table 1.

Table 1. Properties of bituminous Colombian coal "El Cerrejon"

Components	Proximate Analysis (wt. %)				Ultimate Analysis (wt. %)						Energy
	Moisture	Volatile matter	Fixed carbon	Ash	C	H	N	S	O	Ash	LHV (kJ/kg)
Fresh	7.5	34.0	49.9	8.6	70.8	3.9	1.7	0.5	7.20	15.9	25880
Pre-Treated	2.3	33.0	55.9	8.8	65.8	3.3	1.6	0.6	17.6	11.1	21899

The schematic of the flow sheet of the oxy-fuel combustor used in the ASPEN Plus simulation is shown in Figure 1. The coal is pulverized and decomposed in the RYIELD reactor before it is fed into the RGIBBS reactor for combustion. It should be noted that the RYIELD and RGIBBS reactors together represent the oxy-fuel combustor since ASPEN Plus cannot model the entire process with one reactor block. The primary oxidant enters the combustor via the IN-GAS stream. The flue gases and ash are released from the combustor in the ASH and OUT-GAS streams respectively. The dashed lines connecting Q1 and Q2 in Figure 1 indicate the flow of thermal energy (heat) in the system.

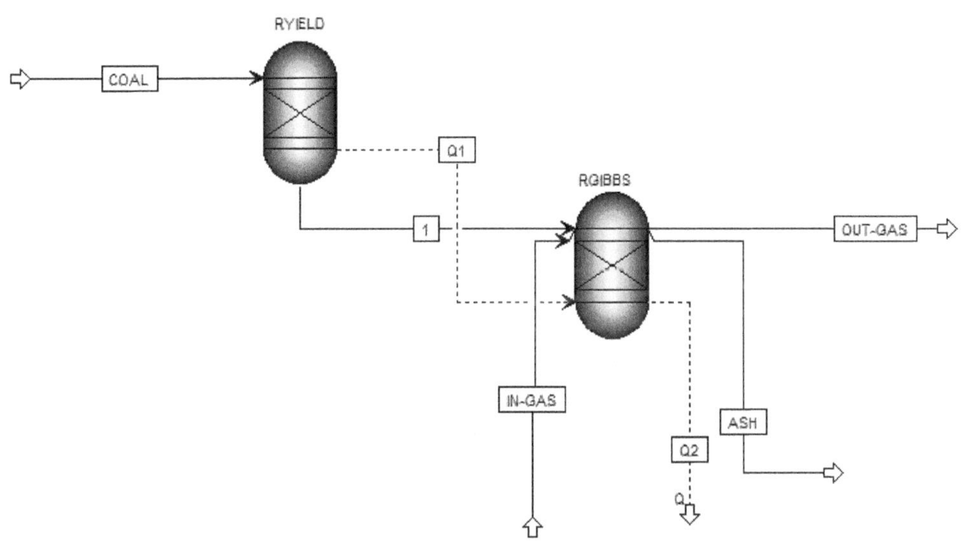

Figure 1. Schematic of the oxy-fuel combustor model in ASPEN Plus

The ASPEN Plus simulations are conducted with both air and oxygen as the primary oxidant. The mass of oxygen required for the stoichiometric ratio is obtained by summing the oxygen requirements for combusting each component of the coal except nitrogen, which is effectively inert at the temperatures considered, minus the mass of oxygen already present in the coal. The ASPEN Plus model uses 100kg of "El Cerrejon" coal for which the stoichiometric mass of oxygen is approximately 212kg. When air is used as the primary oxidant, the same mass of oxygen is used while nitrogen and argon are added such that their relative proportions are correct for air. Considering a ratio of 78% nitrogen, 21% oxygen, and 1% argon, the corresponding mass of air required is 1014kg.

The mass of oxygen in the IN-GAS stream is maintained at the stoichiometric ratio in all cases unless otherwise indicated. Similarly, the temperatures of the RGIBBS reactor and the IN-GAS stream are set at

1000 K and 500 K respectively, unless otherwise indicated; the pressures are set at 1.6 MPa and 1 atm respectively, in line with the work of Gopan et al [2]. The temperature and pressure of the RYIELD reactor are inconsequential since its output is solely based on the proximate and ultimate analysis of the coal. For each case considered, the net energy output of the system corresponding to stream Q2 in Figure 1 is recorded along with the molar composition of the OUT-GAS stream. The parameters that are varied to investigate their effects on the net energy output and flue gas composition are listed in Table 2.

Table 2. Operating parameters considered using ASPEN Plus simulation

Parameter varied	Variation range	Results section
Temperature of RGIBBS reactor	500–1500 K	03.1
Mass of IN-GAS stream	25–175% of stoichiometric O_2	3.2
Temperature of IN-GAS stream	500–1500 K	03.3

3. Results and discussion

3.1 Effect of variation in temperature of RGIBBS reactor

The variation in net energy output of the oxy-fuel combustor model with the RGIBBS reactor temperature is given in Table 3 and Figure 2. According to Figure 2, the relationship between the temperature and the energy output is inversely proportional. The net heat produced by the combustion reaction depends only on the chemical energy stored in the coal; it is independent of temperature. However, as the reactor temperature increases, the temperature of the flue stream increases, which takes away more heat from the reactor. Hence, the net heat output of the reactor decreases with temperature. Furthermore, using pure oxygen as the primary oxidant yields higher energy than using air at all temperatures except at 500 K. At 500 K, there is no energy lost due to heating of the additional components in air to the reactor temperature; therefore the net energy output is the same for both oxygen and air. At higher reactor temperatures, more and more energy is lost due to heating of the nitrogen fraction in air. Consequently, the energy output deficit using air compared to oxygen increases with temperature.

Table 3. Effect of varying RGIBBS temperature on net energy output

Oxidant	500K	750K	1000K	1250K	1500K
O_2	800627 W	774404 kW	745732 kW	715255 kW	683524 kW
Air	800341 W	713490 kW	621540 kW	525577 kW	426270 kW

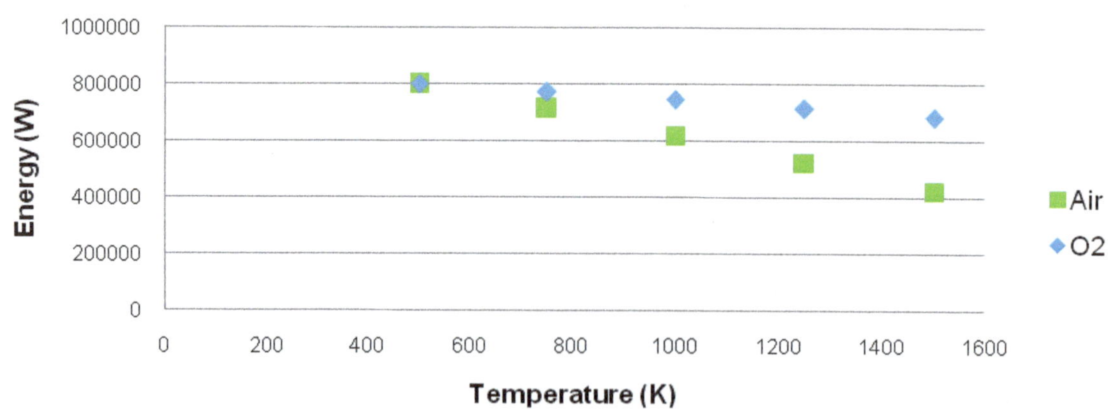

Figure 2. Effect of varying RGIBBS temperature on net energy output

Figure 3 shows the relative distribution of the flue gases in the flue stream of the oxy-fuel combustor model at all temperatures. The most significant products are carbon dioxide and steam. Although the mass of oxygen was prescribed in the stoichiometric ratio, some unused oxygen is observed in the flue stream. The nitrogen originally present in the coal is also present in the flue stream in both cases, although it gets overshadowed by the nitrogen in the air when air is used as the oxidant.

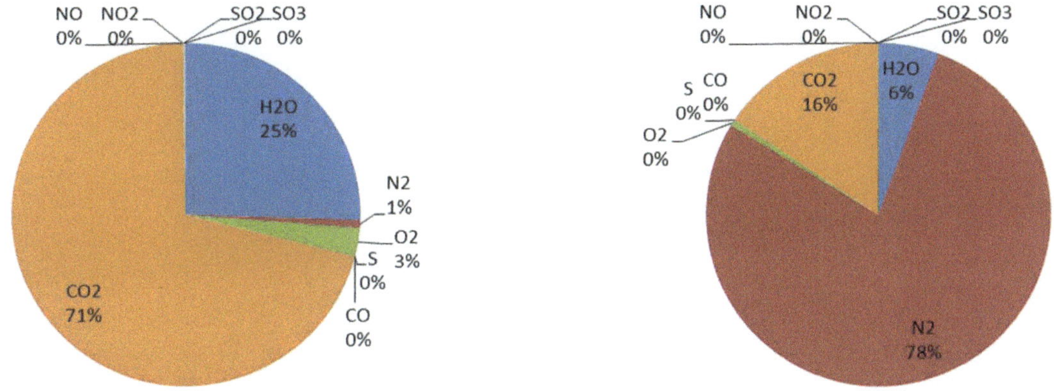

Figure 3. Flue stream composition with oxygen (left) and air (right) using stoichiometric mass of oxygen

NOx and SOx concentrations in the flue stream are negligible according to Figure 3. However, their flow rate variations with temperature, as shown in Figure 4, provide insight into the formation of these pollutants and how they may be reduced. The reaction between nitrogen and oxygen is virtually non-existent at the temperature range considered in the study. The rate of formation of NOx products increases with temperature for both air and oxygen. The formation of SOx products is slightly faster, and there is an inverse relationship in the formation of SO_2 and SO_3. Complete combustion is more favorable at higher reactor temperatures, therefore the molar flow rate of SO_3 decreases and SO_2 increases as the temperature increases.

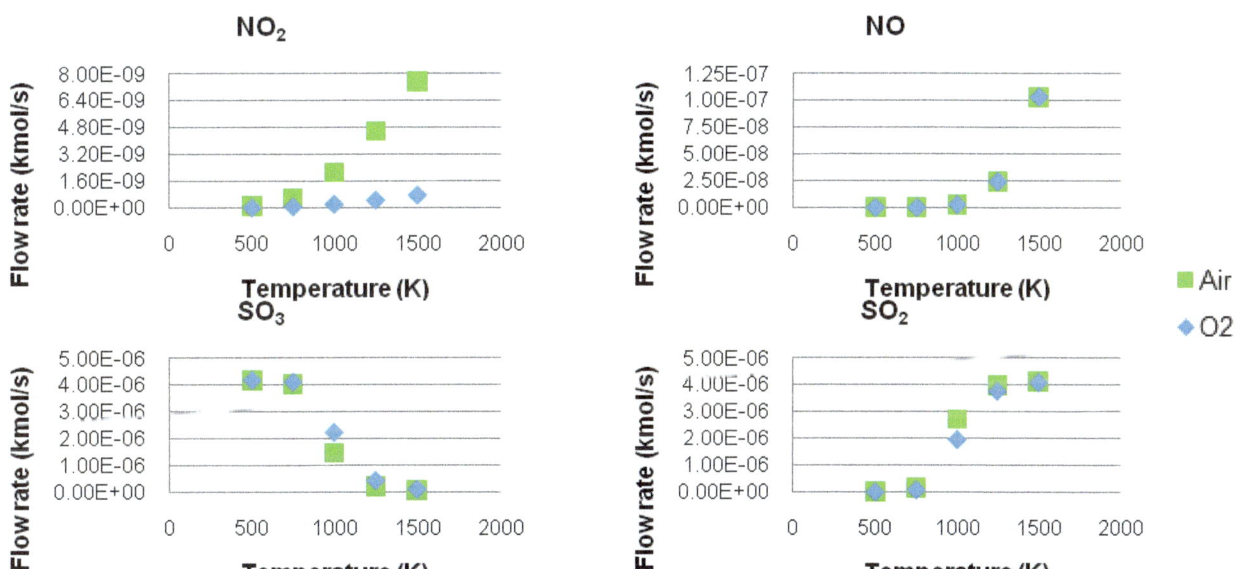

Figure 4. Effect of varying RGIBBS temperature on pollutant formation

3.2 Effect of variation in mass of oxygen in IN-GAS stream

The stoichiometric mass of oxygen required for complete combustion of 100kg of coal is 212kg. In this section, the mass of oxygen in the IN-GAS stream is varied between 25% and 175% of the stoichiometric mass. If air is used as the oxidant, the other components of air are scaled accordingly. The temperature of the oxidant stream and the RGIBBS reactor are maintained at 500 K and 1000 K respectively. Figure 5 shows the effect of the mass of oxygen on the net energy output of the oxy-fuel model. The results in Figure 5 are as expected: the net energy output increases linearly with increasing oxygen mass until the stoichiometric ratio is reached. Beyond the stoichiometric ratio, the net energy output decreases as more energy goes towards heating the excess oxygen to the reactor temperature. Similarly, the difference in the energy obtained using pure oxygen and air also increases as the mass of oxygen increases because the energy must be expended in heating the increased mass of nitrogen since the oxygen and nitrogen are in a fixed ratio in air.

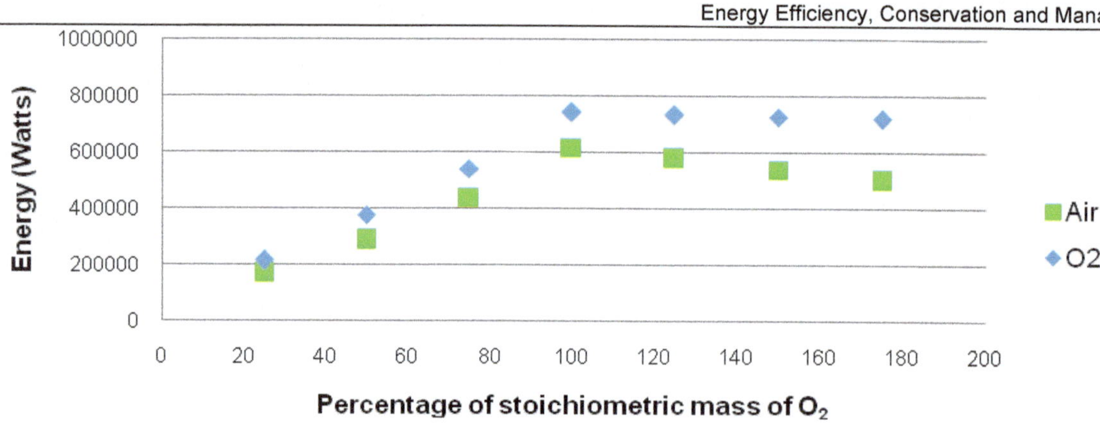

Figure 5. Effect of varying the mass of oxygen on net energy output

The distribution of products in the flue stream by varying the mass of oxygen using pure oxygen and air as oxidant is shown in Figures 6 and 7 respectively. Both figures show that for less than the stoichiometric mass of oxygen, the coal combustion is incomplete as indicated by the high percentage of CO in the flue stream. As the mass of oxygen is increased further, all the CO is converted to CO_2, and the excess oxygen is expelled with the flue stream. Similar to results in section 3.1, when the combustion occurs in air, the large fraction of nitrogen in the air offsets the remaining product fractions; however the trends for incomplete combustion when the oxygen mass is below the stoichiometric ratio are the same.

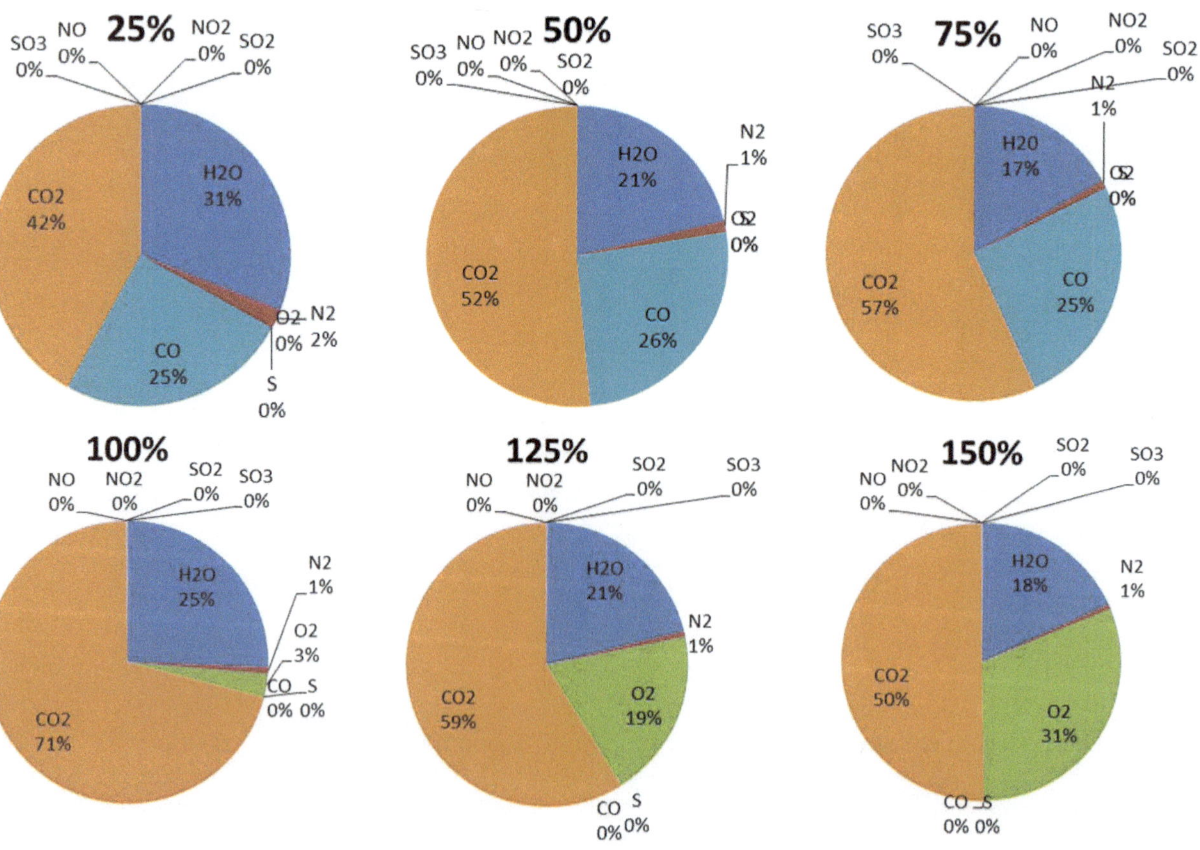

Figure 6. Effect of varying the oxygen mass on flue stream composition using pure oxygen as oxidant

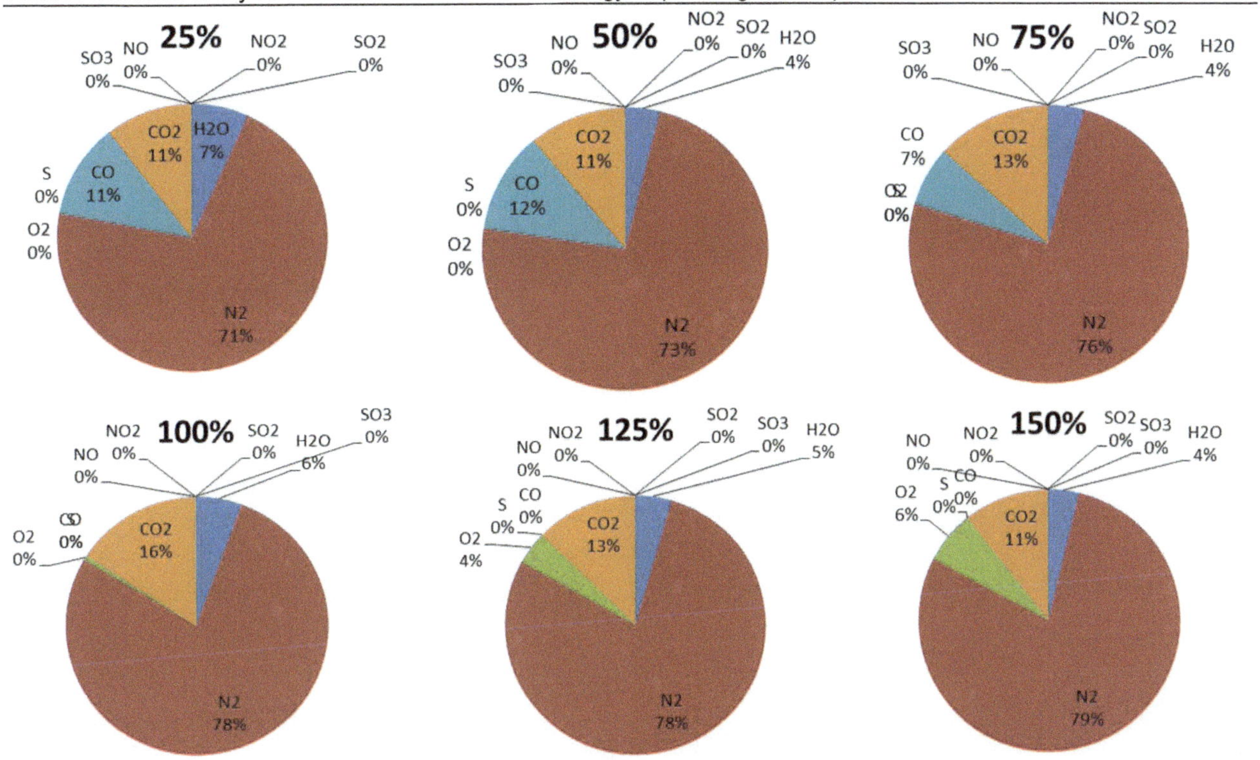

Figure 7. Effect of varying the oxygen mass on flue stream composition using air as oxidant

The detailed flow rates of CO and CO_2, and NOx and SOx products in the flue stream are presented in Figure 8. From Figure 8, it is clear that CO is only formed when the mass of oxygen is less than the stoichiometric ratio (i.e., the combustion is oxygen starved). Also, no NOx and SOx products are formed when the system is oxygen-starved. The concentrations of NO and NO_2 begin to increase when excess oxygen is supplied; this increase is more evident when the oxidant is air because of the larger nitrogen concentration in the reactor. SO_2 and SO_3 also begin to form when excess oxygen is supplied and the inverse relationship in their respective concentrations shown in Figure 4 also applies here as in section 3.1. It should be noted that the NOx and SOx product concentrations remain negligible in relation to other products even at 175% of the stoichiometric mass of oxygen.

3.3 Effect of Variation of temperature of IN-GAS stream

The IN-GAS stream prescribes the chemical composition of the oxidant that is fed into the oxy-fuel combustor. The effect of varying the temperature of a pure oxygen IN-GAS stream on the net energy output of the system is shown in Figure 9. The temperature of the RGIBBS reactor is maintained at 1000 K. Figure 9 shows that the energy output increases linearly as the inlet stream temperature is increased. This is because less energy must be expended on heating the inlet stream to the reactor temperature as the inlet stream temperature increases. The composition of the flue gas stream is identical to that in Figure 3; the details of pollutant formation can be obtained from the data points in Figure 4 which correspond to a reactor temperature of 1000 K.

4. Conclusions

In this paper, ASPEN Plus software was employed to model an oxy-fuel combustor and conduct parametric studies to optimize the energy output of the system. The reactor temperature and mass inflow of oxygen were varied to investigate their effect on the net energy output of the system as well as on the composition of the flue stream of the oxy-fuel combustor. The results from the process simulations suggest that the highest energy output is obtained when the mass of oxygen is at the stoichiometric ratio for the components of the coal. Furthermore, increasing the reactor temperature lowers the energy output since more heat is lost due to the increased temperature of the flue stream. This work provides a good starting point for further studies towards improving the efficiency of an oxy-fuel combustion system.

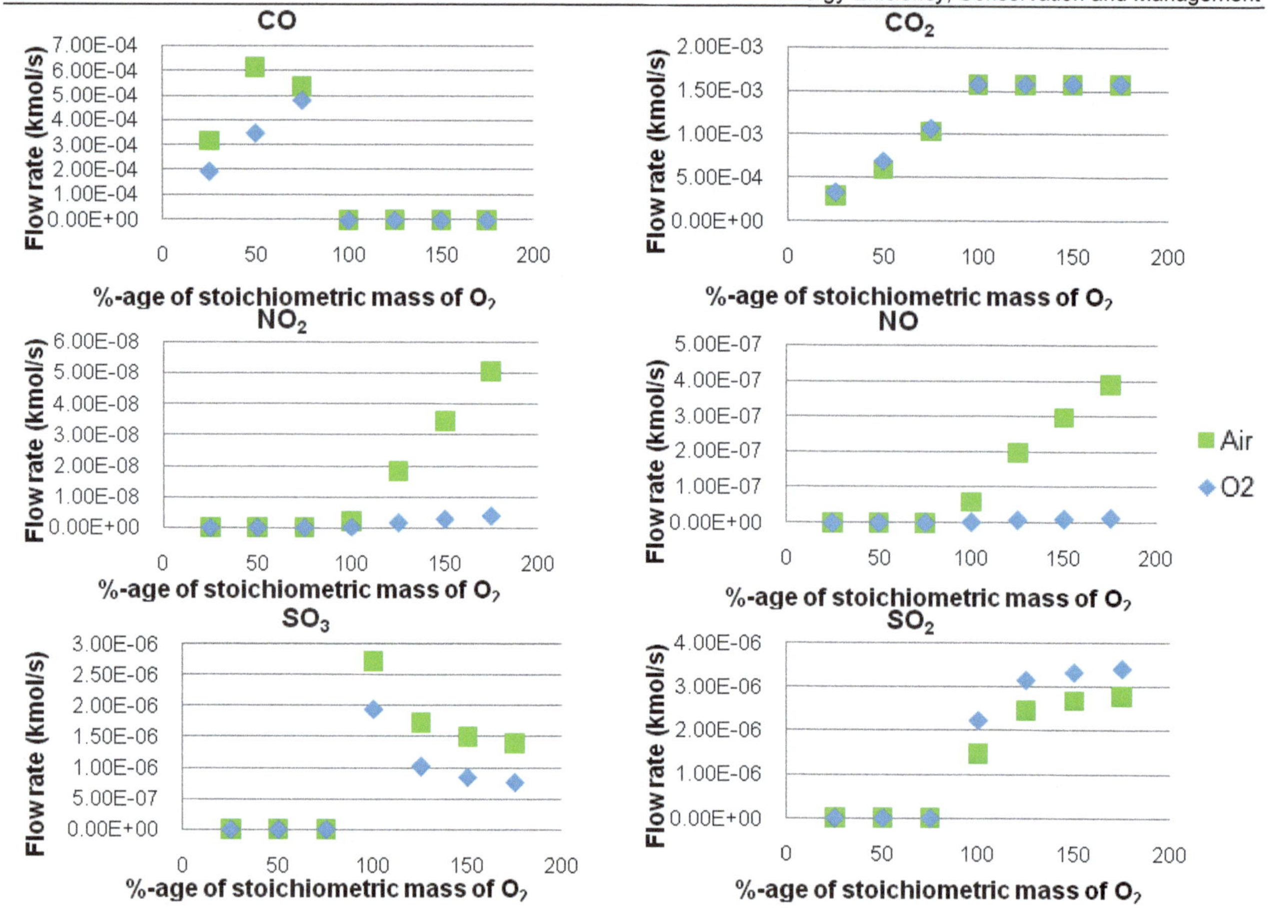

Figure 8. Effect of varying the oxygen mass on pollutant formation

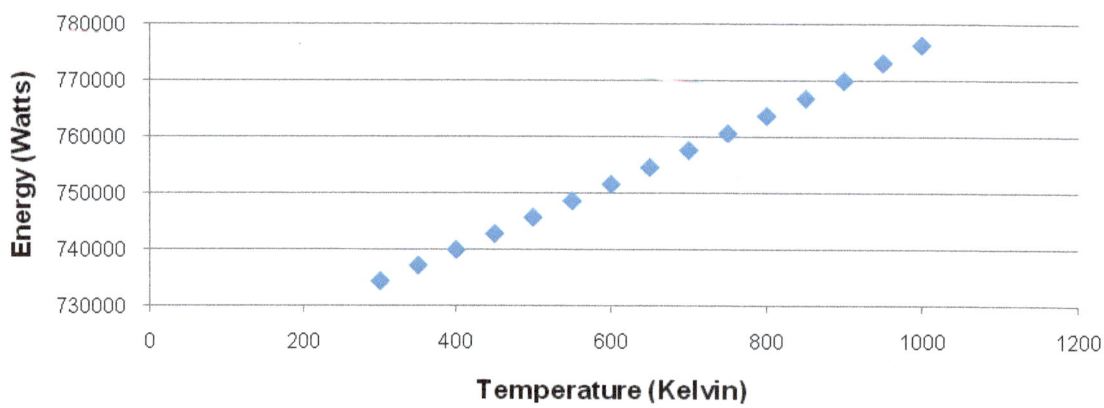

Figure 9. The effect of varying the oxygen stream temperature on net energy output

Acknowledgements

The authors would like to acknowledge the valuable inputs provided by Akshay Gopan of the department of Energy, Environment and Chemical Engineering (EECE) at Washington University in St. Louis. One of the co-authors Suraj K. Puvvada was supported by the STAR program in St. Louis.

References

[1] Zhou L., Zhang Z., Chivetta C., Agarwal R.K. Process simulation and validation of chemical-looping with oxygen uncoupling (CLOU) process using Cu-based oxygen carrier. Energy Fuels,2013, 27(11), 6906-6912.

[2] Gopan, A., Kumfer, B.M., Philips, J., Thimsen, D., Smith, R., Axelbaum, R.L. Process design and performance analysis of a staged, pressurized oxy-combustion (SPOC) power plant for carbon capture. Applied Energy, 2014,125, 179-188.

11

Study of technical feasibility and the payback period of the invested capital for the installation of a grid-connected photovoltaic system at the library of the Technological Federal University of Paraná

Henrique Marin Campos, Ana Katherine Rodríguez Manrique, Bruno Victor Kobiski, Eloy Fassi Casagrande Júnior, Jair Urbanetz Junior

Post-graduation Program on Civil Engineering, Technological Federal University of Paraná, Curitiba, Brazil.

Abstract

This article shows the technical feasibility, and the payback period of the capital invested to install a Grid-connected Photovoltaic (PV) system on the rooftop of the library of the Technological Federal University of Parana (UTFPR), Curitiba campus. The rooftop has 897 square meters, and the photovoltaic modules will be used to supply electricity to four consumption scenarios. It is hoped that with the normative resolution 482 of the National Agency of Electric Energy (ANEEL), published in April 2012, the payback period on the initial investment of the PV system is shorter than when there was no such resolution. It is known that, although the resolution represents a breakthrough for inserting the Grid-connected Photovoltaic power generation, it is still not enough to expand this technology. The high tax of the PV equipment and the absence of incentives for this form of generation still prevent large-scale use. In addition, this article also shows the PV systems installed in Florianópolis (LABSOLAR / UFSC) and Curitiba, such as the Green Office (GO), which is situated at the Technological Federal University of Parana.

Keywords: Sustainability; Grid-connected PV system; Law; Political incentive.

1. Introduction

The ANEEL, through BIG (Database of Information of Generation), presents that the total solar power plants in operation is equal to 2.637 kW. However, it is possible that Grid-connected PV systems increase this quality, because of the Resolution 482/2012 from ANEEL. This law is focused on regulate solar power plants in a range of 100 kW to 1 MW, introducing the net metering system [1, 2].

Currently, in some states of Brazil, the implementation costs related to generate electricity through solar panels for the residential consumers are lower than the taxes of the electrical distribution company. This fact predicts a potential growth on Grid-connected Photovoltaic Power Systems [3].

The Grid-connected Photovoltaic Power System in this study is planned to be situated on the rooftop of the library, at UTFPR, Curitiba. An important concept to this paper is the distributed generation, which means that the generator is located near the consumers, reducing the distance between the source of the energy and its final use.

In the electrical system of Brazil, researchers point out that 15 % of the energy generated in huge blocks is lost in transportation. Among advantages and disadvantages of distributed generation, the mainly favorable points could be: the improvement of the efficiency of energy use and the reduction of the loss in transmission grid; the increase of the partial reliability of power supply in the distributed network; it satisfies the partial increase of loads and reduces the investment of electricity generation facilities; and finally, it uses clean energy such as solar, wind and biomass to reduce the emission of wasted gas during electricity generation [4, 5].

2. Solar radiation

The sun may be regarded as a black body that emits radiation at a temperature of 5700 K. The constant is defined as the solar energy from the sun per unit area in a time interval of 1 second. Recent measurements show that this constant is 1367 $W.m^{-2}$ [6].
The radiation received by the earth is the sum of direct and diffuse radiation, conditioned by cloudiness or other weather conditions. This radiation has photons that can be harnessed and converted into electricity, and the energy delivered by them is at least 1 $kW.m^{-2}$ [7].

3. Photovoltaic systems

Photovoltaic systems are responsible for the conversion of sunlight into electricity. These are divided basically in Isolated Photovoltaic System (IPVS) and Grid-connected Photovoltaic System (GCPVS).

3.1 Isolated photovoltaic systems

IPVSs are common where there is no distributed network energy supply. These systems consist of solar modules, charge controller, battery and inverter. The batteries are responsible to feed the loads that can operate in Direct Current (DC) or Alternate Current (AC). This system configuration is noted in Figure 1 [8].

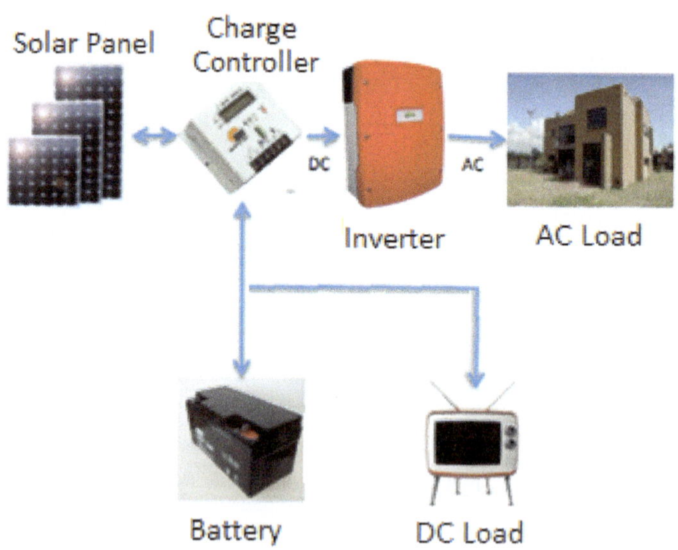

Figure 1. Isolated photovoltaic system configuration. Source: [8]

3.2 Grid-connected photovoltaic systems

The GCPVSs are quite simpler than the SFIs, because they consist of the solar modules, compounding a solar panel and the inverter. This system operates generating electrical energy in parallel to the distributed network. There are two ways to cause an insufficiency on the power generated by the solar panel: the increase of the loads or the low levels of solar radiation. This kind of system has the electric network providing a backup in case of insufficiency on the power generated.
On the other hand, when there is more power generated than the loads consumption, this power is injected in the grid. This system is used in urban areas, since there is availability of electricity supply to consumers in times of low productivity and it is also possible to convert the amount of energy that exceed the loads consumption into a credit to be used by the consumer on his next energy bill, in

accordance to Resolution 482/2012 [1]. Another feature of this system is that if there is a fault in the network, the inverter automatically shuts down the system. Thus, the phenomenon of "islanding" is avoided, giving greater security to network operators by preventing injection of energy from this source [9]. This system configuration is noted in Figure 2.

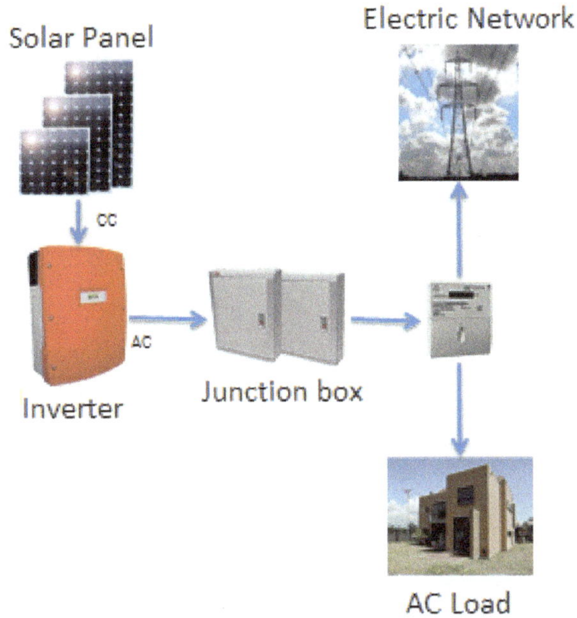

Figure 2. Grid-connected photovoltaic system configuration. Source: [8]

4. Photovoltaic systems in Brazil

4.1 System of Federal University of Santa Catarina

Situated at the Solar Energy Laboratory – LABSOLAR, at the UFSC (Federal University of Santa Catarina), there is the first Grid-connected Photovoltaic System of Brazil, with a total implanted power of 2 kW, provided by 65 solar modules with the amorphous silicon cells (a-Si), 52 being opaque and semi-transparent 13 [10]. In the Figure 3 the system is shown [11]. The system occupies an area of 40,8 m^2 and previously was divided in four circuits with an Wurth 650W inverter. After November 2008, the inverters were replaced by a high efficient one, with 2500 W.

Figure 3. Solar energy laboratory – LABSOLAR. Source: [11]

4.2 System of the green office of UTFPR

The Green Office is a project that aims to show the use of solar energy in a sustainable building, situated at the UTFPR. The whole system is divided into the two configurations explained, the Grid-connected Photovoltaic System and the Isolated Photovoltaic System. The first one consists of ten modules of 210 W, which was built with polycrystalline silicon cells, and a 2000 W inverter [12, 13]. In Figure 4 the system is shown.

Figure 5 shows the general single line diagram of this installation. The electrical panel of the Green Office is called QFL-V-05-TR and its energy consumption is monitored by the Power meter 2.

Figure 4. Green office photovoltaic system. Source: [13]

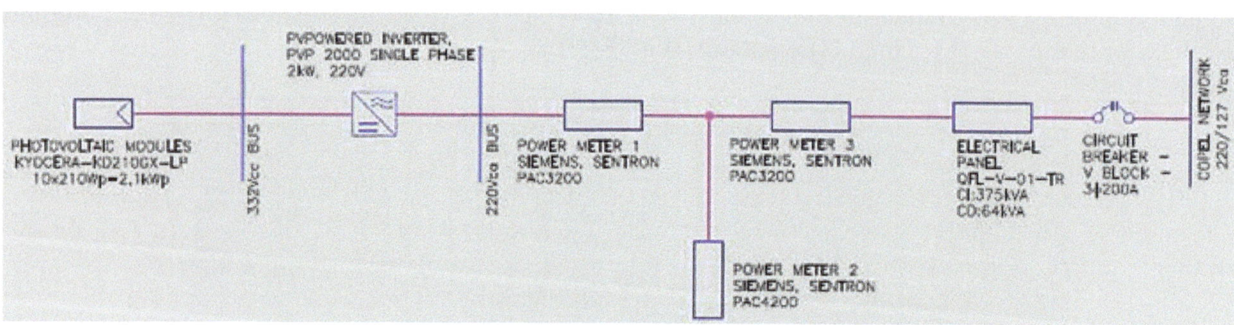

Figure 5. General single line diagram of green office. Source: [13]

This system generated 5,95 MWh from December 14, 2011 to June 18, 2014.

The IPVS has a total implanted power of 870 Wp and is composed of two arrays, one consists of two modules, totalizing a power of 174 Wp, and the other consists of eight modules, totalizing a power of 696 Wp. Figure 6 presents these two arrays.

5. Design of the photovoltaic system

5.1 Scenarios for the photovoltaic power generation

To design the GCPVS, four different scenarios of load consumption were considered, as shown in Table 1. The load description, which was the base for the calculations, is shown in "Appendix".

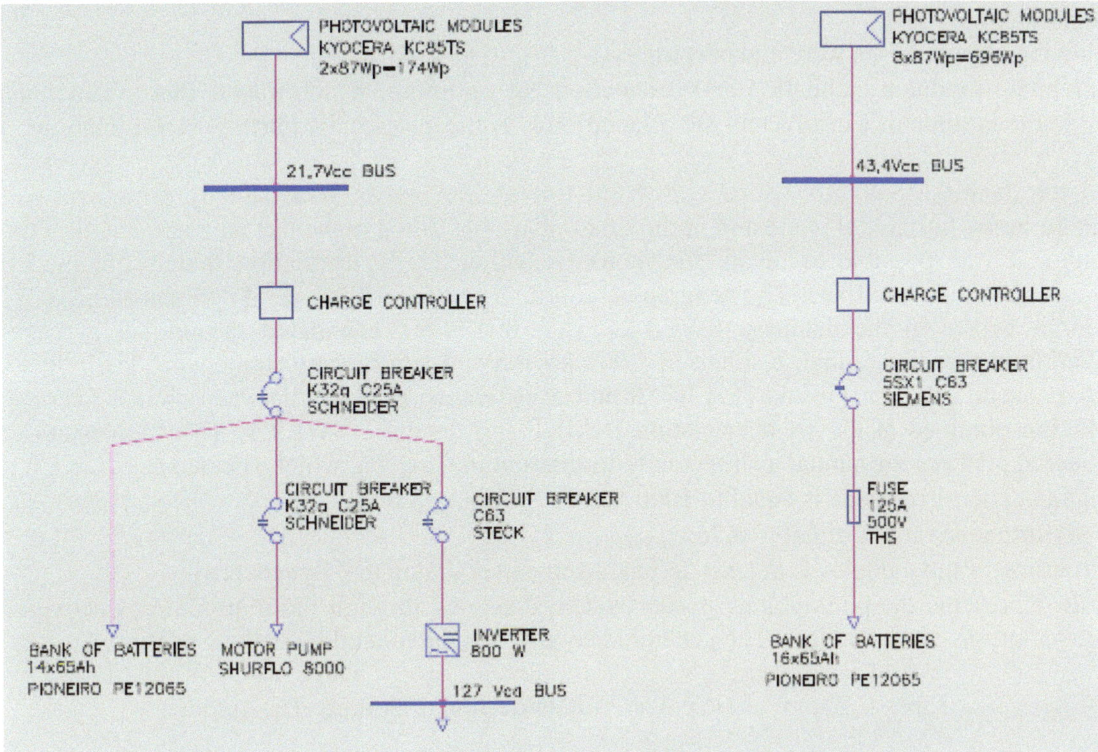

Figure 6. Single line diagram of SFI which consists of two arrays. Source: [13]

Table 1. Scenarios of energy consumption. Source: The authors

Scenario	Load descriptions	Diary consumption (Mon-Fri) (kWh)[2]	Diary consumption (Sat) (kWh)[3]	Photovoltaic power (kWp)[4]
1	Lighting and TUGs[1] at the 1st floor	142,98	67,9	46,6
2	Total loads at the 2nd floor and electrical sockets for the BWC at the 1st floor	96,47	45,8	31,4
3	Electrical sockets for the computers at the 2nd floor	37,44	17,8	12,2
4	TUGs and electrical sockets for the BWC at the 1st floor	21,35	10,1	7,0

[1]: TUGs: Electrical sockets for general use

[2]: The diary consumption was calculated in order to design the photovoltaic power. The period considered to have solar radiation was from 8 to 18 h, which occurs from Mondays to Fridays.

[3]: In this case the period considered was from 8 to 12:45, because on Saturdays it is when the library is working.

[4]: The formula is given by the Equation 1, considering the diary consumption from Monday to Friday, which is the critical case.

$$P_{FV} = \frac{EG}{H_{TOT} PR} \quad (1)$$

where P_{FV} is the photovoltaic power system installed (Wp); E is the load consumption (Wh/day); G is the irradiance under Standard Test Conditions (STC) (1.000 W/m²); H_{TOT} is the solar radiation incident on the surface of the photovoltaic modules (Wh/m².dia); PR is the performance ratio, equal to 0,75.

For this paper, the load consumption was considered to be constant during the days. This point is justified because there is no energy metering specific for the library, so the real load profile is not predictable.

5.2 Requirements

The following requirements were considered:

- Photovoltaic modules inclination and orientation are optimum, which means that the inclination is equal to the latitude of Curitiba and the orientation is to the geographic north [14]. Inclination: -25,43° [15];
- From the database SWERA (Solar and Wind Energy Resources Assessment), which provides the radiation in the horizontal plane and in the tilted plane (the tilted plane has an angle equal to the local latitude), it was possible to obtain the monthly average daily irradiation that is obtained by the photovoltaic module surface. The geographic coordinates of Curitiba, -25,43° S and -49,27° W, were the input values in the database SWERA. From this, were calculated the annual average daily irradiation in Curitiba, which resulted in 5,001 kWh/m². According to Montenegro [16], 80% of the result is used, although Fusano [17] has found a difference of 5,5% between annual average daily irradiation obtained at the weather station INMET and database SWERA. To conclude, this paper considered 5,5% of the annual average daily irradiation in Curitiba, which is equal to 4,726 kWh/m²;
- The global solar irradiance is equal to 1000 W/m², assuming STC;
- The performance ratio is equal to 0,75;
- The lifetime of the modules is at least 25 years and can reach up to 35 years [18].

The Table 2 presents the photovoltaic power system designed through the requirements above and the diary consumptions at the Table 1. The generated energy was calculated by using Equation 1.

Table 2. Photovoltaic power system designed. Source: The authors

Scenarios	1	2	3	4
Photovoltaic power (kWp)	46,6	31,4	12,2	7,0
Generated energy (kWh)	143,0	96,5	37,4	21,4

5.3 Photovoltaic power system costs

In the Table 3 is presented the investment cost for photovoltaic power systems, according to EPE [19]. Table 4 presents the costs applied to the grid-connected photovoltaic power systems of the library.

Table 3. Photovoltaic power systems investment costs (R$/Wp). Source: [19]

Power	Modules	Inverters	Instalation and services	Total
Residencial (4-6kWp)	4,88	1,25	1,53	7,66
Residencial (8-10kWp)	4,42	1,09	1,38	6,89
Comercial (100kWp)	3,81	0,92	1,18	5,91
Industrial (≥1.000kWp)	3,50	0,66	1,04	5,20

Table 4. Photovoltaic power systems total costs. Source: The authors

Scenario	Item	Power (Wp)	Unitary cost (R$/Wp)	Partial cost	Total cost (C_T)
1	Module	46555	3,81	R$177.374,5	R$ 275.139,97
	Inverter	46555	0,92	R$42.830,59	
	Instalation & services	46555	1,18	R$54.934,88	
2	Module	31409	3,81	R$119.669,67	R$ 185.629,32
	Inverter	31409	0,92	R$ 28.896,61	
	Instalation & services	31409	1,18	R$ 37.063,05	
3	Module	12190	3,81	R$ 46.445,71	R$ 72.045,71
	Inverter	12190	0,92	R$ 11.215,24	
	Instalation & services	12190	1,18	R$ 14.384,76	
4	Module	6952	4,42	R$ 30.725,93	R$ 47.896,30
	Inverter	6952	1,09	R$ 7.577,21	
	Instalation & services	6952	1,38	R$ 9.593,16	

5.4 Study of the payback period of the invested capital

The study of the payback period of the invested capital was based on the following requirements.

- It was calculated the time required to recover the capital invested in the project implementation;
- the energy tax price is included on the category horossazonal green, A4 group whose supply voltage is between 2.3 and 25 kV;
- the Normative Resolution n° 482 of the National Electric Energy Agency (ANEEL) proposes the insertion operation of microgeneration, which reaches up to 100 kW of power distribution systems. According to this resolution, the generated active power that exceeds the consumption creates an energy credit to be used primarily at the same time of generation. Therefore, this credit is valid primarily for the hours of sunshine, which are included in the off-peak period, which is when there is photovoltaic generation (off-peak period is between 21:01 and 17:59). These credits are valid for up to 3 years.
- the calculations related to the payback period of the invested capital depart weekly analysis, which is divided as follows: Monday to Friday, when all the energy generated by the system is absorbed by the loads, resulting in energy savings that can be treated in financial terms; Saturday, when part of the energy generated is absorbed by the loads, as is valid from Monday to Friday, and the rest of the energy flows of the dependencies UTFPR or is injected into the network so that it can also be assessed financial return; Sundays, when all energy generated is injected into the network, or into the UTFPR electrical installation that is not located in the library. The behavior of the loads in practice, present variations throughout the day, but were not considered in this paper, due to not having a study of the behavior of the library loads. In Table 5 the energy taxes from the *Companhia Paranaense de Energia* (COPEL) are shown [20].

Table 5. COPEL energy taxes. Source: [20]

Consumption (R$/kWh)		Demand (R$/kWh)		
Peak	Off-peak	Peak	Off-peak	Exceeding
R$ 1,00493	R$ 0,22597	R$ 8,25000	R$ 8,25000	R$ 16,50000

Firstly, the calculations considered an annual decrease of productivity equal to 0,5%. [16]. Secondly, three conditions were pointed out.

- Condition 1: The university must pay ICMS, PIS and COFINS taxes, which represents 35,56% of the energy injected in the electric network [16].
- Condition 2: The university must pay ICMS, PIS and COFINS taxes just for the consumption.
- Condition 3: The university must pay ICMS, PIS and COFINS taxes just for the consumption and the exceeding energy that is injected in the electric network is sold for the double price of the energy bought from the network.

5.4.1 Results of the payback period of the invested capital for the three conditions

The payback period of the invested capital was calculated from the Equation 2. The annual financial returns were summed until the result was the amount of capital invested in the project.

$$\sum_{i}^{n} RFA_{Ti} \tag{2}$$

where i is the year; RFA_T the annual financial return of the year i. Logically, the weekly financial return was calculated, as shown in the Equations 3, 4 and 5.

$$RFS_{C1} = E_{P_1} T_{C_{FP}} 5 + \left(E_{P_2} T_{C_{FP}}\right) + \left(E_{i_1} + E_{i_2}\right) T_{C_{FP}} 0.64 \tag{3}$$

$$RFS_{C2} = E_{P_1} T_{C_{FP}} 5 + \left(E_{P_2} T_{C_{FP}}\right) + \left(E_{i_1} + E_{i_2}\right) T_{C_{FP}} \tag{4}$$

$$RFS_{C3} = E_{p_1}T_{C_{FP}}5 + \left(E_{P_2}T_{C_{FP}}\right) + \left(E_{i_1} + E_{i_2}\right)T_{C_{FP}}2 \quad (5)$$

where RFS_{C1} is the weekly financial return to the condition 1 (R\$); RFS_{C2} is the weekly financial return to the condition 2 (R\$); RFS_{C3} is the weekly financial return to the condition 3 (R\$); E_{P1} is the diary saved energy, which means the energy absorbed from the photovoltaic power system, without demanding the network, from Monday to Friday (kWh); E_{P2} is the diary saved energy during the Saturday (kWh); E_{i1} is the diary injected energy, which means the amount of energy generated that exceeds the loads consumption, during the Saturday (kWh); E_{i2} is the diary injected energy, during the Sunday (kWh); Tc_{FP} is the energy tax for the period of photovoltaic generation (R\$/kWh).
The equations 6, 7 and 8 show the calculation of the monthly financial return, while the Equations 9, 10 and 11 shows the calculation of the annual financial return.

$$RFM_{C1} = RFS_{C1}4 \quad (6)$$

$$RFM_{C2} = RFS_{C2}4 \quad (7)$$

$$RFM_{C3} = RFS_{C3}4 \quad (8)$$

$$RFA_{C1} = RFM_{C1}12 \quad (9)$$

$$RFA_{C2} = RFM_{C2}12 \quad (10)$$

$$RFA_{C3} = RFM_{C3}12 \quad (11)$$

where RFM_{C1} is the monthly financial return to the condition 1; RFM_{C2} is the monthly financial return to the condition 2; RFM_{C3} is the monthly financial return to the condition 3; RFA_{C1} is the annual financial return to the condition 1; RFA_{C2} is the annual financial return to the condition 2; RFA_{C3} is the annual financial return to the condition 3.
The annual financial returns were summed until the result was the amount of capital invested in the project, as shown in the Tables 6, 7, 8 and 9. Table 10 sum up the payback period of the invested capital for the three conditions.

Table 6. Payback period of the invested capital: scenario 1 of energy consumption. Source: The authors

Year	Annual financial return accumulated (RFA_{C1}) (R\$)	Annual financial return accumulated (RFA_{C2}) (R\$)	Annual financial return accumulated (RFA_{C3}) (R\$)
1	10.004,62	10.856,04	13.221,10
2	19.960,53	21.657,80	26.372,43
3	29.868,00	32.405,55	39.454,31
4	39.727,25	43.099,56	52.467,10
5	49.538,53	53.740,10	65.411,14
22			275.268,05
28		284.313,18	
30	279.911,24		

6. Conclusions

Since there is no compensation, like deductions in income tax and fiscal incentives, regarding the installed photovoltaic power, as well as because of the same unitary installation costs for this power range, the payback period of the invested capital was the same for the scenarios 1, 2 and 3. On the other hand, for a smaller installed photovoltaic power system (7 kW), the prices are higher, so there is an increase in the payback period of the invested capital for scenario 4.
The payback period of the invested capital for the implementation of the library grid-connected photovoltaic power system in UTFPR is smaller than the lifetime of the modules, for scenarios 1, 2, 3

and 4 in the three conditions studied [18]. It is important to clarify that in this article were suggested three conditions for achieving results, but the number of these conditions could increase as more systems will have been installed and the rates that could be charged in an interconnection to the network of Copel will have been set. The expectative is that the rates that will be charged in GCPVSs will become clearer, considering that there will be more systems in operation in Parana.

Table 7. Payback period of the invested capital: scenario 2 of energy consumption. Source: The authors

Year	Annual financial return accumulated (RFA_{C1}) (R$)	Annual financial return accumulated (RFA_{C2}) (R$)	Annual financial return accumulated (RFA_{C3}) (R$)
1	6.749,84	7.324,27	8.919,91
2	13.466,82	14.611,92	17.792,74
3	20.151,11	21.863,13	26.618,72
4	26.802,88	29.078,08	35.398,09
5	33.422,28	36.256,96	44.131,07
22			185.715,67
28		185.421,07	
30	188.848,40		

Table 8. Payback period of the invested capital: scenario 3 of energy consumption. Source: The authors

Year	Annual financial return accumulated (RFA_{C1}) (R$)	Annual financial return accumulated (RFA_{C2}) (R$)	Annual financial return accumulated (RFA_{C3}) (R$)
1	2.619,72	2.842,67	3.461,96
2	5.226,69	5.671,12	6.905,65
3	7.820,97	8.485,43	10.331,15
4	10.402,63	11.285,67	13.738,56
5	12.971,72	14.071,91	17.127,98
22			72.079,22
28		74.447,73	
30	73.295,09		

Table 9. Payback period of the invested capital: scenario 4 of energy consumption. Source: The authors

Year	Annual financial return accumulated (RFA_{C1}) (R$)	Annual financial return accumulated (RFA_{C2}) (R$)	Annual financial return accumulated (RFA_{C3}) (R$)
1	1.493,88	1.621,02	1.974,17
2	2.980,50	3.233,93	3.937,92
3	4.459,87	4.838,78	5.891,30
4	5.932,05	6.435,60	7.834,36
5	7.397,07	8.024,44	9.767,15
26			48.073,28
32		48.046,38	
35	48.188,18		

Table 10. Payback period of the invested capital to the conditions 1, 2 and 3. Source: The authors

Scenario	Time to the condition 1	Time to the condition 2	Time to the condition 3
1	29 years and 6 months	27 years and 1 month	22 years
2	29 years and 6 months	27 years and 1 month	22 years
3	29 years and 6 months	27 years and 1 month	22 years
4	34 years and 10 months	31 years and 11 months	25 years and 11 months

It may be mentioned, too, that the legislation is incomplete, although it represents a breakthrough with the introduction of net metering concept that before its publication, in April 2012, was unknown in

Brazil. The proposed on the condition 3 of this study is the possibility of commercialization between the consumer and the electric network, through the sale of the exceeding energy that is injected in the network. This practice is called Feed-in tariff and represented huge increases on the installed photovoltaic power in countries like Germany, which is a country that invests extensively in solar energy. In this case, the consumer becomes also a minigenerator of energy, and that energy generated in case of surplus is sold to the dealership for a premium rate 2,5 to 3 times the normal rate. This variation is a function of the location of the installation and the power installed [21]. This practice is already current in Japan as well, where a new Feed-in scheme started on July 2012 and brought an increase of 33% on the installed photovoltaic power comparing that year to the previous one. This new Feed-in scheme is aimed to non-residential segments, like large scale photovoltaic projects, in the industrial and commercial sectors. Researches pointed out that the payback period for a 100 kW solar photovoltaic plant in Japan is 8,05 years, which is quite less in comparison to the 14,65 years obtained in Germany [22].

Brazilian legislation related to the implementation of solar photovoltaic energy as a form of distributed generation is very different when compared to Japan and Germany for example. According to the law N. 10848/2004 and Decree N. 5163/2004, it is not allowed to sale the exceeding energy to the electric network. So, it represents a barrier for the economic feasibility of grid-connected photovoltaic systems. However, there are other public policies that might encourage the expansion of grid-connected photovoltaic systems, such as financing of equipment, deductions in income tax and fiscal incentives [23].

It is also important to mention that this research is a simplified economic analysis, although its results are consistent with the literatures that still conclude the economic unfeasibility on grid-connected photovoltaic systems. However, this and many other researches are limited because it does not take into account the negative externality costs that occur for other decentralized energy sources, like fossil fuel. This source pollutes the air and contribute to the increase on greenhouse gases emission, so it has a societal cost, because affects everyone's life. The evaluation of the impacts of externalities should be developed and it would be an important and favorable step for decision making on renewable energies [24].

To conclude, the power tariff in the tariff structure horossazonal green of COPEL, which is what fits UTFPR, is still cheap so the payback times become longer when compared to others Brazilian states [16]. If the rate increases will be possible to obtain a shorter payback period on the invested capital. It is worth mentioning that this study is not considering a possible reduction in demand due to the installation of GCPVS, which would imply savings in electricity bills, reducing the time of return on invested capital. It is a suggestion for future work, to make a study of the possible demand reduction in UTFPR with the installation of a GCPVS and to perform such a study, it is necessary to access the mass memory of the Power Metering in operation at the university.

Appendix

The loads and its descriptions are shown in reference to the electrical panels QFL-L-01-PR, QFL-L-01-SG and QFL-L-01-TC, which are installed on the first, second and third floor, respectively, at the library.

QFL-L-01-PR

Load descriptions	Installed power (W)	Demand factor	Demanded power (W)	Time of usage (h)	Diary energy consumed in average (kWh)
Lighting	15744	0,8	12595,2	13,75	173,18
Ventilation	3876	1	3876	13,75	53,30
TUGs	17030	0,1	1703	13,75	23,42
Electrical sockets for computers	12480	0,9	11232	13,75	154,44
Electrical sockets for bathroom	4320	0,1	432	13,75	5,94

QFL-L-01-SG

Load descriptions	Installed power (W)	Demand factor	Demanded power (W)	Time of usage (h)	Diary energy consumed in average (kWh)
Lighting	4912	0,8	3929,6	13,75	54,03
Ventilation	1020	1	1020	13,75	14,03
TUGs	3770	0,1	377	13,75	5,18
Electrical sockets for computers	4160	0,9	3744	13,75	51,48
Electrical sockets for bathroom	1440	0,1	144	13,75	1,98

QFL-L-01-TC

Load descriptions	Installed power (W)	Demand factor	Demanded power (W)	Time of usage (h)	Diary energy consumed in average (kWh)
Lighting	5040	0,8	4032	13,75	55,44
Ventilation	1224	1	1224	13,75	16,83
TUGs	4030	0,1	403	13,75	5,54
Electrical sockets for computers	4680	0,9	4212	13,75	57,92
Electrical sockets for bathroom	960	0,1	96	13,75	1,32

Acknowledgements

Thanks to the Technological Federal University of Parana that provides the load descriptions of the library, as well as to the Green Office from UTFPR and Labsolar from UFSC that represents important initiatives on researching about solar photovoltaic energy. Thanks to CAPES for the financial resources given that made possible this research.

References

[1] ANEEL. Resolução normativa n° 482, de 17 de abril de 2012. Disponível em: <http://www.aneel.gov.br/cedoc/ren2012482.pdf>. Acesso em: 4 abril 2014.

[2] ANEEL: Banco de Informações de Geração. Capacidade de geração do Brasil. Disponível em: <http://www.aneel.gov.br/aplicacoes/capacidadebrasil/capacidadebrasil.cfm>. Acesso em: 05 maio 2014.

[3] SOLARBUZZ. Emerging PV Markets Report: Latin America & Caribbean. Report Sample. Dezembro 2012.

[4] PLATAFORMA ITAIPU. Geração distribuída: solução para a eficiência energética. Disponível em: <http://www.plataformaitaipu.org/plataforma/geracao-distribuida>. Acesso em: 02 maio 2014.

[5] Bo, Wang, Ka, Lan. Analysis of the distributed generation system and the influence on power loss. IEEEXPLORE: 2011.

[6] Andújar Sagredo, Rodrigo. Estudio técnico-económico de una planta solar de alta temperatura en una central de ciclo combinado. Proyecto fin de carrera. Universidad Pontificia Comillas Escuela técnica superior de ingeniería (ICAI). Madrid, 2004.

[7] FUNDACIÓN TERRA. "Perspectiva ambiental: Energía fotovoltáica". En: Suplemento de perspectiva escolar. Versión en español en formato digital (2000).

[8] Rodriguez, K, Cadena, A, Aristizabal, J. Estudo da viabilidade técnica e econômica de uma instalação fotovoltaica na "Sabana de Bogotá", 2011.

[9] Santos, Isis Portolan, Urbanetz Junior, J., Rüther, Ricardo. Energia solar fotovoltaica como fonte complementar de energia elétrica para residências na busca da sustentabilidade, 2007.

[10] Urbanetz, J. J. Sistemas Fotovoltaicos Conectados a Redes de Distribuição Urbanas: Sua Influência na Qualidade da Energia elétrica e Análise dos Parâmetros que Possam Afetar a Conectividade. Universidade Federal de Santa Catarina. Florianópolis, p. 189. 2010. Tese (Doutorado em Eng. Civil).

[11] FOTOVOLTAICA UFSC. 2 kWp - Primeiro sistema do Brasil integrado à arquitetura e interligado à rede elétrica pública. Disponível em: <http://www.fotovoltaica.ufsc.br/projetos/detalhe/2kwp-primeiro-sistema-do-brasil-integrado-a-arquitetura-e-interligado-a-rede-eletrica-publica>. Acesso em: 04 maio 2014.

[12] Urbanetz Junior, J., Casagrande Junior, E. F.. Sistema Fotovoltaico Conectado à Rede Elétrica do Escritório Verde da UTFPR. Congresso brasileiro de planejamento energético. Curitiba, agosto 2012.

[13] Campos, Henrique M., Oliveira, Allan R., Amarante, Joao G.. Estudo da eficiência energética do Escritório Verde da Universidade Tecnológica Federal do Paraná, câmpus Curitiba. 2013. 247 f. Trabalho de conclusão de curso (Graduação) – Curso de Engenharia Industrial Elétrica. Universidade Tecnológica Federal do Paraná, Curitiba, 2013.

[14] CEPEL; CRESESB. Grupo de Trabalho de Energia Solar (GTES). Manual de engenharia para sistemas fotovoltaicos. Rio de Janeiro, 2004, 206 p.

[15] DB-CITY. Geografia Curitiba. Disponível em: <http://pt.db-city.com/Brasil--Paran%C3%A1--Curitiba>. Acesso em: 16 maio 2014.

[16] Montenegro, A. Avaliação do retorno do investimento em sistemas fotovoltaicos integrados a residências unifamiliares urbanas no brasil. 2013. 175 f. Dissertação (Mestrado em Engenharia Civil) – Programa de Pós-Graduação em Engenharia Civil, Universidade Federal de Santa Catarina, Florianópolis, 2013.

[17] Fusano, Renato H. Análise dos índices de mérito do Sistema Fotovoltaico Conectado à Rede do Escritório Verde da UTFPR. 2013. 94 f. Trabalho de conclusão de curso (Graduação) – Curso de Engenharia Industrial Elétrica. Universidade Tecnológica Federal do Paraná, Curitiba, 2013.

[18] Deambi, S. Solar PV Power: A global perspective. India: The Energy and Resources Institute, 2011.

[19] EPE. Análise da inserção da geração solar na matriz elétrica brasileira. Rio de Janeiro, maio de 2012.

[20] COPEL. Taxas e tarifas. Disponível em: <http://www.copel.com/hpcopel/root/nivel2.jsp?endereco=%2Fhpcopel%2Froot%2Fpagcopel2.nsf%2F5d546c6fdeabc9a1032571000064b22e%2F0a363cf546237cc203257488005939ce>. Acesso em: 30 junho 2014.

[21] GERMANY. Act revising the legislation on renewable energy sources in the electricity sector - Act implementing Directive 2001/77/EC of the European Parliament and of the Council of 27 September 2001 on the promotion of electricity produced from renewable energy sources in the internal electricity market. 2004.

[22] Muhammad-Sukki, Firdaus, Abu-Bakar, Siti H., Munir, Abu B., Yasin, Siti H. M., Ramirez-Iniguez, Roberto, McMeekin, Scott G., Stewart, Brian G., Sarmah, Nabin, Mallick, Tapas K., Rahim, Ruzairi A., Karim, Md E., Ahmad, Salman, Tahar, Razman M. Feed-in tariff for solar photovoltaic: The rise of Japan. Renewable Energy, Philadelphia, v. 68, p. 636-643, 2014.

[23] Neto, Giovani Z., Costa, Wagner T., Vasconcelos, Vinicius B. A resolução normativa n ° 482/2012 da ANEEL: possibilidades e entraves para a microgeração distribuída. Congresso brasileiro de energia solar. Recife, abril 2014.

[24] Kwan, Calvin L., Kwan, Timothy J. The financials of constructing a solar PV for net-zero energy operations on college campuses. Utilities Policy. Philadelphia, v. 19, p. 226-234, 2011.

12

Improve of produced gas quality by using air/steam in fluidized bed gasifier

Salami N, Skala Z

Energy institute, Faculty of Mechanical Engineering, Brno University of Technology, Technická 2896/2, 616 69 Brno, Czech Republic.

Abstract
The aim of this work is to determine the best operating parameters of system air- steam gasification in in fluidized bed gasifier,which achive the best gas quality. To accomplish this task, many experiments have been performed to study the effect of reactor temperature(T101), steam to biomass ratio $(\frac{S}{B})$ and temperature of provided steam (Tf1), low heating value(LHV), gas yield, carbon conversion efficiency and gasifier efficiency.

Keywords: Gasification; Gasifying agent; Air; Steam.

1. Introduction

The renewable energy sources, as solar, wind, hydraulic and biomass energies, have been used since many centuries and their applications continued until the "industrial revolution", at which time, because of the low price of petroleum, they were abandoned [1].

Biomass energy is an important source of energy for majority of the world's population. The use of biomass energy is expected to increase in the near future, with growth in population. In developing nations, biomass is an even more important resource providing as much as 35 % of the energy needs in some areas of the globe, particularly in isolated areas where it is often the only resource available [3].

Biomass is available for exploitation for conversion to the bio-fuels as well as for power generation applications. There are various conversion technologies that can convert biomass resources into power, heat, and fuels. In view of this a variety of processes exists for biomass conversions [5]. The most used of these are thermal conversions, bio-chemical and chemical conversions and direct combustion. The thermal conversion processes consist of fast and slow pyrolysis and biomass gasification. Gasification is a process for converting carbonaceous materials to a combustible or synthetic gas (H_2, CO, CO_2, CH_4)[1]. In general, gasification involves the reaction of carbon with air, oxygen, steam, carbon dioxide, or a mixture of these gases at 700 °C or higher to produce a gaseous product that can be used to provide electric power and heat or as a raw material for the synthesis of chemicals, liquid fuels, or other gaseous fuels such as hydrogen [5].

2. Chemical reactions in the gasification process

Table 1 includes the reactions that take place in a gasifier during the gasification process [1].

Table 1. Reactions that take place in a gasifier during the gasification process

	NO		heat of reaction $\Delta H_r^0\left(\frac{kJ}{mol}\right)$
The combustion reaction	1	$C+O_2 \rightarrow CO_2$	+393,5
	2	$2H_2+O_2 \rightarrow 2H_2O$	+482,3
partial combustion	3	$C+\frac{1}{2}O_2 \rightarrow CO$	+110,5
gasification reaction	4	$C+H_2O \leftrightarrow CO+H_2$	-131,3
	5	$C+2H_2O \leftrightarrow CO_2+H_2$	-90,2
	6	$C+CO_2 \leftrightarrow 2CO$	-172,4
Methanation reaction	7	$C+2H_2 \leftrightarrow CH_4$	+74,8
	8	$2CO+2H_2 \rightarrow CH_4+CO_2$	+247,3
	9	$CO+3H_2 \rightarrow CH_4+H_2O$	+206,1
	10	$CO_2+4H_2 \rightarrow CH_4+H_2O$	+165,0
	11	$CO+3H_2 \leftrightarrow CH_4+H_2O$	+205,1
	12	$CH_4+2O_2 \rightarrow CO_2+2H_2O$	+801,0
	13	$2CO+O_2 \rightarrow 2CO_2$	+576,3
Water gas reaction	14	$CO+H_2O \leftrightarrow CO_2+H_2$	+41,1

3. Gasification medium

Thermochemical gasification of biomass is a well-known technology that can be classified depending on the gasifying agent: air, steam, steam–oxygen, air– steam, .., etc[2].

3.1 Air gasification

The using of air as a gasifying agent is not complex way, however the produced gas is with lower calorific value primarily approximately (3.5-7.8$\frac{MJ}{Nm^3}$), with little amout of hydrogen and carbon monoxide and high amount of nitrogen.The produced gas by this method it suitable for boiler and engine applications but not for uses that require its transportation through pipelines [6]. Air gasification is widely used compared to oxygen and steam due to its economical and operational advantages, However this technology produces a gas with a low heating value 4 ($\frac{MJ}{Nm^3}$) and an 8-11 vol.% H_2 content only [7].

3.2 Steam gasification

Steam gasification needs an external heat source, if it is used alone as gasifying agent [7]. Provide steam will enhance gas quality, which it enhance hydrogen content and heating value. The high temperature will enhance devolatilization process of biomass to produce gas [8]. Steam will react with carbon monoxide to produce hydrogen and carbon dioxide it is called the water-gas shift reaction Equation 14, Table 1. However the excessive increase of steam provided, will be lowered the gas quality [8].

3.3 Steam-air gasification

Using a mixture of steam and air as a gasifying agent will enhance syngas quality. Oxygen in the air will help to provide the needed energy because the exothermic nature of burning biomass. Reducing the Nitrogen content of the product gas will increase the heating value of the gas [7].

In the present work, it have been studied the effect of steam and air gasification in fluidized bed on syngas quality and It has been chosen the optimal parameters of air steam gasification system to achieve the best quality [7].

4. Experimental study

4.1 Experimental unit biofluid 100

Experiments are carried out at fluidized bed atmospheric gasifier with stationary fluidized bed called Biofluid 100. The experimental set-up, shown in (Figures 1 and 2), consists of six main parts: (i) a fluidized bed reactor of atmospheric pressure, (ii) biomass feeding section, (iii) steam, air providing and preheating section, (iv) gas metering, cleaning and sampling section, (v) temperature control section, (vi) gas offline analysis section. A mixture of air and steam was used as the fluidizing agent and introduced into the reactor. Air was supplied by a compressor and was heated to 200 °C in a preheater. The steam of 150 °C was produced in a steam generator and its mass flowrate was measured by a steam flowmeter then the steam heat in reheater to 450 °C Blower-compressed air is delivered to the gasifier, to under its grate, as primary air ensuring partial oxidization of fuel and maintaining the fluidized bed. Moreover, air can be supplied at two other levels as secondary air and tertiary air. The parameters of the gasifier are as follows:

- Output (in generated gas) 100 (kW)
- Input (in fuel) 150 (kW)
- Fuel consumption max. 40 $\left(\frac{\text{kg}}{\text{h}}\right)$
- Air flow max. 50 $\left(\frac{\text{m}^3}{\text{h}}\right)$
- Air temperature 200 (°C)
- Output (steam generator) 18 (kW)
- Steam temperature (output steam generator 150 °C, then it will be heated to 450 °C)
- Steam flow 18 $\left(\frac{\text{kg}}{\text{h}}\right)$

Measured quantities: T 101-104…temperatures in the gasifier, T105...temperature inside the cyclone Tf1...temperature of the incoming primary air (the temperature of the mixture of incoming primary air and steam), T107...gas temperature at jacket outlet, F 1-3…air flows, F107...gas flow,P107…outlet gas pressure, Ppal... tank pressure, DPfv...fluidized bed pressure difference.

Figure 1. Atmospheric fluidized bed gasifier Biofluid 100

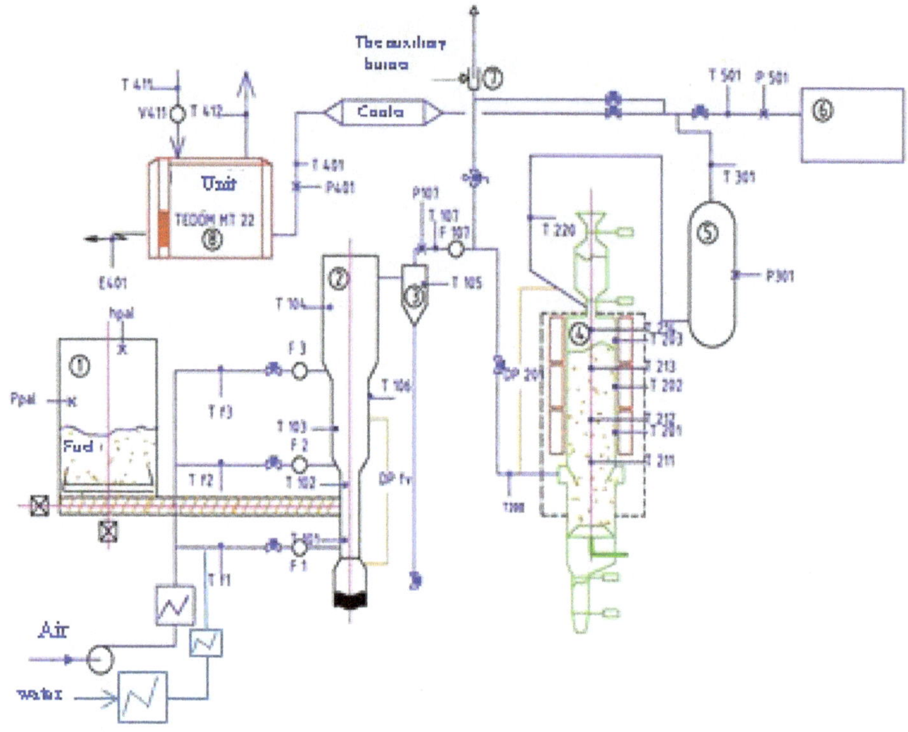

Figure 2. Simplified layout of the gasifier connections

4.2 Fuel material

The fuel material used in this experiment is pine wood chips obtained from a local timber mill were used as the feedstock. The analyses of the feedstock were reported in Table 2.

Table 2. The analyses of the feedstock by RWTUV Praha, Spol.sr.o. Laboratories and test Olomoucka 7/9 656 66 Brno [9]

Specified	Sample in delivered stat	Water-free sample	Sample inflammable matter
Total water %	11.00	-	-
Ashes %	0.47	0.53	-
Caloric value LHV at 25C $kJ.kg^{-1}$	16426	18760	18860
Hydrogen H %	5.46	6.13	6.16
Carbon C %	44.56	50.07	50.34
Nitrogen N %	0.11	0.12	0.12
Oxygen O_d %	38.38	43.13	43.36
Chlorine Cl %	<0.01	<0.01	<0.01
Volatile sulphur S_{vk} %	<0.01	<0.01	<0.01
Sulphated ash S_A %	<0.01	<0.01	-
Total sulphur S_t %	0.01	0.01	-

5. Results and discussion

5.1 Steam to biomass ratio S/B, reactor temperature T101

The first goal is determination the optimum temperature of the reactor T101 and optimal ratio of steam to biomass. To achieve this goal, many experiments have been done at different reactor temperatures and different values of steam to biomass ratio. Reactor temperature was varied from 770 to 861 (°C) in 20 (°C) increments. Steam rate was varied from 0 to 20 $\left(\frac{kg}{h}\right)$, thus steam to biomass ratio varied from 0 to 0.85 $\left(\frac{kg_{steam}}{kg_{biomass}}\right)$ Steam temperature was about Tf1 =261 (°C), equivalence ratio ER about 0.29, air flow rate avaried from 14 to 24 $\left(\frac{kg}{h}\right)$ and biomass flow rate also avaried from 15 to 26 $\left(\frac{kg}{h}\right)$. Samples for mutual comparison are selected at similar gasification conditions, for every reactor temperature separately.The results of this testes were reported in Figures 3 to 11.

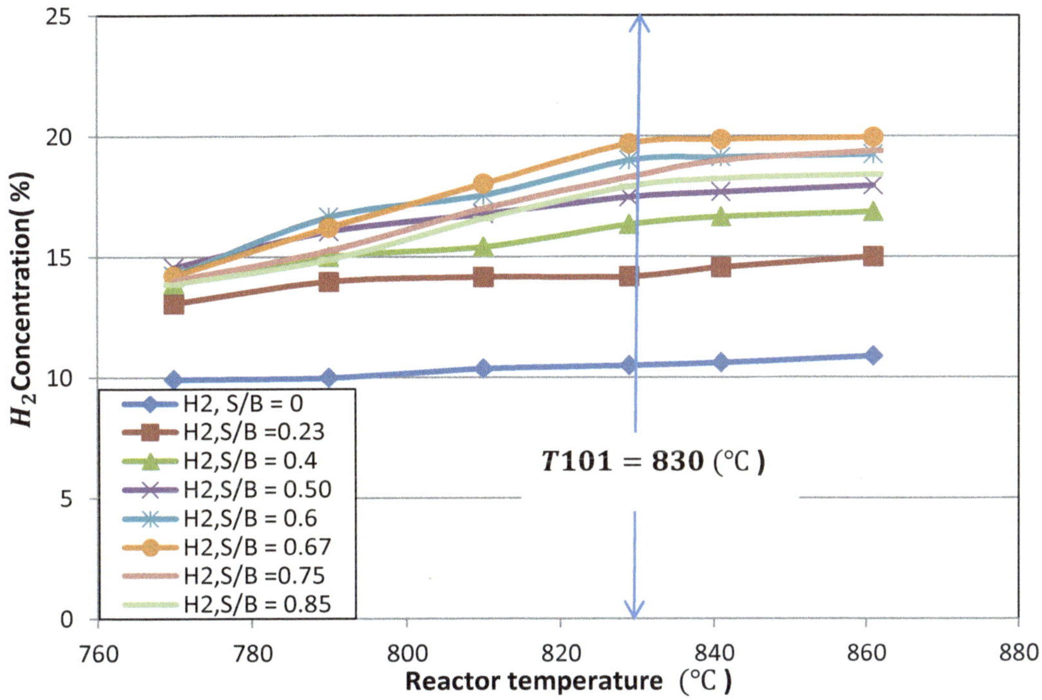

Figure 3. The effect of S/B and T101 on hydrogen content in produced gas

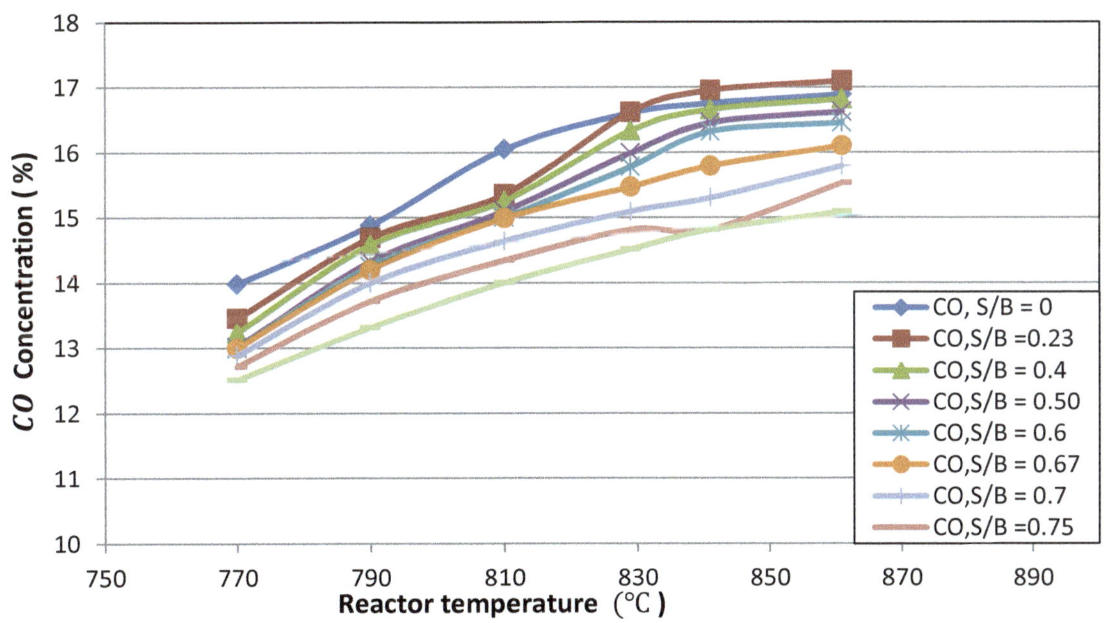

Figure 1. The effect of S/B and T101 on Carbon monoxide content in produced gas

It has been calculated each of Low heating value, gas yield, Carbon efficiency, gasifier efficiency and Equivalence ratio depend on conditions and results of experiment for every gas samples due to:

- Low heating value LHV $\left(\frac{MJ}{m^3}\right)$ has been calculated by this flowing equation [7]:

$$LHV = (30\,CO + 25.7H_2 + 85.4CH_4 + 151.3\,C_nH_m).4.2/1000 \quad (15)$$

where: CO, H_2, CH_4 and C_nH_m are the gases concentrations of the product gas.

- gas yield ($V_{gas}, \frac{m^3}{h}$),which has been calculated by this equation[5]:

$$V_{gas} = \frac{0.79}{N_2}.F_1 \tag{16}$$

where: Air contains 79 % nitrogen by volume.
N_2 (%): Content of nitrogen in produced gas.
$F1\left(\frac{m^3}{h}\right)$: primary air flow rate, it has been measured during the experiment.

- Carbon efficiency η_c(%), which has been calculated by equation [5]:

$$\eta_c = \frac{\sum \text{mass of carbout in gas produce}}{\sum \text{mass of carbon in biofuile}} X100$$

$$\eta_c = \frac{(CO+CO_2+CH_4+2.C_2H_2).M_{gas}\ X\ 12}{C} X100 \tag{17}$$

where: CO, CO_2, CH_4 and C_2H_6 (%): are the gas concentrations in produced gas
C (%): The carbon content in the ultimate analysis of biomass (Table 2).

- Gasifier efficiency η_{geff} (%), which has been calculated by flowing equation [6]:

$$\eta_{geff} = \frac{\sum HHV_{gas}}{HHV_{biomass}} X100 \tag{18}$$

where:
$HHV_{biomass}\left(\frac{MJ}{kg_{biomass}}\right)$: High heating value of biomass
$\sum HHV_{gas}\left(\frac{MJ}{kg\ biomass}\right)$: High heating value of produced gas

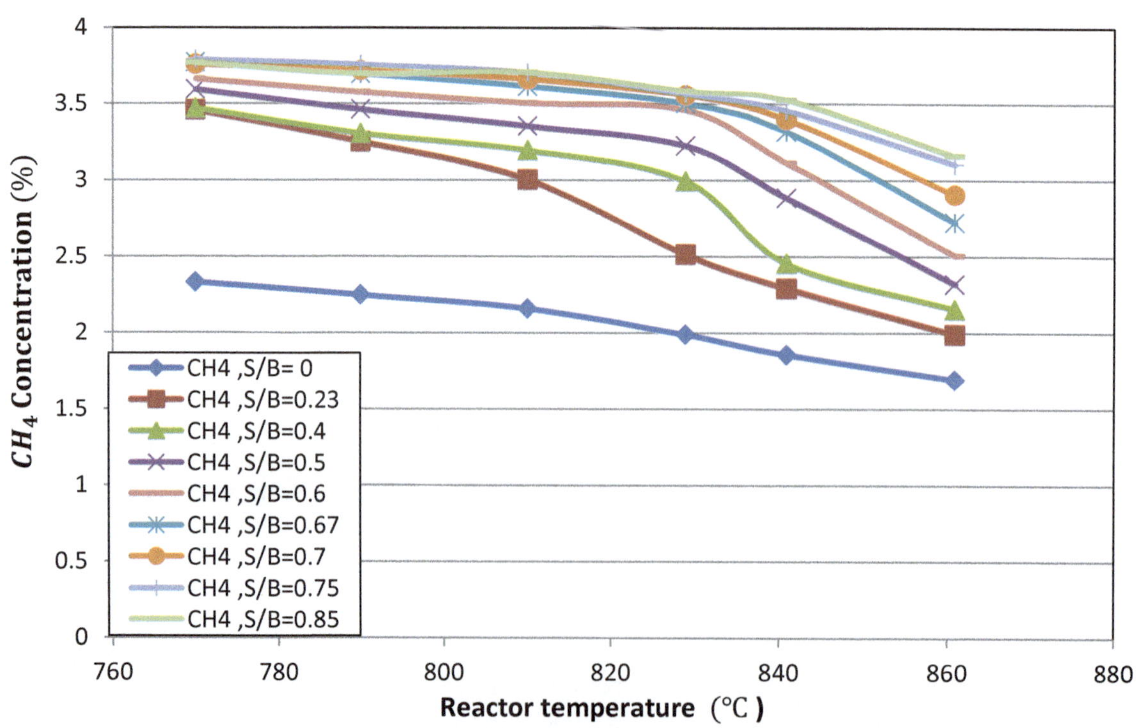

Figure 2. The effect of S/B and T101 on methane content in produced gas

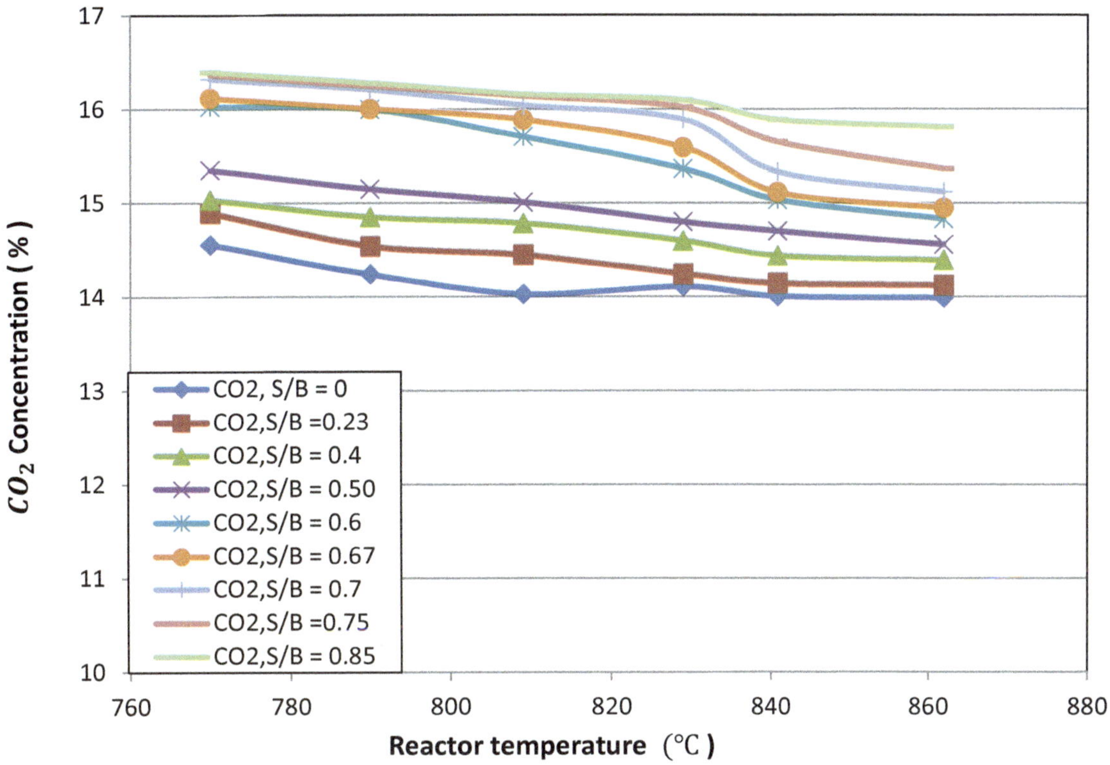

Figure 3. The effect of S/B and T101 on carbon dioxide content in produced gas

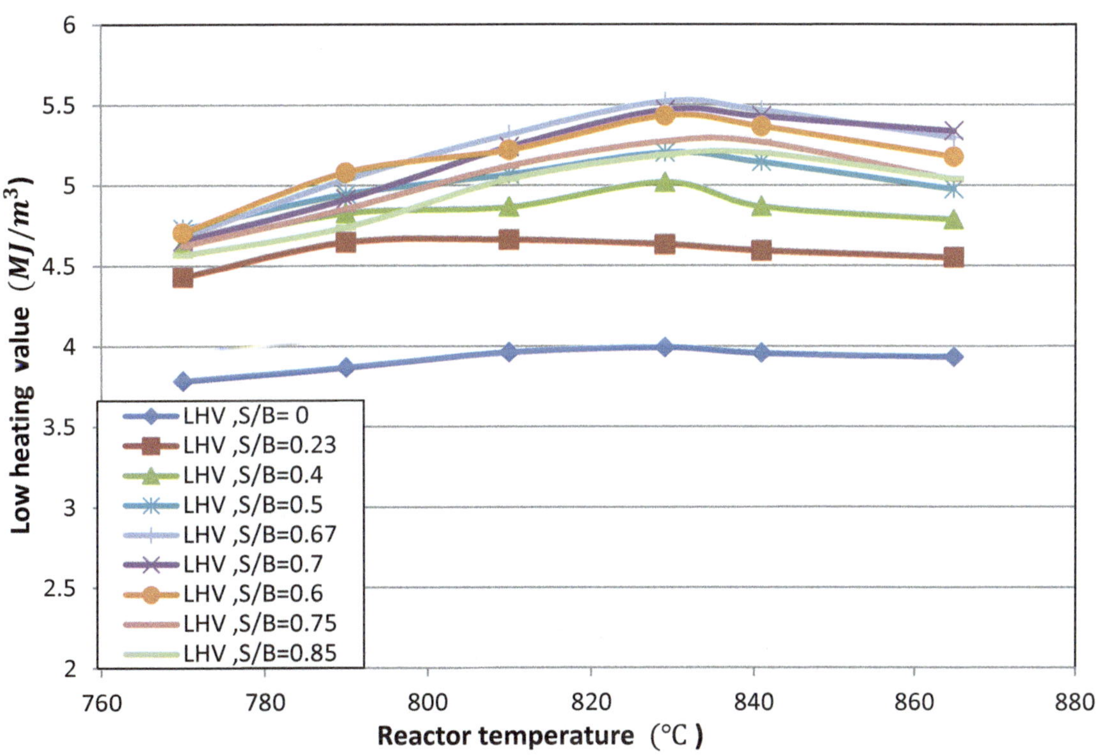

Figure 4. The effect of S/B and T101 on low heating value of produced gas

5.2 The optimal temperature of the steam

The objective of these experiments is to determine the temperature of the feed steam that achieves the best properties of produced gas. It has been depended on the results of experiments that have been carried out previously. These experiments has been done at constant reactor temperature T101= 829 (°C) and different values of the steam temperature Tfl: (180,200,220,240 and 260 °C), which Tfl is Temperature of steam and air mixture. Primary air during the experiment about F1= 20-21$\left(\frac{m^3}{h}\right)$, feed flow rate was about B = 23$\left(\frac{kg}{h}\right)$. Steam to biomass ratio value was 0.68 ($\frac{kg\ Steam}{kg\ Biomass}$). Equivalence ratio was about 0.29.Samples for mutual comparisons are selected at similar gasification conditions, for every reactor temperature separately.

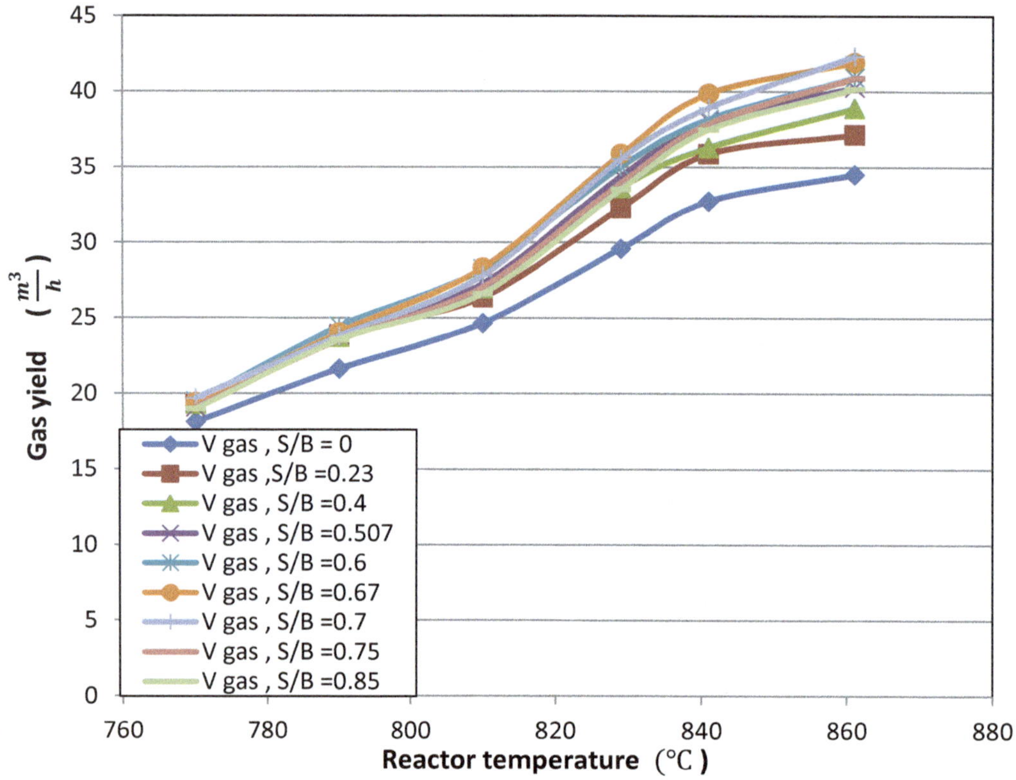

Figure 5. The effect of S/B and T101 on the gas yield

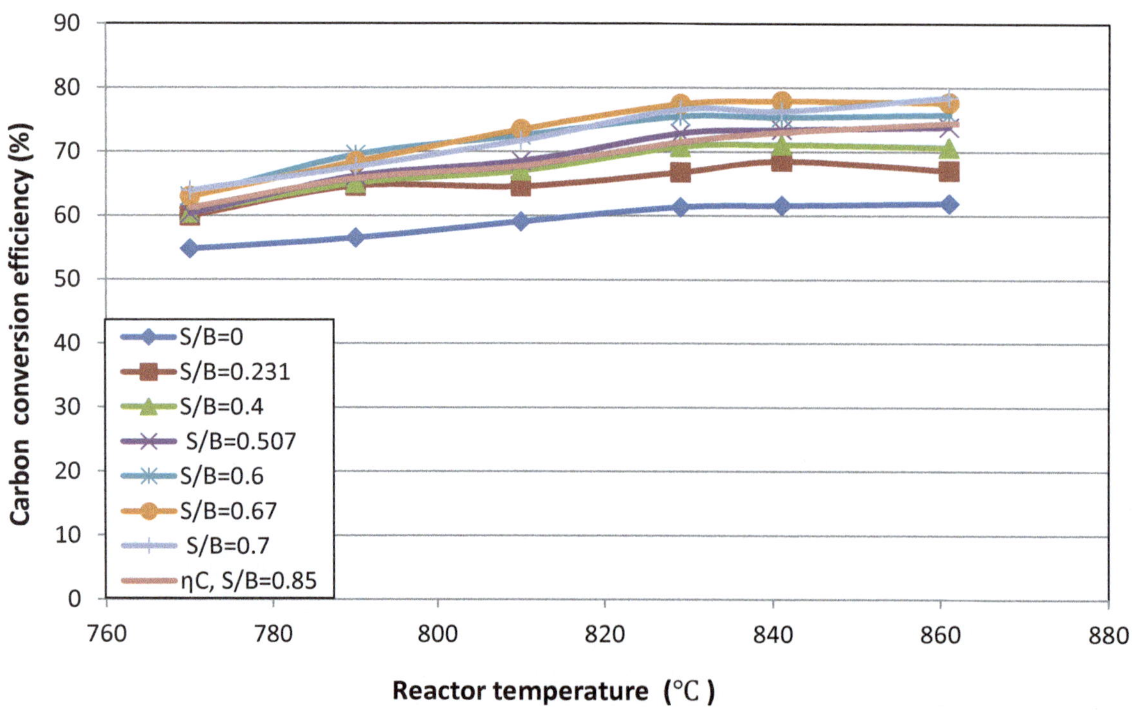

Figure 9. The effect of $\frac{S}{B}$ and T101 on carbon conversion efficiency

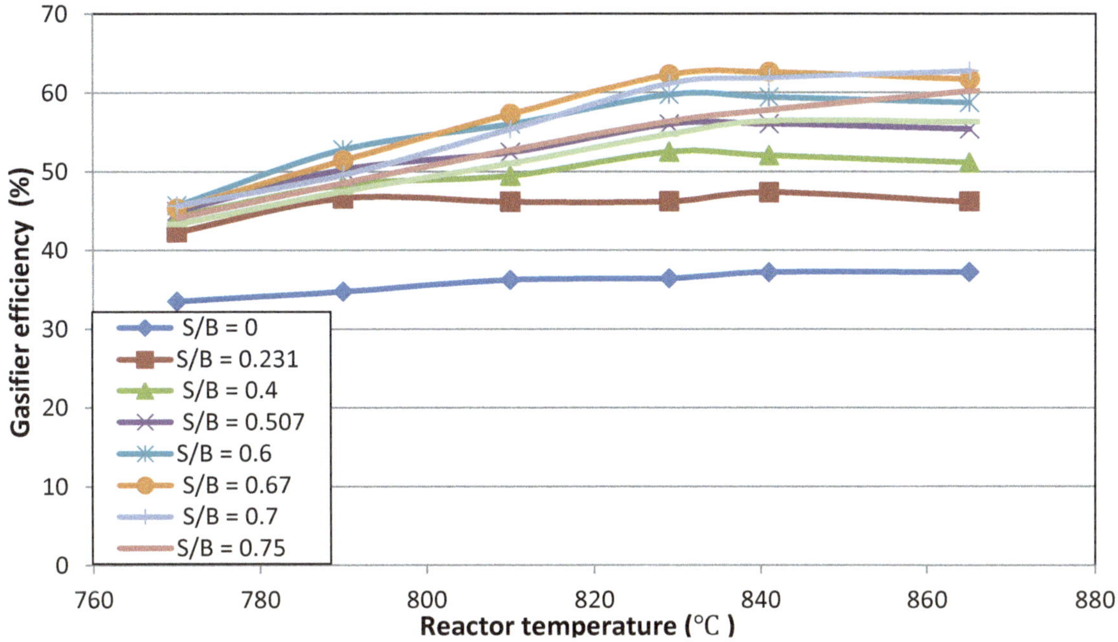

Figure 10. The effect of $\frac{S}{B}$ and T101 on gasifier efficiency

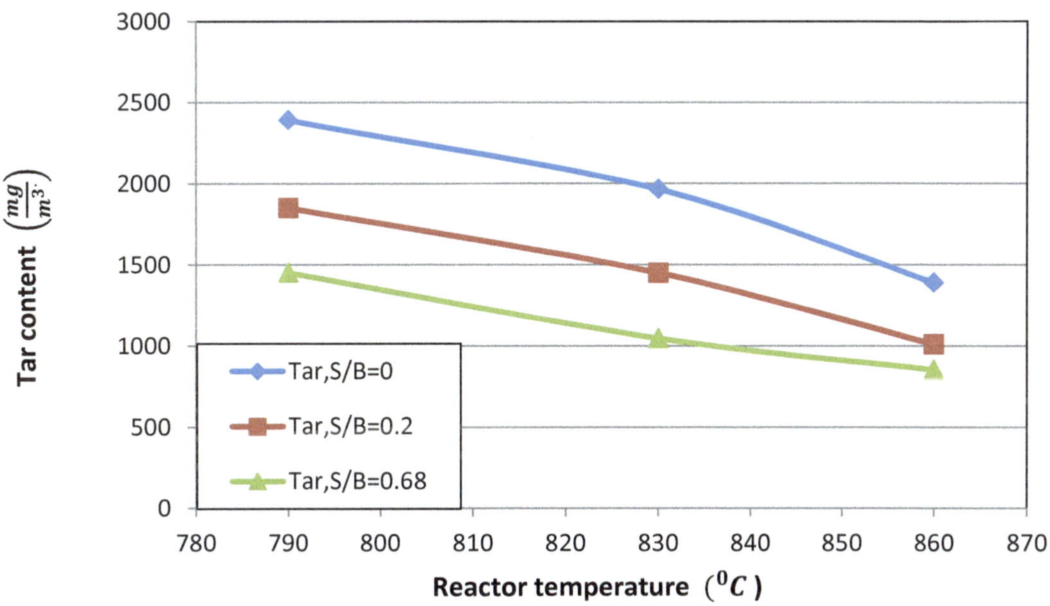

Figure 11. The effect of $\frac{S}{B}$ and T101 on tar content

It has been calculated each of Low heating value, gas yield, Carbon efficiency and gasifier efficiency, as it have been discussed in the previous paragraph for each gas samples.The results of samples analysis were reported in Table 3 and Figures 12 and 13.

5.3 The evaluation of experiments and discuss the results

- From Figures 3 to 11, it can be observed, that the increasing in reactor temperature T101 lead to increase in hydrogen, carbon monoxide, gas yield and carbon conversion efficiency, and decrease in methane and tar content and carbon dioxide. But Low heating value and gasifier efficiency at the first increase with T101, until reach to temperature 829 °C, they start decreasing by increasing T101. And this due to:
  - By depending on Le Chatelier's principle, higher temperatures improve the reactants in exothermic reactions and improve the products in endothermic reactions [7]. Therefore the

endothermic reactions (19) and (20) will be enhanced with increase of temperature, which leads to increase of hydrogen concentration and decrease of CH_4 concentration.

$$CH_4 + H_2O \rightarrow CO + 3H_2 \ - 206 \text{ KJ} \tag{19}$$

$$CH_4 + 2H_2O \rightarrow CO_2 + 4H_2 - 165 \text{ KJ} \tag{20}$$

- The Boudouard reaction Equation 6, it can be seen that this reaction will be improved by high temperature, so activity so the carbon dioxide decreasing and carbon monoxide increasing and this lead to enhance gas yield [4].
- The methane formation reaction Equation 7 is improved by low temperature therefore decrease methane by increasing temperature.
- By increasing temperature this due to increasing of combustible gases (CO, H_2), but the heavier hydrocarbons as (tar, ethane, ethane and methane), which have high low heating value have been cracked at high temperature and produce carbon and hydrogen, that conversion to combustible gases and non-combustible gas therefore low heating value decreases at high temperature.

Table 3. The effect of Tf1 on $\left(\text{LHV}, V_{gas}, \eta_C \text{and } \eta_{geff}\right)$

TF1 (°C)	180	200	220	240	260
Gas LHV$\left(\frac{MJ}{m^3}\right)$	5.180	5.23	5.35	5.42	5.52
Gas yield $\left(\frac{m^3}{h}\right)$	34.12	33.894	35.35	35.38	35.00
Carbon conversion efficiency (%)	71.9	73.05	75.66	76.19	77.46
Gasifier efficiency (%)	55.62	56.56	58.9	60.36	62.30
B = 23 $\left(\frac{kg}{h}\right)$; T101= 829 °C; S = 15.6 $\left(\frac{kg}{h}\right)$; $\frac{S}{B}$ = 0.68, ER=0.29,F1= 21.1$\left(\frac{m^3}{h}\right)$					

- From Figures 3 to 11, it can be observed, that the increasing in values of ratio of steam to biomass lead to increases each of hydrogen and methane, carbon dioxide, gas yield, low heating value, conversion efficiency and gasifier efficiency and decreased tar, but to decreases carbon monoxide. However, the excessive increase in the provided steam lead to the reduction of the concentration of hydrogen and thus leads to the lower of the value of each (low heating value, gas yield). The best ratio of steam to biomass, which achieve the best gas quality increases by reactor temperature, it can be calculated by equation (28) for our experimental conditions (Tf1=261 °C and ER =29), it was $\frac{S}{B} = 0.67\left(\frac{\text{kg steam}}{\text{kg biomass}}\right)$ at T101 =829°C and this due to:
  - The water–gas shift reaction, Equation 14, will be more activate with steam and it leads to increase in the ratio of hydrogen and carbon dioxide to carbon monoxide in the gas [5].
  - The water–gas reaction Equation 4 will be more activity reaction by steam [5] so it enhance gas yield.
  - The excessive increase of steam at steam temperature 260 °C (according to our experiments) leads to a reduction reaction temperature, so it leads to decreasing in hydrogen, low heating value and gas yield during the experiment after a certain value of steam to biomass ratio.
- From Figure 11, it can be seen, that tar content decreases by increases T101 and S/B, where tar content decreased from 2390 to 1390 $\left(\frac{mg}{m^3}\right)$ by increased temperature from 770 to 861°C when used only air, but tar content decreased from 1450 to 853 $\left(\frac{mg}{m^3}\right)$ by using steam and air mixture at $S/B = 0.67\left(\frac{kg_{steam}}{kg_{biomass}}\right)$ in the same range of increasing temperature. And this due to:
  - Steam converts high molecular weight hydrocarbons of tar into smaller gas products including H_2, CH_4, CO and CO_2 [8].Also cracking of the heavier hydrocarbons as (tar, ethane, ethane and

methane) will be done by high temperature and produce carbon and hydrogen, part of carbon converts to carbon monoxide [5]. Therefore Tar content will be decreased by increasing each of reactor temperature T101 and steam to biomass ratio S/B.

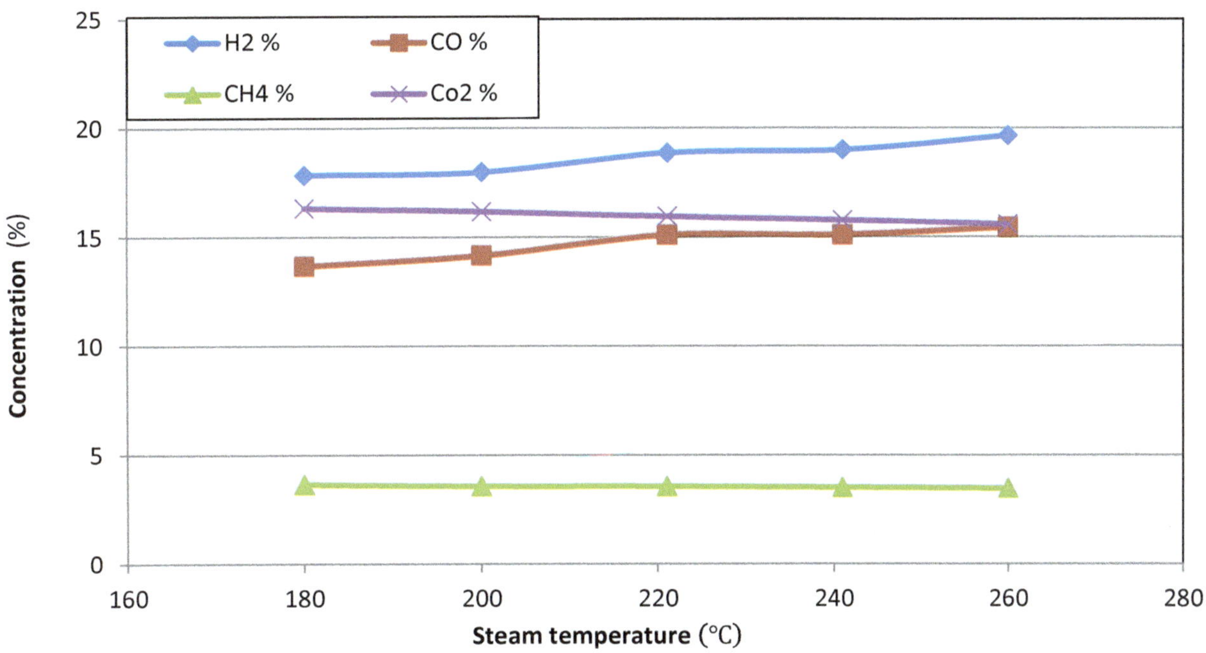

Figure 12. The effect of Tf1 on the basic components of the produced gas at $\left(\frac{S}{B} = 0.68, T101 = 830\text{ °C}, ER = 0.29\right)$

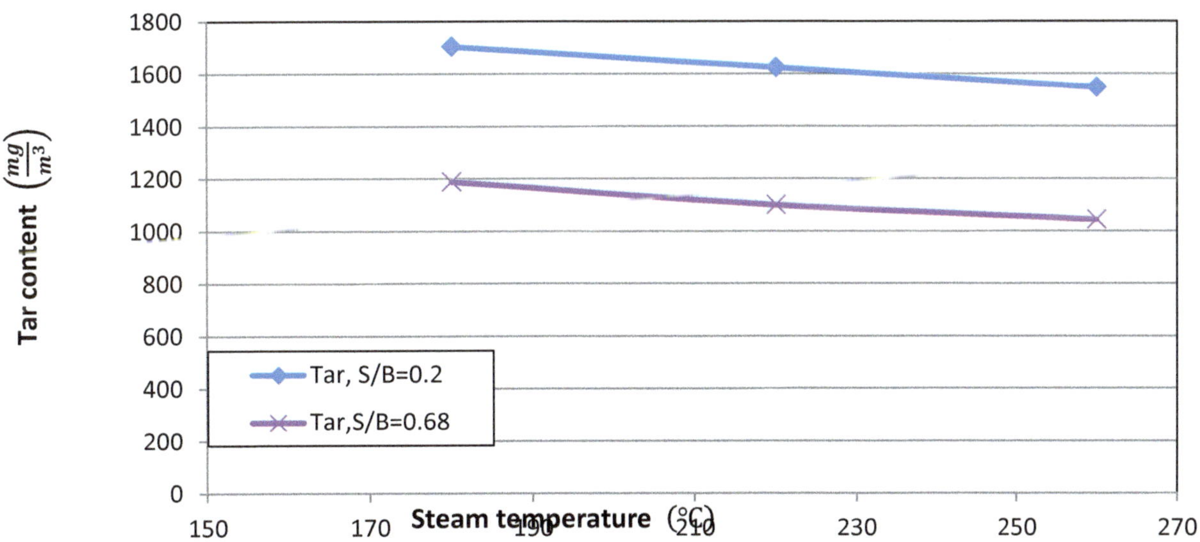

Figure 13. The effect of Tf1 on tar content

- Figure 12 and Table 3 show, that the effect of steam temperature Tf1 on the quality of gas. Steam temperature is important for the biomass gasification process.The increasing in temperature Tf1, which is the temperature of a mixture of steam and air at the inlet of the reactor, enhances the reaction temperature and thus improves the production of endothermic reactions according to Le Chatelier's principle. Therefore the Equations 19 and 20, the water-gas reaction Equation 4, Boudouard reaction Equation 6, were improved with rising temperature of steam but the methane formation reaction equation 7 was favored by low temperature. This explains, that the increase in concentration of hydrogen and carbon monoxide in produced gas by increasing steam temperature, while the concentration of methane and carbon dioxide decrease with rising steam temperature Figure 12, also increase each of low heating value, gas yield, carbon conversion efficiency and gasifier efficiency with steam temperature, as it is clear in Table 3. From Figure 13 show that tar content decreases with increasing Tf1. By depending on results discussion, it can be seen, that the best value of steam temperature is the higher value, it can be.

6. Conclusion

From the results of the experiments and discussion, it has found, that by using of the mixture of steam and air, the gas quality will be improved, and the parameters, which achieved the best quality of produced gas at the experimental conditions ($\mathbf{ER} = \mathbf{0.29}$.).are $\boldsymbol{T101} = \mathbf{830}$°C, $\mathbf{S/B} = \mathbf{0.68}$ and $Tf1$., is the higher value, where $\boldsymbol{H_2}$ increased from 10.3 to 19.67 %, $\boldsymbol{CH_4}$ from 2 to 3.5 %, $\boldsymbol{LHV}$ from 3.9 to 5.55 $\left(\frac{MJ}{m^3}\right)$ and tar content decreased from 1970 to 1050 $\left(\frac{mg}{m^3}\right)$ by using only air and by using mixture of steam and air at the best conditions respectively.

References

[1] DEMIRBAS, Ayse Hilal a Imren DEMIRBAS. Importance of rural bioenergy for developing countries. Energy Conversion and Management. 2007, roč. 48, č. 8, s. 2386-2398. ISSN 01968904.DOI:10.1016/j.enconman.2007.03.005.Dostupnéz: http://linkinghub.elsevier.com/retrieve/pii/S0196890407000763.

[2] BASU, Prabir. Combustion and gasification in fluidized beds. Boca Raton: CRC, 2006, 473 s. ISBN 08-493-3396-2.

[3] GEK PROJECT. Gasifier Experimenters Kit: pushing wood gas beyond the Imbert [online]. [Cit. 2012-09-02]. Available from http://gekgasifier.com/.

[4] ARENA, Umberto. Process and technological aspects of municipal solid waste gasification. A review. Elsevier: Waste Management. roč. 35, č. 4, 625-639.

[5] GIL, Javier, Jose CORELLA, María P AZNAR a Miguel A CABALLERO. Biomass gasification in atmospheric and bubbling fluidized bed: effect of the type of gasifying agent on the product distribution. Elsevier: Biomass and Bioenergy. 2009, roč. 17, s. 389-403.

[6] Handbook biomass gasification. Enschede: BTG Biomass Technology Group, 2005, xxii, 378 s. ISBN 90-810-0681-9.

[7] LV, Z.H XIONG, J CHANG, WU, Y CHEN a J.X ZHU. An experimental study on biomass air-steam gasification in a fluidized bed. Bioresource Technology. 2004, roč. 95, č. 101, 95–101.

[8] Sadaka, S.S., Ghaly, A.E., Sabbah, M.A., 2002a. Two phase biomass air–steam gasification model for fluidized bed reactors. Part I: model development. Biomass Bioenergy 22, 439-462.

[9] Test report n.861/05-698/Ves laboratory RWUV Praha, spol.sr.o Laboratories and test Olomoucka 7/9 656 66 Brno 30.11.2005.

13

Estimating the annual range of global illuminance on a vertical south facing building facade

Tijo Joseph, Animesh Dutta

School of Engineering, University of Guelph, Guelph, Ontario, Canada.

Abstract
Towards assessing the daylighting potential for a campus building and in consideration of the recommended strategy of maximizing window exposure on south-facing walls in northern latitudes, the range of global illuminance on a south facing vertical surface at the building location was estimated over an annum, under both clear and cloudy sky conditions, using a calculation methodology proposed by the Illuminating Engineering Society of North America. The illuminance is observed to be a variable over the day with the daily variation estimated to range as high as 35KLx, over the year and under different sky conditions. Overall, it is estimated that the dynamic variation of global illuminance on a south facing façade,specific to the study location, ranges from 14KLx to 100Klx.

Keywords: Daylighting; Global illuminance; Illuminance on vertical facade; Illuminance under clear or cloudy skies; Illuminance on south facing surface.

1. Introduction

The goal of daylighting for a building is to use natural light, when available from the sun, to serve the lighting needs within the building [1]. The key advantages with daylighting are two-fold – one is a decrement in electricity usage which otherwise would have been required to power artificial lighting sources and the second is the edge daylighting offers over artificial lighting from a human comfort perspective [2-5]. Therefore, assessing the potential of daylighting for a building is an important aspect of building energy studies. Artificial lighting in buildings, particularly in the commercial sector, can account for a significant portion of its electricity usage. It is reported that lighting in office buildings can account for as high as 50% of the electricity consumption, while in general, artificial lighting can use up 25 to 40% of the energy supplied to buildings [2, 6]. In the United States for example, lighting is estimated to account for 10% of the total electricity usage in the country [7]. Field survey and simulation studies estimate that using daylighting in place of artificial lighting can contribute to energy savings ranging anywhere from 20 to 70% [8-10].

Daylighting as a strategy to improve the comfort level of occupants in a building is underpinned by a number of vantages that natural light offers. For one, natural light best befits human vision through its graduated build-up and build-down during sun rise and sun set, plus, it also achieves a better color rendering score [2, 4]. Studies have established that natural lighting is more conducive to a productive and healthy working environment than is artificial lighting and cases have been made on this basis to legislate daylighting performance for buildings [11, 12]. Daylighting, on a par lighting level basis with artificial lighting, contributes less heat to the lit area which in turn can impact the cooling load generally reducing it [13]. The key however is to achieve a proper integration of the daylighting and artificial lighting systems in order to reap the benefits of a reduced electricity off take [13].

The entry of daylight into a building is typically through side fenestration like windows or through roof fenestration like skylights. In the northern hemisphere, the south facing facades of a building offer the most daylight entry and in addition, also offer the most control on ingress of direct sunlight using shading techniques [1]. The south facade is thus a top priority when considering daylighting strategies for a building located in the northern hemisphere. For most buildings, the amount of illuminance received on its vertical facades is an important daylighting design consideration and this knowledge becomes more relevant in context of an increasing adoption of features like curtain walling in modern buildings [2, 14].As part of daylighting studies for a campus building located in Canada, this paper seeks to estimate the annual range of global illuminance on a south facing vertical facade.

2. Building location and preliminary sun path study

The building under study has an orientation in the north-east direction, a site altitude of 338m and latitude and longitude references as 43° 31'53.03" and 80°13'34.17" respectively. A preliminary sun path study, using Autodesk's Vasari 3D sun path diagram generator, is performed in order to visualise the range within which the sun moves in reference to the building over a year. The result is presented in Figure 1. As is evident, the south facing facades present the best opportunity for daylighting.

Figure 1. One year sun path study between 8am to 8pm [Courtesy – *Autodesk® Vasari*]

3. Factors influencing daylight availability at a location

When light rays from the sun reach the earth's atmosphere, close to 20% of the light is absorbed and another 25% is reflected back. [2].What is left reaches the earth partly as direct sunlight and partly as diffused light [2]. The atmosphere, which is largely made up of clouds, aerosols, water vapor and other particulate matter, is responsible for the scattering effect of sunlight which results in diffused light(skylight). In essence, the daylighting potential for a building can be contributed by sunlight, skylight or reflected sunlight or skylight from the ground surface or obstructions in proximity to the building. The amount of daylight available at a location depends on various factors. This includes, the site latitude, the site longitude, local meteorological conditions, local air quality, time of the day and time of the year and not least, characteristics of the location's immediate surroundings including the presence of nearby trees or buildings which can act as obstructions [1,2].

The apparent position of the sun with reference to any location on earth, which is one of the key factors determining the solar radiation received at a site, can be defined by two parameters – the altitude angle and the azimuth angle [8]. The solar altitude angle is the vertical angle of the sun above the horizon, while the solar azimuth angle is the angle of the sun on the horizontal from the due south. The altitude and azimuth angles can in turn be determined using the site latitude and longitude values, the solar declination angle and the solar time with the last two parameters being variables ranging between limits over a day and over a year respectively. The solar declination angle varies across seasons from +23.5 degrees during summer solstice to -23.5 degrees during winter solstice. This variation is presented in Table 1. A sun path chart for the building location, plotting the sun's elevation angle (altitude angle) and azimuth angle for different times of the day, is given in Figure 2.

Table 1. Seasonal declination angle change

Jun-22	23°27'	(Summer Solstice)
May 21/Jul 24	20°	
Apr 16/Aug 28	10°	
Mar 21/Sept 23	0°	(Autumn & Spring Equinox)
Feb 23/Oct 20	(-) 10°	
Jan 21/Nov 22	(-) 20°	
Dec-22	(-) 23°27'	(Winter Solstice)

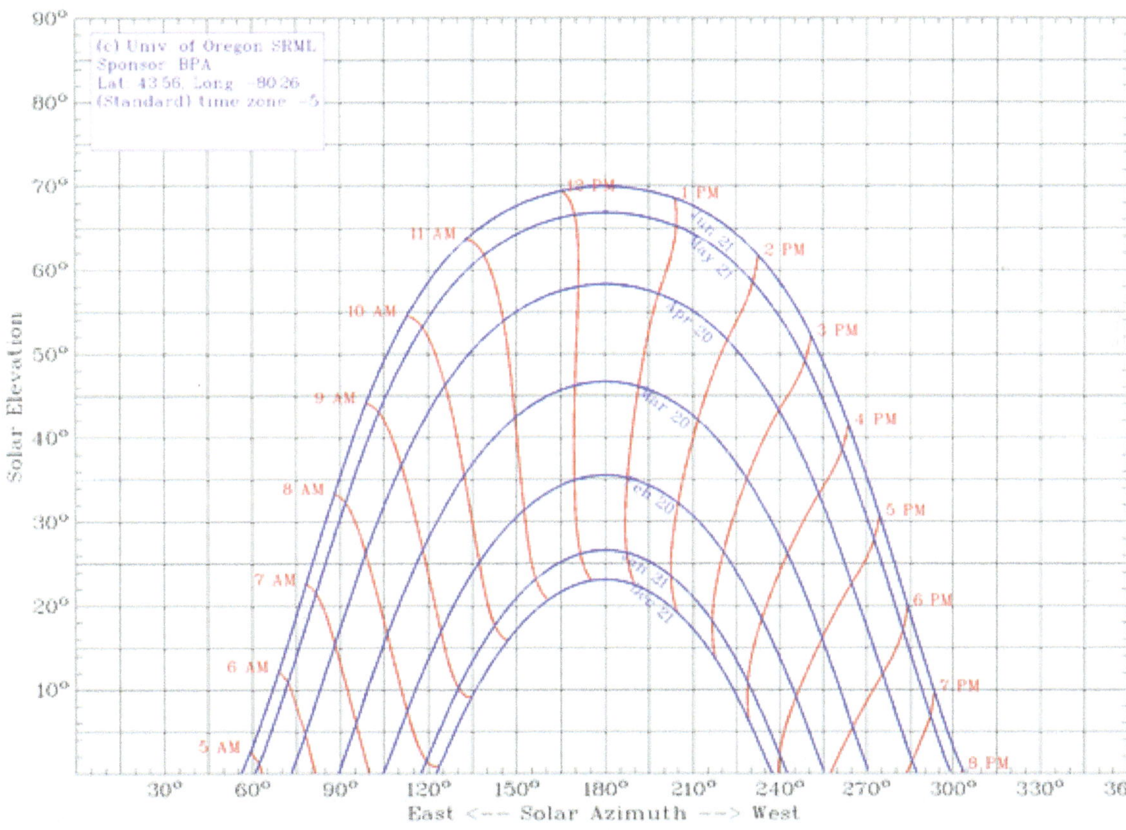

Figure 2. Sun path chart [Courtesy – *http://solardat.uoregon.edu/SunChartProgram.html*]

4. Global illuminance on a south facing vertical surface

Extensive reference is made to the IESNA Lighting Handbook [15] for the calculation methodology to be followed and the specific parameters required in estimating the illuminance for a south facing vertical surface. The relevant set of equations proposed by IESNA is presented in this section.

$$t_{solar} = t_{std} + EoT + \frac{12(M_{std} - Long)}{\pi} \tag{1}$$

where t_{solar} is the solar time in decimal hours, t_{std} is the standard time or daylight time in decimal hours, EoT is the equation of time correction applied in decimal hours considering the earth's elliptical orbit around the sun and the variable declination angle over the seasons, M_{std} is the local standard meridian reference in radians and $Long$ is the site longitude in radians.

$$\delta = 0.4093\sin\left[\frac{2\pi\ (J-81)}{368}\right] \tag{2}$$

where δ is the solar declination angle in radians and J is the Julian date ranging from 1 to 365.

$$Ang_{alt} = \arcsin\left[\sin lat\ .\sin\delta - \cos lat\ .\cos\delta\ .\cos\left({}^{\pi t_{solar}}/_{12}\right)\right] \tag{3}$$

where Ang_{alt} is the solar altitude angle in radians, and lat is the site latitude in radians.

$$E_{xt} = E_{sc}\left[1 + 0.034\cos\frac{2\pi\ (J-2)}{365}\right] \tag{4}$$

where E_{xt} is the extraterrestrial solar illuminance in KLx after correction is applied accounting for the earth's elliptical orbit and E_{sc} is the solar illumination constant which is the direct solar illuminance on a sun facing surface for a clear day and given as 128KLx.

$$Ang_{azm} = \arctan\left\{\frac{-\left[\cos\delta\ .\sin\left({}^{\pi t_{solar}}/_{12}\right)\right]}{-\left[\cos\ lat\ .\sin\ \delta + \sin\ lat\ .\cos\ \delta\ .\cos\left({}^{\pi t_{solar}}/_{12}\right)\right]}\right\} \tag{5}$$

where Ang_{azm} is the solar azimuth in radians.

$$m = \frac{1}{\sin\ Ang_{alt}} \tag{6}$$

where m is the optical air mass with no applicable dimensional unit and which varies as a function of the angle of the sun with respect to the earth's surface [2]

$$E_{dni} = E_{xt}\ .e^{-cm} \tag{7}$$

where c is the atmospheric extinction coefficient and assigned values 0.21 for clear and 0.8 for partly cloudy sky and E_{dni} is the direct normal solar illuminance in KLx.

$$E_{dhi} = E_{dni}\ .\sin Ang_{alt} \tag{8}$$

where E_{dhi} is the direct horizontal solar illuminance in KLx.

$$E_{khi} = A_i + Bsin^{C_i} Ang_{alt} \tag{9}$$

where E_{khi} is the horizontal diffuse sky illuminance due to unobstructed skylight in KLx, A_i is the sunrise/sunset illuminance in KLx and assigned values 0.8 for clear and 0.3 for partly cloudy sky, B is the solar altitude illumination coefficient in KLx with values 15.5 for clear and 45 for partly cloudy sky and C_i is solar altitude illuminance exponent with values 0.5 for clear and 1 for partly cloudy sky.

$$Ang_{z} = Ang_{azm} - Ang_{e} \tag{10}$$

where Ang_z is the solar-elevation azimuth angle measured in the horizontal place between the normal to the vertical face of study and the south in radians and Ang_e is the elevation azimuth in radians.

$$Ang_{i} = \arccos(\cos Ang_{alt}\ .\cos Ang_{z}) \tag{11}$$

where Ang_i is the incident angle in radians and represents the angle between the normal to the vertical surface under study and the direction to the sun.

$$E_{dvi} = E_{dni} .\cos Ang_i \tag{12}$$

where E_{dvi} is the direct vertical solar illuminance in KLx.

$$E_{kvi} = A_i + B\cos Ang_i^{C_i} \tag{13}$$

where E_{kvi} is the diffuse vertical illuminance in KLx.

$$E_{gi} = \rho_g.(E_{dhi} + E_{khi}) \tag{14}$$

where ρ_g is a ground reflectivity coefficient or albedo with a nominal value of 0.2 typically assigned and E_{gi} is the illuminance reflected off the ground in KLx.

The incident global solar radiation is then the sum of direct beam radiation, sky radiation, and the ground-reflected radiation and accordingly, the global vertical illuminance on a south facing surface is given by:

$$Ev_{south} = E_{dvi} + E_{kvi} + E_{gi} \tag{15}$$

where Ev_{south} is the total illuminance on a south facing vertical surface in KLx.

5. Results and discussion

Based on the calculation methodology proposed by IESNA, the global illuminance on a vertical south facing surface at the building location, under both cloudy and clear sky conditions, is estimated for Julian days 1 to 365. The resultant plots are presented in Figures 3 to 5.The plots show that illuminance is a variable over time, over the year and under different sky conditions. At 12 solar noon, when the illuminance is at its peak over the day, under clear sky conditions, the global illuminance value on the vertical face ranges from a low of around 65KLx to a high of around 100KLx as seen in Figure 3. On the other hand, when the sky condition is cloudy, this range drops to between 14KLx and 21KLx as observed in Figure 3. It is thus evident that the sky condition plays a significant role in determining the global illuminance value. Figures 4 and 5 depict the annual illuminance trends for solar time 9am and solar time 3pm respectively. These plots show that even for the same time of the day, over the year, as the earth revolves round the sun, the illuminance level on a vertical face varies with the range of variation as high as 35KLx. This daily and seasonal trends need to be factored in when conducting solar design studies for buildings [16, 17].

In order to ascertain the fraction of time the building location is typically exposed to overcast or clear sky cover, reference is made to the Kitchener Airport weather station cloud cover report accessed through www.weatherspark.com. This report is based on historical records from 1994 to 2012. This station is selected for its nearest proximity to the building location among available weather station records and it is within reason to assume that these records are representative of local cloud cover trends over the building. Notably as seen in Figure 6, the sky condition is observed to be overcast for close to 50% of the time during winter. As established earlier, under overcast sky conditions, the level of vertical illuminance on a surface in the south cardinal direction is relatively much lower thus emphasizing the relevance of sky condition data in support of daylighting studies.

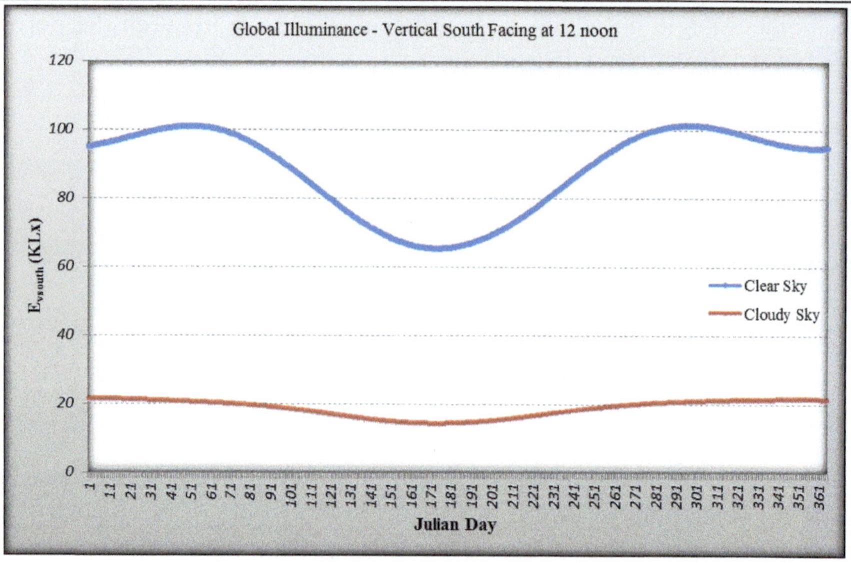

Figure 3. Global illuminance versus Julian day under clear and cloudy skies at 12noon

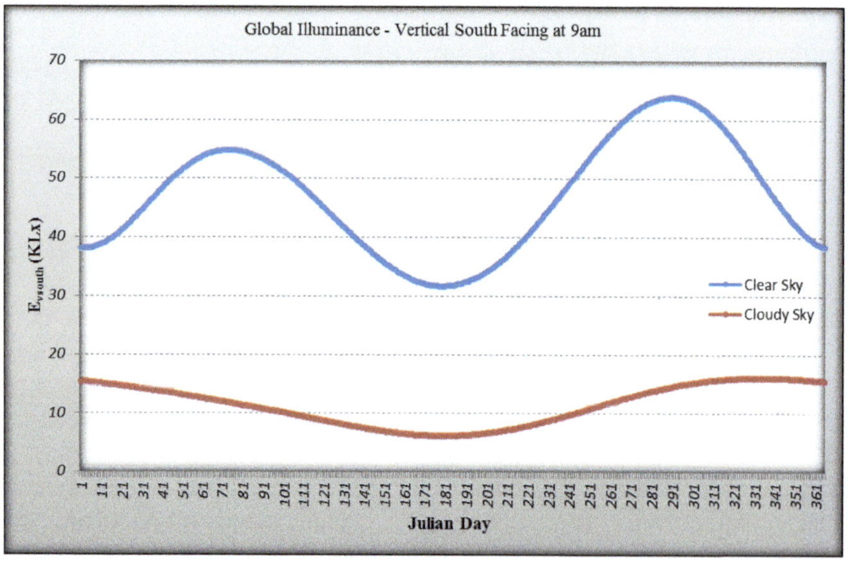

Figure 4. Global illuminance versus Julian day under clear and cloudy skies at 9am

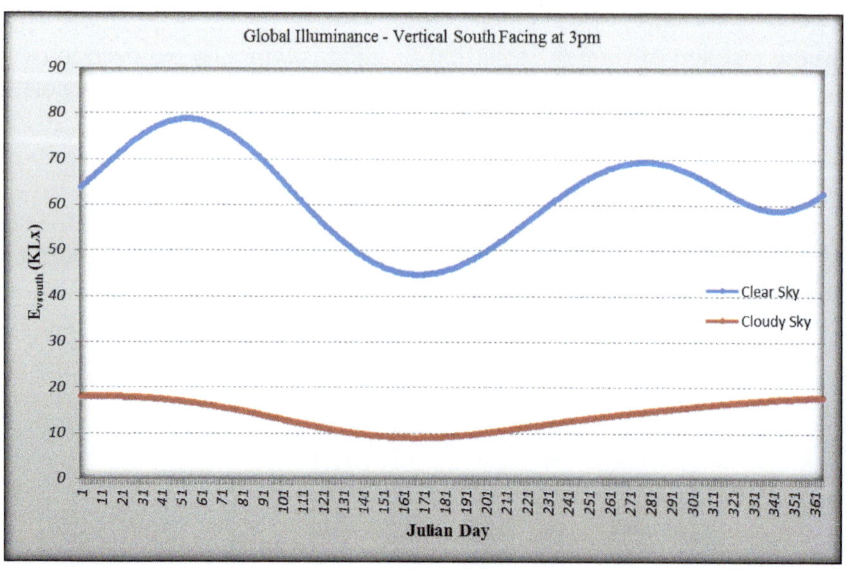

Figure 5. Global illuminance versus Julian day under clear and cloudy skies at 3pm

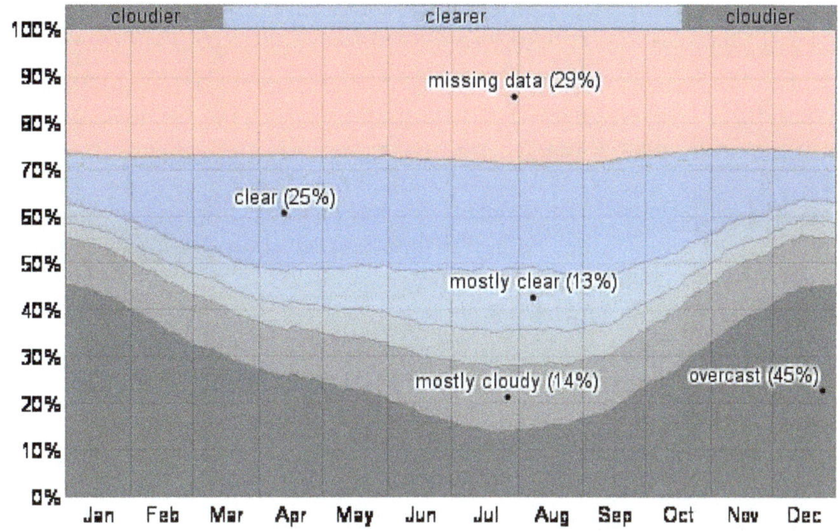

Figure 6.Cloud cover report - Waterloo region [Courtesy *–http://weatherspark.com/#!graphs;ws=28280*]

8. Conclusions

The following describes the key conclusions of this study:

- Using an IESNA defined calculation methodology, the global illuminance on a vertical south facing building surface is shown to be a variable across a day, over a year as well as under different sky conditions.
- Specific to the building location in this study, the global vertical illuminance, covering all dynamic scenarios (daily, annual and under varying sky conditions),is observed to range from a low of 14KLx to a high of 100KLx. Over the course of a day, the global vertical illuminance variation can be as high as 35KLx.
- By estimating the varying global illuminance level on building surfaces specific to a building's geographical location, and along with knowledge about the fenestration transmission properties and shading systems of the building, can aid in the creation of an effective daylighting scheme and also support feasibility studies investigating strategies such as incorporation of light transport systems.

References

[1] Enermodal Engineering Limited. (2002). Daylighting Guide for Canadian Commercial Buildings. Canada: Public Works and Government Services Canada.

[2] McNicholl, A., & Lewis, J. O. (Eds.). (1994). Daylighting in Buildings. Energy Research Group, University College Dublin for the European Commission Directorate-General for Energy (DGXVII).

[3] Li, D. H. W., Lau, C. C. S., & Lam, J. C. (2004). Predicting daylight illuminance by computer simulation techniques. Lighting Research and Technology, 36(2), 113-128.

[4] Alrubaih, M. S., Zain, M. F. M., Alghoul, M. A., Ibrahim, N. L. N., Shameri, M. A., &Elayeb, O. (2013). Research and development on aspects of daylighting fundamentals. Renewable and Sustainable Energy Reviews, 21, 494-505.

[5] Shehabi, A., DeForest, N., McNeil, A., Masanet, E., Greenblatt, J., Lee, E. S., &Milliron, D. J. (2013). US energy savings potential from dynamic daylighting control glazings. Energy and Buildings, 66, 415-423.

[6] Krarti, M. (2011). Energy audit of building systems: an engineering approach. U.S.: Taylor & Francis.

[7] Energy Information Administration. (2012). Annual Energy Outlook 2012 with Projections to 2035. Catalogue No. DOE/EIA-0383.Washington, D.C.: U.S. Department of Energy.

[8] Muneer, T. (2004). Solar Radiation and Daylight Models (2nd ed.). Oxford: Elsevier Butterworth-Heinemann.

[9] Reinhart, C. F., Mardaljevic, J., & Rogers, Z. (2006). Dynamic daylight performance metrics for sustainable building design. Leukos, 3(1), 1-25.

[10] Ihm, P., Nemri, A., &Krarti, M. (2009). Estimation of lighting energy savings from daylighting. Building and Environment, 44(3), 509-514.

[11] Heschong, L. (2002). Daylighting and human performance. ASHRAE journal, 44(6), 65-67.

[12] Boubekri, M. (2004). An argument for daylighting legislation because of health. Journal of the Human-Environment System, 7 (2), 51–56.

[13] Li, D. H. W., Lam, J. C., & Wong, S. L. (2005). Daylighting and its effects on peak load determination. Energy, 30(10), 1817-1831.

[14] Li, D. H. (2010). A review of daylight illuminance determinations and energy implications. Applied Energy, 87(7), 2109-2118.

[15] IESNA. (2000). Daylighting. In M.S.Rea (Ed.), The IESNA Lighting Handbook; Reference & Application (pp. 335 – 378). New York: Illuminating Engineering Society of North America.

[16] Pérez-Burgos, A., de Miguel, A., & Bilbao, J. (2010). Daylight illuminance on horizontal and vertical surfaces for clear skies. Case study of shaded surfaces. Solar Energy, 84(1), 137-143.

[17] Li, D. H., & Lam, J. C. (2000). Measurements of solar radiation and illuminance on vertical surfaces and daylighting implications. Renewable energy, 20(4), 389-404

14

Mechanical behaviour of membrane electrode assembly (MEA) during cold start of PEM fuel cell from subzero environment temperature

Maher A.R. Sadiq Al-Baghdadi

Fuel Cell Research Center, International Energy and Environment Foundation, Najaf, P.O.Box 39, Iraq.

Abstract
Durability is one of the most critical remaining issues impeding successful commercialization of broad PEM fuel cell transportation energy applications. Automotive fuel cells are likely to operate with neat hydrogen under load-following or load-levelled modes and be expected to withstand variations in environmental conditions, particularly in the context of temperature and atmospheric composition. In addition, they are also required to survive over the course of their expected operational lifetimes i.e., around 5,500 hrs, while undergoing as many as 30,000 startup/shutdown cycles.
Cold start capability and survivability of proton exchange membrane fuel cells (PEM) in a subzero environment temperature remain a challenge for automotive applications. A key component of increasing the durability of PEM fuel cells is studying the behaviour of the membrane electrode assembly (MEA) at the heart of the fuel cell. The present work investigates how the mechanical behaviour of MEA are influenced during cold start of the PEM fuel cell from subzero environment temperatures. Full three-dimensional, non-isothermal computational fluid dynamics model of a PEM fuel cell has been developed to simulate the stresses inside the PEM fuel cell, which are occurring during fuel cell assembly (bolt assembling), and the stresses arise during fuel cell running due to the changes of temperature and relative humidity. The model is shown to be able to understand the many interacting, complex electrochemical, transport phenomena, and stresses distribution that have limited experimental data.

Keywords: PEM fuel cells, Cold start; Subfreezing temperature; Durability, Hygro-thermal stress, CFD.

1. Introduction

Many problems arise when a PEM fuel cell is frozen, most of these are primarily caused by internal ice formation. Internal ice crystallization causes physical tearing of the polymer membrane, clogging of internal fuel passages, dehydration of the membrane, and separation of the Gas Diffusion Layer (GDL) from the Membrane Electrode Assembly (MEA). Automotive fuel cells are likely to operate with neat hydrogen under load-following or load-levelled modes and be expected to withstand variations in environmental conditions, particularly in the context of temperature and atmospheric composition (Figure 1). Prior to the commercialization of the PEM fuel cells, however, successful development of a rapid and reliable cold-start capability from the subzero environment temperature becomes a prerequisite. In order to maintain good proton conductivity, the water content in the membrane must be kept at a certain level to ensure that the membrane remains fully hydrated. However, the water in the pores of the catalyst layer, gas diffusion layer and the membrane may freeze once the cell temperature decreases below the freezing point. Thus, methods which could enable the fuel cell startup without or with slight performance degradation at subfreezing temperature need to be studied. In addition, repeated freeze/thaw cycles would lead to the irreversible performance decay and device damage. As is well known the densities of water and ice at 0 °C are 999.8 kg/m^3 and 916.8 kg/m^3, respectively, and there is about a 9% volume expansion when water freezes at this temperature. As a result, water freezing will definitely generate unbalanced stresses in fuel cell and the stress disappears when the volume gets smaller with the melting of ice. The repeated generation and disappearance of the unbalanced stress in the fuel cell with the phase change of ice will cause damage on the structure and performance of the components to some extent. For further investigation, the damage is classified into three parts: impact on membrane, impact on catalyst coated membrane (CL), and impact on gas diffusion layer (GDL).

Figure 1. Toyota PEM fuel cell hydrogen vehicle (FCHV) covered in snow

In order to achieve a high performance, the sulfonated fluoroethylene membrane should be well hydrated. Hence, when the ambient temperature decreases below, the existence of ice may have a significant impact on the membrane. Since the ion conductivity between the membrane and catalyst is low, the interface between the membrane and CL has attracted more and more attention. After a long time of operation, the interfaces will change significantly and the cathode CL will also be dissociated seriously.

The CL of PEM fuelcell generally comprises three components: Pt/C electrocatalyst, PTFE, and polymer resins, such as Nafion®. These three components in the CL form three networks, namely, water and proton transport channels, electronic transport channel, and the gas transport channel. Only the gas transport channel composed of PTFE is hydrophobic, and water is present in the networks composed of Pt/C electrocatalyst and Nafion® resins. When the water in the CL freezes and expands, the hydrophilic network undoubtedly will be the most directly affected. Since these networks interact with each other, the hydrophobic network will inevitably suffer from the physical effect of the hydrophilic network.

The physical structure of GDL may be directly damaged by the formation of ice. Once the water in the hydrophilic porous structure in the diffusion layer freezes, the volume expands. As a result, the hydrophobicity of micro-pores becomes large. After repeated freeze/thaw cycling, the hydrophilic and hydrophobic network structure could possibly be damaged due to the generated stress.

To further understand the fundamental cold-start mechanisms of PEM fuel cells, much research has been expended in this area in recent years [1].

Oszcipok et al. [2] conducted isothermal potentiostatic cold-start measurements of a single cell. It was shown that the product water initially increased the membrane humidity at subfreezing temperatures, and after the membrane humidity reached its maximum, the product water would flood the catalyst layer and the gas diffusion layer (GDL) and became frozen. Ice formation would subsequently lead to strong current density decay and cell degradation. They also carried out a mathematic curve fitting and statistic regression analyses, and showed that dryer membrane and high gas flow rates would benefit the PEM fuel cell cold-start operations.

To obtain direct pictures of ice formation in a PEM fuel cell during cold starts from subfreezing temperatures, Ge and Wang [3] developed a transparent cell with a silver mesh used as the cathode GDL and conducted visualization experiments on liquid water transport and ice formation using this cell. It was concluded that the freezing-point depression of water in the cathode catalyst layer should be less than 2 C and its role in cold-start practice should thus be negligible. They later further narrowed down the value range to -1 ± 0.5 C.

Yan et al. [4] experimentally investigated the effect of subzero temperatures on PEM fuel cell performance and cold-start capability. They cooled down the cell temperature during operation, measured polarization curves at subzero temperatures, and found that when the environmental temperature was -15 C the cell cathode temperature was below zero and cell operation stopped. For the study of cold-start capability, they varied the startup temperature, gas purge process, cell insulation, air stoichiometry, and preheat of the reactant gas and tested whether startup succeeded or failed. In this study, they concluded the preheat of reactant gas also affected the results because the hotter gases could remove more water vapor from the cell. However, it is unlikely because the specific heat of the dry gas is so small. The gas is cooled down to cell temperature as soon as it enters the cell, no matter how much the gas is preheated.

The Membrane-Electrode-Assembly (MEA) is the core component of PEM fuel cell and consists of membrane with the gas-diffusion layers including the catalyst attached to each side. The fuel cell MEA durability plays a vital role in the overall lifetime achieved by a stack in field applications. Within the MEA's electrocatalyst layers are three critical interfaces that must remain properly intermingled for optimum MEA performance: platinum/carbon interface (for electron transport and catalyst support); platinum/Nafion interface (for proton transport); and Nafion/carbon interface (for high-activity catalyst dispersion and structural integrity). The MEA performance shows degradation over operating time, which is dependent upon materials, fabrication and operating conditions [5-8].

Mechanical stresses which limit MEA durability have two origins. Firstly, this is the stresses arising during fuel cell assembly (bolt assembling). The bolts provide the tightness and the electrical conductivity between the contact elements. Secondly, additional mechanical stresses occur during fuel cell running because PEM fuel cell components have different thermal expansion and swelling coefficients. Thermal and humidity gradients in the fuel cell produce dilatations obstructed by tightening of the screw-bolts. Compressive stress increasing with the hygro-thermal loading can exceed the yield strength which causes the plastic deformation. The mechanical behaviour of the membrane depends strongly on hydration and temperature [9]. Due to water management issues, operating conditions need to be carefully chosen in order to properly operate fuel cells. Because of the gas consumption along the feeding channels and water production at the cathode, internal cell humidification is highly inhomogeneous. Consequently, operating fuel cells are very often close to critical operating conditions, such as flooding and drying, at least locally [10].

An operating fuel cell has varying local conditions of temperature, humidity, and power generation (and thereby heat generation) across the active area of the fuel cell in three-dimensions. Nevertheless, except of ref. [9], no models have yet been published to incorporate the effect of hygro-thermal stresses into actual fuel cell models to study the effect of these real conditions on the stresses developed in membrane and gas diffusion layers. In addition, as a result of the architecture of a cell, the transport phenomena in a fuel cell are inherently three-dimensional, but no models have yet been published to address the hygro-thermal stresses in PEM fuel cells with three-dimensional effect. Suvorov et al. [11] reported that the error introduced due to two-dimensional assumption is about 10%. Therefore, in order to acquire a complete understanding of the mechanical behaviour of the MEA during the cell startup process, mechanical response under transient hygro-thermal stresses should be studied under real cell operating conditions and in real cell geometry (three-dimensional).

2. Model description

Three-dimensional CFD model of a PEM fuel cell was used with stress model has been developed, validated, and discussed in detail by the current author in his previous paper [9]. This model is modified for transient PEM fuel cell simulations. In brief, the model is based on the computational fluid dynamics method and considers multi-phase, multi-component flow inside the gas flow channels and the porous media of a PEM fuel cell with straight flow channels. The full computational domain consists of cathode and anode gas flow channels, and the membrane electrode assembly as shown in Figure 1. The model includes the transport of gaseous species, liquid water, protons, energy, and water dissolved in the ion-conducting polymer. Water transport inside the porous gas diffusion layer and catalyst layer is described by two physical mechanisms: viscous drag and capillary pressure forces, and is described by advection within the gas channels. Water transport across the membrane is also described by two physical mechanisms: electro-osmotic drag and diffusion. Water is assumed to be exchanged among three phases; liquid, vapour, and dissolved, and equilibrium among these phases is assumed. This model takes into account convection and diffusion of different species in the channels as well as in the porous gas diffusion layer, heat transfer in the solids as well as in the gases, and electrochemical reactions. The model reflects the influence of the operating parameters on fuel cell performance to investigate the in situ total displacement and degree of the deformation of the polymer membrane of PEM fuel cells. A unique feature of the model is to incorporate the effect of hygro and thermal stresses into actual three-dimensional fuel cell model. In addition, the temperature and humidity dependent material properties are utilize in the simulation for the membrane.

3. Results and discussion

Results for a sixty seconds from the start of operation of a PEM fuel cell operates from subzero startup temperature (-20 C) are discussed in this section. The cell operates at nominal current density of (1.2 A/cm^2), inlet gas temperature and initial cell temperature of (-20 C), surrounding temperature of (-20 C), cathode and anode pressure of (3 atm), and air and fuel stoichiometric flow ratio of (2). The geometrical and operational parameters, electrode and membrane parameters, and the material properties for the fuel cell components used in this model are taken from ref. [9].

The temperature distribution inside the fuel cell has important effects on nearly all transport phenomena, and knowledge of the magnitude of temperature increases due to irreversibilities might help preventing membrane failure.

The transient distributions of the cell temperature during the cold-start process are presented in Figure 3. The cell temperature continuously increases with time because of waste heat released during a cold-start from −20 C. In general, the temperature at the cathode side is higher than at the anode side, due to the reversible and irreversible entropy production. Naturally, the maximum temperature occurs, where the electrochemical activity is highest, which is near the cathode side inlet area. The temperature peak appears in the cathode catalyst layer, implying that major heat generation takes place in this region.

The temperature distribution profile remains similar at each operating instant, although the overall cell temperature is increasing as the operation time progresses. These results indicate that a similarity solution for the temperature distribution could be assumed during the transient fuel cell startup process.

The durability of proton exchange membranes used in fuel cells is a major factor in the operating lifetime of fuel cell systems. Figure 4 shows stress distribution (contour plots) and total displacement (deformed shape plot, scale enlarged 140 times) inside the cell on the y-z plane at x=10 mm during the cold-start process. The figure illustrates the effect of stresses on the cell MEA. Because of the different thermal expansion and swelling coefficients between gas diffusion layers and membrane materials with non-uniform temperature distributions in the cell during operation, hygro-thermal stresses and deformation are introduced. The non-uniform distribution of stress, caused by the temperature gradient in cell MEA, induces localized bending stresses, which can contribute to delaminating between the membrane and the gas diffusion layers. It can be seen that the total displacement and the degree of the deformation in MEA are directly related to the temperature, where the temperature is highest in the centre of the channel and coincide with the highest reactant concentrations. In addition, the deformation that occurs in MEA under the land areas is much smaller than under the channel areas due to the clamping force effect. This result may explain the occurrence of cracks and pinholes in the membrane during cell startup and shutdown cycles, especially at high load conditions and subzero environment temperatures.

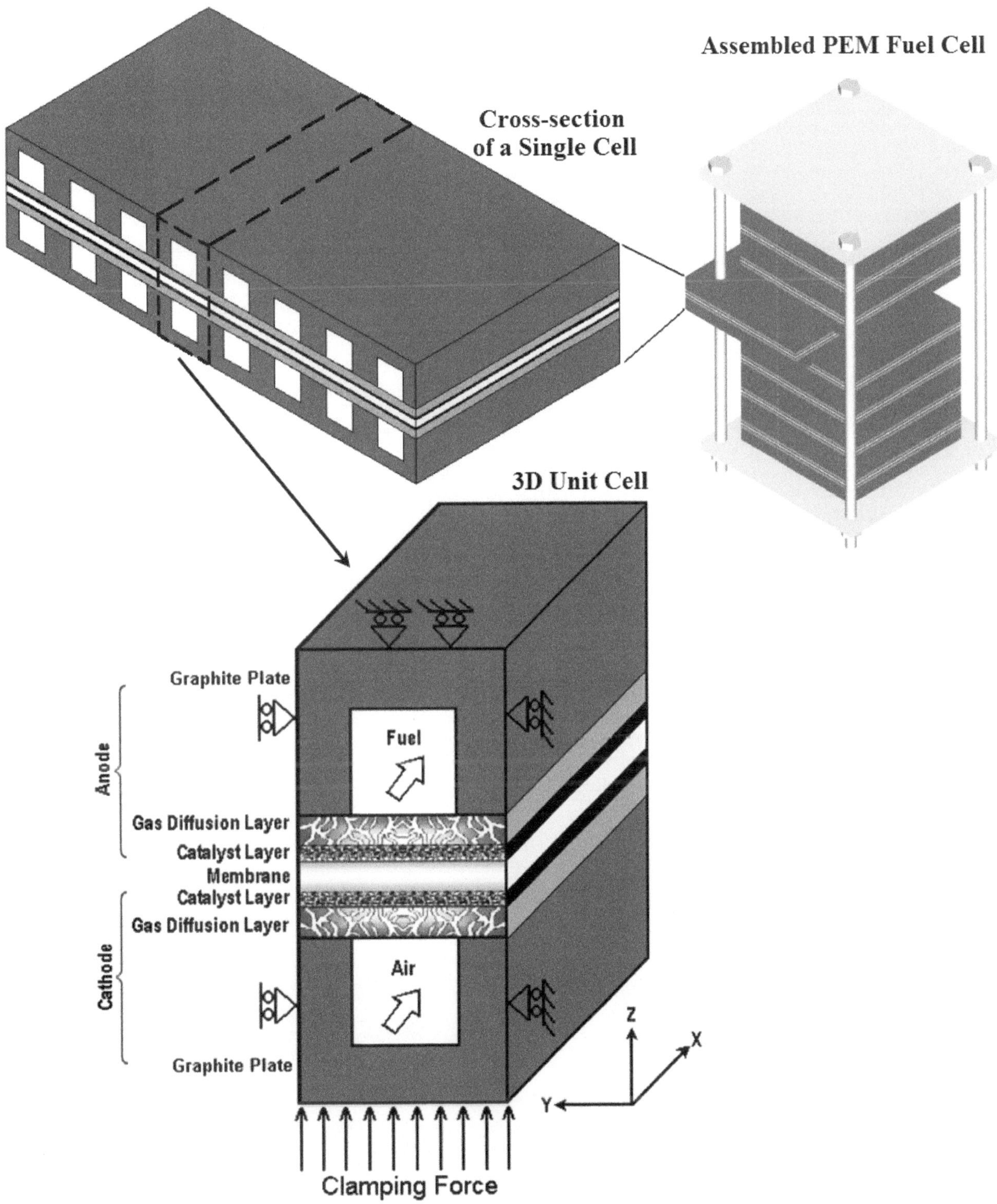

Figure 2. Three-dimensional computational domain

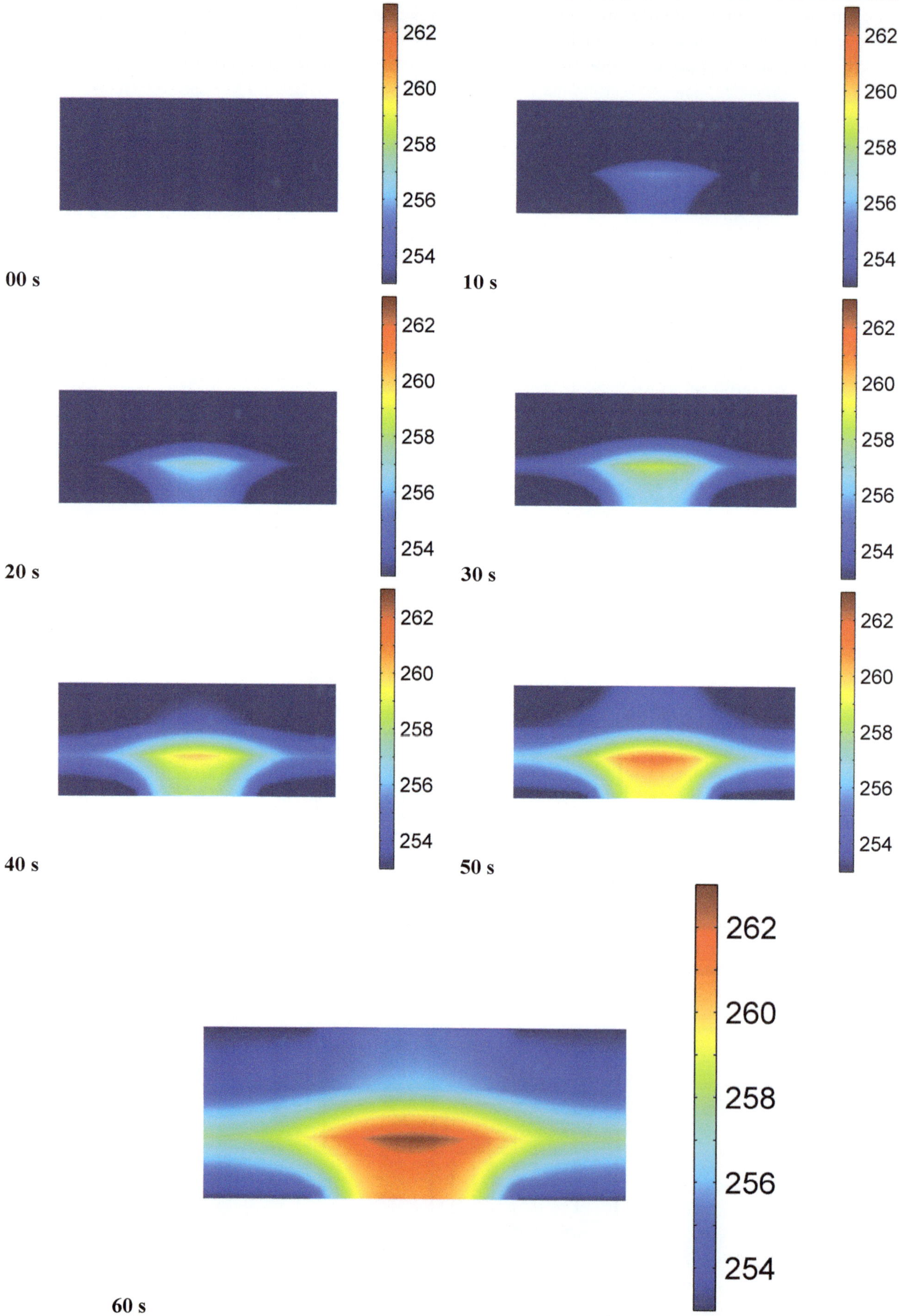

Figure 3.Temperature distribution [K] in MEA on the y-z plane at x=10 mm during cold-start of the cell from subzero temperature

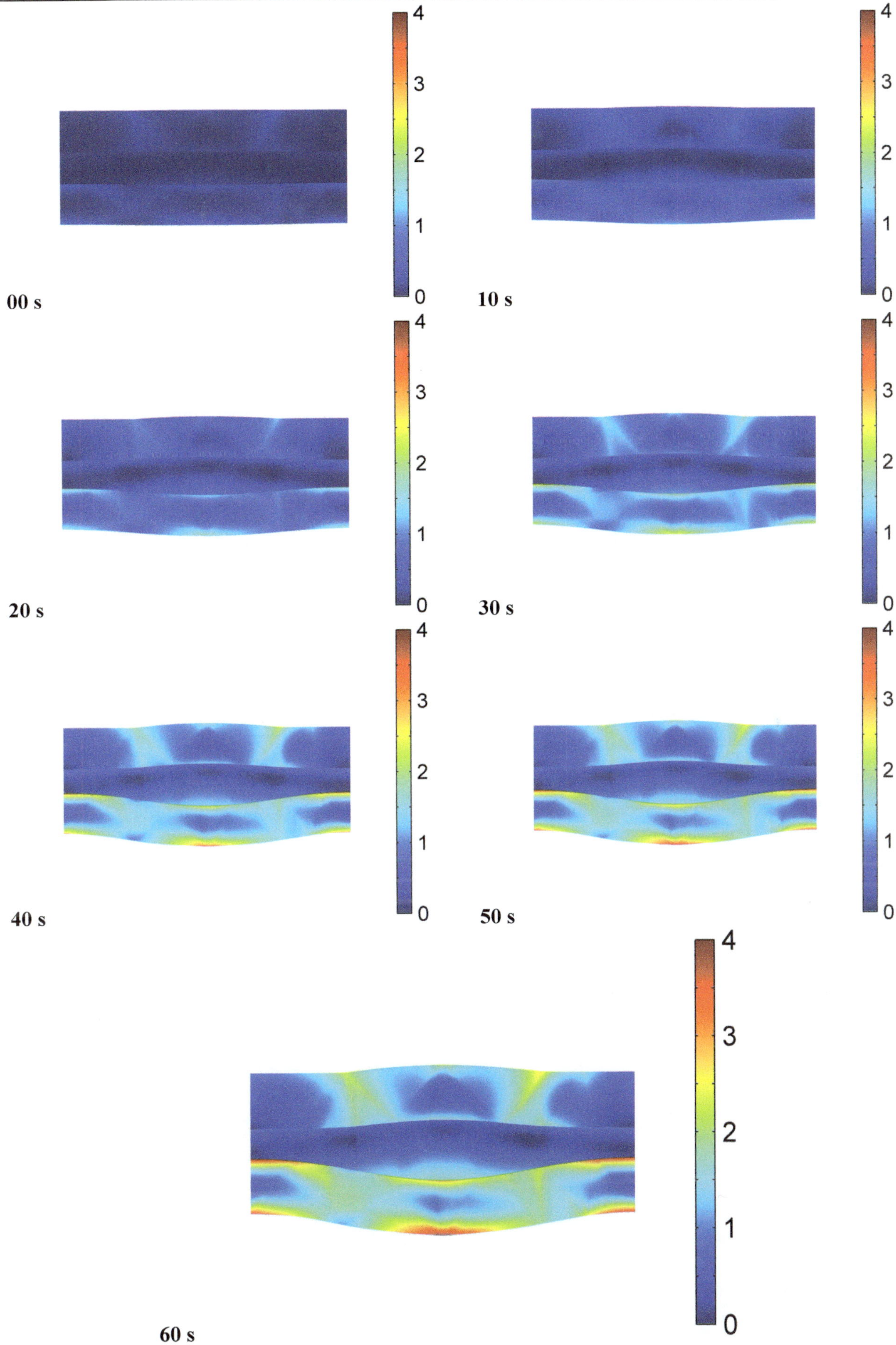

Figure 4. Stresses [MPa] and deformation (deformed shape plot, scale enlarged 140 times) in MEA on the y-z plane at x=10 mm during cold-start of the cell from subzero temperature

4. Conclusion

The behaviour of the cell during operation has been studied and investigated under cold start real operating conditions. The results show that the non-uniform distribution of stresses, caused by the temperature gradient in the cell, induces localized bending stresses, which can contribute to delaminating between the membrane and the gas diffusion layers. These stresses may explain the occurrence of cracks and pinholes in the membrane during cell startup and shutdown cycles over a long period.

In conclusion, the development of physically representative models that allow reliable simulation of the processes under realistic conditions is essential to the development and optimization of fuel cells, improve long-term performance and lifetime, the introduction of cheaper materials and fabrication techniques, and the design and development of novel architectures. In order to achieve the cold startup of PEM fuel cell rapidly without any performance degradation, novel and high performance PEM fuel cell materials, advanced PEM fuel cell system, and appropriate startup strategies are demanded.

References

[1] Hua Meng, Bo Ruan. Numerical studies of cold-start phenomena in PEM fuel cells: A review. Int. J. Energy Res. 2011; 35:2-14.

[2] Oszcipok M, Riemann D, Kronenwett U et al. Statistic analysis of operational influences on the cold start behavior of PEM fuel cells. Journal of Power Sources 2005; 145: 407-415.

[3] Ge S, Wang CY. In situ imaging of liquid water and ice formation in an operating PEFC during cold start. Electrochemical and Solid-State Letters 2006; 9: A499-A503.

[4] Qiangu Yan, Hossein Toghiani, Young-Whan Lee, Kaiwen Liang, and Heath Causey. Effect of sub-freezing temperatures on a pem fuel cell performance, startup and fuel cell components. Journal of Power Sources 2006, 160(2): 1242-1250.

[5] Haile SM. Fuel cell materials and components. Acta Materialia 2003; 51:5981-6000.

[6] Cooper JS. Design analysis of PEMFC bipolar plates considering stack manufacturing and environmental impact. Journal of Power Sources 2004; 129:152-169.

[7] Yu PT, Gu W, Makharia R, Wagner FT, Gasteiger A. The Impact of Carbon Stability on PEM Fuel Cell Startup and Shutdown Voltage Degradation. 210th ECS Meeting, Cancun, Mexico, Durability – Fuel Starvation and Start/Stop Degradation, October 29-November 3 2006.

[8] Larminie J, Dicks A. Fuel Cell Systems Explained, Second Edition. John Wiley & Sons 2003. Chichester.

[9] Maher A.R. Sadiq Al-Baghdadi. Prediction of deformation and hygro-thermal stresses distribution in PEM fuel cell vehicle using three-dimensional CFD model. International Journal of Energy and Environment IJEE, 2012; 3(4): 485-504.

[10] F. Brèque, J. Ramousse, Y. Dubé, K. Agbossou, P. Adzakpa. Sensibility study of flooding and drying issues to the operating conditions in PEM Fuel Cells. International Journal of Energy and Environment IJEE, 2010; 1(1):1-20.

[11] Suvorov, A.P.; Elter, J.; Staudt, R.; Hamm, R.; Tudryn, G.J.; Schadler, L.; Eisman, G. Stress relaxation of PBI based membrane electrode assemblies. Int. J. Solids and Structures, 2008; 45(24): 5987-6000.

15

Indoor tests on the effect of wind speed on still performance

Abdul Jabbar N. Khalifa, Marwa AW. Ali

Al-Nahrain University, College of Engineering, Mechanical Engineering Department, Jadiriya, P.O. Box 64040, Baghdad, Iraq.

Abstract

Wind speed is an important parameter that affects the productivity and efficiency of solar stills. The literature shows conflicting opinions about the effect of wind speed on the total yield of fresh water from solar stills. One reason behind such discrepancy could be attributed to the uncontrolled effect of some of the meteorological parameters. This study reports an investigation on the effect of wind speed on the performance of basin type stills carried out indoor using a fan to generate airflow analogous to the outdoor wind and heaters to provide uniform heat flux to the basin. The tests were conducted for four different wind speeds of 1.14, 2.06, 2.92 and 4.01 m/s in addition to tests with stationary air. It was found that increasing wind speed will definitely increase the yield of solar stills and high wind speeds may give less improvement in productivity than moderate wind speeds.

Keywords: Solar still; Wind speed; Indoor test; Productivity; Heat flux.

1. Introduction

The performance of solar stills is affected by three different sets of parameters namely, design parameters (such as glass inclination and basin water depth), operational parameters (such as salinity of water, feed water preheating) and climatic parameters (such as wind velocity and solar radiation intensity). In contrast to the first two groups, the latter cannot be controlled as they are related to the conditions of the weather. However, some of these variables can be controlled to some extent if indoor tests are used. Hence, the performance of solar stills can be evaluated by using heaters beneath the still to simulate the energy from the sun and to provide the necessary heat for evaporation. It is possible to control the level of the input power by regulating the input voltage to the heaters and/or the number of heating elements used. Fan(s) may also be used to create a condition analogous to the outdoor wind on the body and condensing surface(s) of the still. Such measures allow some control on the input power and wind speed and thus should reduce the effect of outdoor conditions on the accuracy of the experimental results.

Wind speed is an important parameter that affects the productivity and efficiency of solar stills. The magnitude and trend of wind effect on several configurations of solar stills have been investigated in the literature. Table 1 summarizes the majority of the experimental and theoretical studies on this issue. It can be noticed from the table that most of the studies on wind effect is theoretical. One would expect that the results from these theoretical studies agree as they are all based on the same fundamental relations of heat and mass transfer. However, discrepancy among the results of these studies is evident even for the similar configurations of stills. The verdict about the effect of increasing wind speed on productivity varies among the theoretical investigations from negative [1-5] by up to 13% [5], insignificant [6, 7], to positive [8-13] by up to 50% [13]. Moreover, some investigations propose a certain optimum value of wind speed at which the maximum yield is obtained [14-17].

Table 1. Summary of studies on how wind speed affects solar still performance

Author(s)	Configuration	Conclusion about the effect of wind speed
	Outdoor/ Theoretical Studies	
Elsherbiny and Fath [1]	Single slope still	Increasing the wind speed cause a small reduction in still productivity
Nafey et al. [5]	Single slope still	Increasing the wind speed from 1 to 9 m/s decreases the productivity by 13%
Fath and Hosny [6]	Single slope still with enhanced evaporation and additional condenser	Wind speed from 0-5 m/s has insignificant influence on the productivity
Fath et al. [7]	Naturally circulated humidifying/dehumidifying solar still	Increasing wind speed from 0-5 m/s has little effect on the productivity
Zurigat and Abu-Arabi [13]	Regenerative single slop still	Increasing the wind speed from 0-10 m/s can increase the productivity by more than 50%
Mamlook and Badran [8]	Asymmetric greenhouse type with mirrors on its inside walls	Increasing the wind speed from 3.1-5 m/s increase the productivity by 15%
Elsafty et al. [2]	Still that uses parabolic reflector-tube absorber	increase the wind speed from 0-5 m/s, decreases the productivity marginally
Madhlopa and Johnstone [3]	Single slope still with evaporator and condenser chambers	Wind speed (range 2-6 m/s) reduce the productivity marginally
Tiwari et al. [9]	Passive and active stills integrated with a flat plate collector	Productivity increase continuously with increasing wind speed (range 0-5 m/s)
El-Sebaii [14, 15, 16]	Basin type stills	There is a critical mass/depth of basin water (45 kg/4.5 cm) beyond which productivity increases as wind velocity increases up to a typical velocity of 10 and 8 m/s on typical summer and winter days, respectively.
	Outdoor/ Combined Theoretical and Experimental Studies	
Morse and Read [4]	Double slope solar still	The influence of wind on productivity is unimportant, (range 2.24-8.94 m/s). The difference in productivity is only 3 %
Toure and Meukam, [10]	Single basin solar still	The wind speed has little effect on productivity. Increasing wind speed from 0-9 m/s increases the production by 10%
Abdenacer and Nafila [11]	Green-house effect solar still	Productivity increases with wind velocity
Dimri et al. [12]	Active solar still coupled to two flat plate collectors	As wind speed increases from 1-5 m/s, the productivity increases
Kumar and Tiwari [17]	Shallow basin solar still	The exergy efficiencies increase rapidly for wind speed up to 2 m/s and decreases further
	Outdoor/ Experimental Studies	
Farid and Hamad [18]	Single basin solar still	There is a significant decrease in efficiency with the increase of wind velocity
Badran [20]	Single slope solar still	Productivity increases by 35% with increasing wind speed from 2.7-5 m/s
Younis [19]	Single slope solar still	Productivity decreases slightly as wind speed increases (range 0-8 m/s)

Indoor Studies		
Maalej [21]	single slope basin still, large size fan High and medium intensity bulb projectors are used to give three intensity levels of 1009.4, 709.77 and 346.99 W/m²	The increase in wind velocity from 0-1.6 m/s yields a reduction of 2% in the still performance (range 0-3.35 m/s)

Experimental studies on the effect of wind speed on still performance are scarce in the literature. Again, discrepancy among these studies is noticed. While Farid & Hamad [18] and Younis et al. [19] found a negative effect of wind speed on productivity, Badran [20] reported a positive effect by up to 35%.

As far as indoor investigations on the effect of wind speed on stills is concerned, Maalej [21] carried out an experimental and theoretical study by simulating the external breeze on his still using a large size fan to generate a uniformly distributed air flow analogous to wind outdoor. A reduction of only 2% in the still performance was reported when the wind velocity is increased from zero to 1.6 m/s.

The literature survey shows conflicting opinions about the effect of wind speed on the total yield of fresh water from solar stills. This parameter is sensitive and must be studied to reach a conclusion about its effect. One reason behind the discrepancy among the conclusions of the different studies may be attributed to the uncontrolled effect of some of the meteorological parameters. Accordingly, the aim of the present research work is to investigate the effect of wind speed on the performance of the basin type still by thorough tests inside the laboratory during which the power input and wind speed are simulated by a suitable heating system and a variable-speed fan respectively, and thus minimizing the effect of some of the factors that affect the accuracy of the experimental results.

2. Experimental work

A single sloped basin type still is constructed from a galvanized steel sheet 0.8 mm thick to form 0.7 by 0.7 m basin. A glass cover 4 mm thick is placed on the top of the basin at a slope equal to the latitude angle of the location of 33° from horizontal [22] to serve as a condensation surface. Silicon rubber sealant is applied on the glass cover to insure tightness of the still. The bottom and sides of the still is insulated with a sheet of glass wool 10 mm thick (thermal conductivity of 0.04 W/m °C) to increase heat retention. The whole construction is contained in a rigid plywood structure. A schematic of the experimental still showing the main components and locations of the temperature measurements is given in Figure 1.

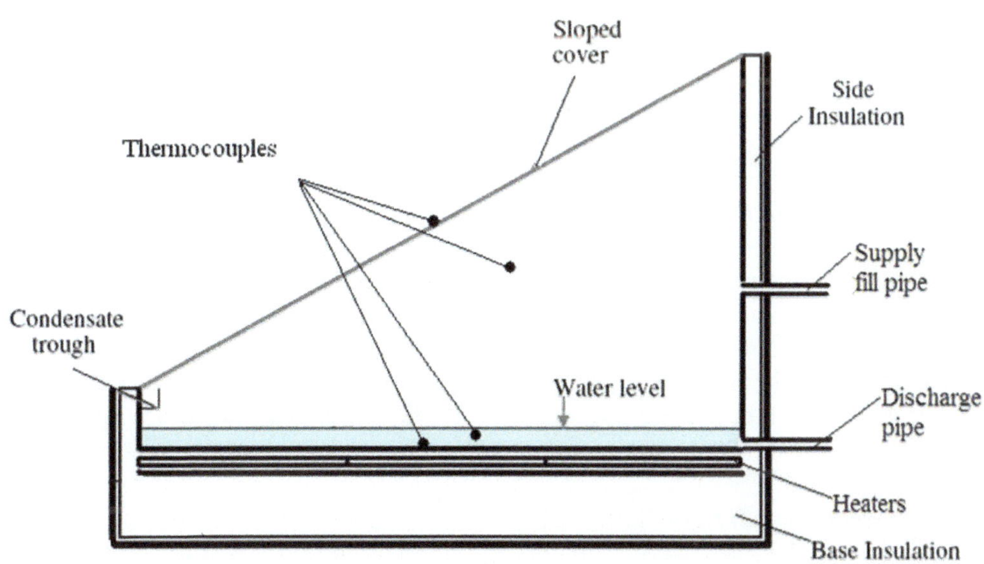

Figure 1. A schematic of the experimental still showing the main component and locations of temperature measurements

The heat input to the still from solar radiation in outdoor condition is simulated by a set of three electric heaters placed beneath the basin as shown in Figure 2. These heaters, which measures 66 by 21.5 cm each, are connected in series to supply a uniform heat flux to the basin area of the still. Each heater is manufactured by winding a wire that has a resistance of 2.5 Ω/m on a 0.4 mm thick sheet of mica which is then sandwiched between two similar sheets of mica. The whole construction is finally contained in a 1-mm-thick plate envelope of galvanized iron to prevent deformation of the mica sheets as they get heated. A variac (input 380 V, output 0-450 V, 50-60 Hz) is used to regulate the input voltage to the heaters and thereby the input power. The three heaters that are connected in series supply a maximum total power of 1000 W. The input voltage and current are constantly monitored to insure a stable power input to the heaters at the desired setting.

Figure 2. Three separate heaters connected in series at the base of the still

A graduated flask is used to collect the distilled water at half-hourly intervals for continuous five hours. All test are carried out during the same hours each day from 8:30 am to 1:30 pm to insure that the tests are carried out at an almost the same surrounding temperature in the laboratory. Obviously, apart from this point, local time does not mean much in indoor tests. The surrounding temperature during the tests were at a minimum of 15 °C and a maximum of 18 °C with an average value of around 16.6 °C.

Calibrated K- type thermocouples are embedded at various locations in the still (Figure 1) to measure the temperatures of the exterior surface of the glass cover, the vapor enclosed inside the still, the brine and the basin plate, in addition to the surrounding temperature. A12-channel temperature recorder data logger (Model BTM-4208SD) with a resolution of 0.1 °C and a sampling rate that can be varied between one second to one hour is used to record these measurements.

To simulate the wind speed in the indoor tests, a 280 W fan is situated in front of the still to create a condition analogous to the outdoor wind. The speed of the fan is controlled by varying the input voltage to the fan motor using an additional variac. Four different wind speeds are created, namely 1.14, 2.06, 2.92, 4.01 m/s. Tests at stationary air are also carried out. The speed of the wind is measured by placing a digital anemometer at a right angle to the sloped cover of the still. During these tests, the water depth in the still basin is kept at 4 cm and the output power of the heaters is set at 459 W (equivalent to 937 W/m^2).

3. Basic heat and mass transfer relations

The operation of a solar still is governed mainly by convection and radiation. A very small amount of energy is also lost to the ground (or atmosphere) by conduction through the base. Within the still, convective heat transfer occurs simultaneously with evaporative mass transfer while radiative heat transfer occurs in the inside and outside regions along with other modes.

Dunkle [23] presented the following relations that account for the heat transfer across the bulk of the humid air inside the still by free convection, which is then released at the glass cover,

$$q_{cb} = 0.884\left[(T_b - T_g) + \frac{(P_b - P_g)(T_b + 273)}{268.9 \times 10^3 - P_b}\right]^{1/3} (T_b - T_g) \tag{1}$$

Or,

$$q_{cb} = h_{cb}\,(T_b - T_g) \tag{2}$$

where q_{cb} is the heat transfer by convection between the brine surface and the glass cover (W/m²), h_{cb} the convective heat transfer coefficient between the brine surface and the glass cover (W/m² K) and T_b & T_g the temperatures of the brine and glass respectively. The partial pressures P_b and P_g of the saturated vapor at the brine and glass cover temperatures (N/m²) respectively are calculated from,

$$P_b = exp\left[25.317 - \frac{5144}{(T_b + 273)}\right] \tag{3}$$

$$P_g = exp\left[25.317 - \frac{5144}{(T_g + 273)}\right] \tag{4}$$

The heat transfer by evaporation (q_{eb}) from the brine surface to the glass cover (W/m²) is,

$$q_{eb} = 16.273 \times 10^{-3}\, q_{cb}\, \frac{P_b - P_g}{T_b - T_g} \tag{5}$$

While the mass transfer rate is,

$$m_e = \frac{q_{eb}}{L} \tag{6}$$

where L is the latent heat of vaporization of brine (kJ/kg)
The water surface and the glass cover may be considered as infinite parallel planes in stills with small cover slopes and large dimensions, hence,

$$q_{rb} = \frac{\sigma\,(T_b^4 - T_g^4)}{[\epsilon_b^{-1} + \epsilon_g^{-1} - 1]} \tag{7}$$

where σ is the Stefan–Boltzmann constant and ϵ_b & ϵ_g are the emissivities of brine (=0.96) and glass (=0.88) respectively.
Absolute values of the total energy transfer rate are obtained by the addition of equations 2, 5 and 7.
Due to the small thickness of the glass cover, the temperature in the glass may be assumed uniform. The external radiation losses from the glass cover at T_g (K) to the surrounding surfaces at T_a (K) (T_{sky} in the outdoor tests) is expressed as,

$$q_{ra} = \epsilon_g\, \sigma\, [T_g^4 - T_a^4] \tag{8}$$

And the convection heat loss to the surrounding is given as,

$$q_{ca} = h_{ca}\,(T_g - T_a) \tag{9}$$

Hence,

$$q_a = q_{ra} + q_{ca} \tag{10}$$

The external convection coefficient h_{ca} is a function of air velocity V,

$$h_{ca} = 5.7 + 3.8V \tag{11}$$

The instantaneous thermal efficiency of still is the ratio of the thermal energy utilized to get a certain amount of distilled water (q_{eb}) to the input wattage per unit basin area at a given time interval $[W(t)]$, both being in W/m^2,

$$\eta_i = \frac{q_{eb}}{W(t)} \tag{12}$$

4. Results and discussion

The time variation of the half-hourly accumulated yield of the still for different wind speeds during the test hours is shown in Figure 3. It can be seen that the yield increases steadily with time, which is obvious due to the heat input to the still. The still with no wind exhibits the lowest yield. An appreciable increase of 44.7% in yield during the test period is noticed when a wind speed of 1.14 m/s is applied. Increasing the wind speed further to 2.06, 2.92 and 4.01 m/s will cause further, but modest, increase of 9.1% and 11.6% and 5.5% respectively from the 1.14 m/s wind level. Figure 4 shows the variation of the yield per square meter of basin area against the wind speed. It can be seen here that the best result is obtained with a wind speed of 2.92 m/s that caused an increase of 61.6% in the yield in comparison to the no wind case. However, this improvement is not significantly larger than those of the other wind speeds tested as shown earlier. One may conclude that wind speed will definitely increase the yield of solar stills; however, increasing the wind speed to higher levels will not increase the yield much further. In the contrary, high wind speeds may reduce the improvement by a certain amount as with the case of the 4.01 m/s in this study.

To explore the reasons behind the productivity variation with wind speed, we look first at the effect of wind speed on some of the system temperatures. Figure 5 shows the time variation of brine and glass temperatures for different wind speeds. It can be seen that the glass temperature is noticeably reduced when a wind speed of 1.14 m/s is applied due to the increase in heat loss from the glass cover to the surrounding air. Further increase in wind speed will cause further, but modest, reduction in the glass temperature. A comparable behavior in brine temperature with wind speed can be noticed in the figure. However, the brine temperature is less sensitive to wind speed variation due to the higher heat capacity of the brine bulk compared to that of the glass cover and the direct contact of the latter with the moving air.

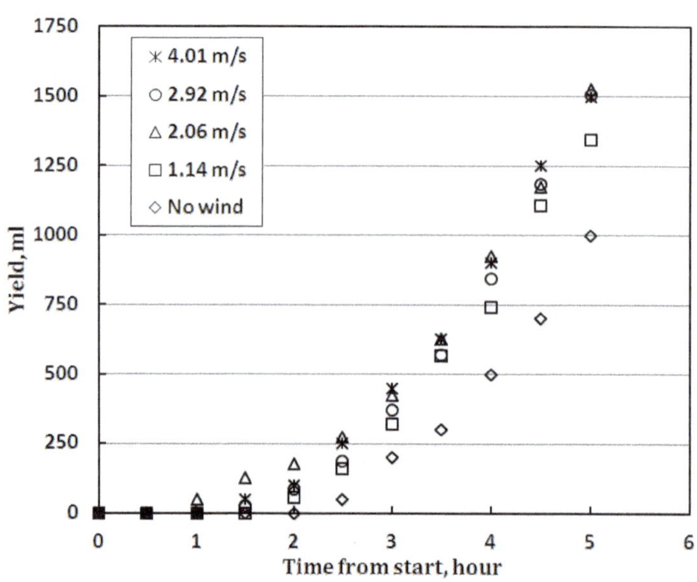

Figure 3. Variation of the half-hourly yield of the indoor still with time for different wind speeds

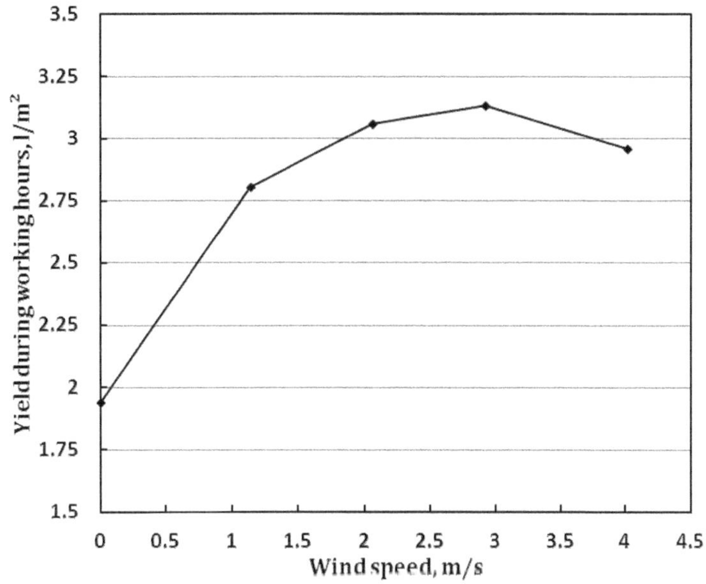

Figure 4. Variation of productivity during working hours with wind speed

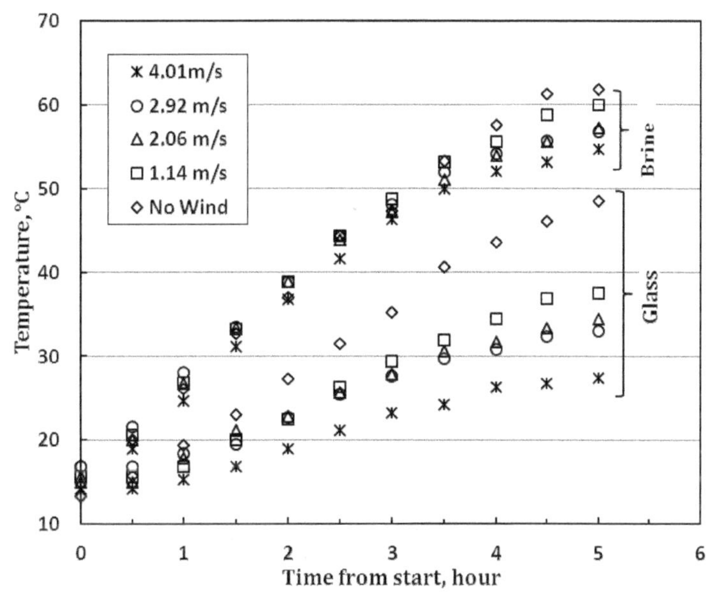

Figure 5. Variation of brine and glass temperatures with time for different wind speeds

The time variation of the brine-glass temperature difference, which is the driving potential of all heat fluxes inside the still, is plotted in Figure 6 for various wind speeds. Obviously, this difference is expected to vary in a similar manner to those of brine and glass temperatures. Consequently, applying 1.14 m/s wind speed increased the brine-glass temperature difference remarkably but further increase in wind speed will only cause modest additional increase in this difference.

It was shown in equations 1 to 5 that the brine-glass temperature difference is the driving potential of heat transfer by convection, radiation and evaporation inside the still. The productivity of the still is strongly related to the heat transport by evaporation. Figure 7 shows the time variation of this heat flux (W/m^2) for the various wind speeds tested. Modest wind speed will cause the evaporative heat flux to increase remarkably due to the increase in the brine-glass temperatures as shown earlier; again, further increase will cause only slight increase in the evaporative flux. As the experimental results for the various wind speeds in Figure 7 is too close, the total energy transported by evaporation during the working period (area under each curve in $W.h/m^2$) is calculated and plotted in Figure 8 against the wind speed. The trend of productivity variation with wind speed shown in Figure 4 and that of energy flux in Figure 8 is evident.

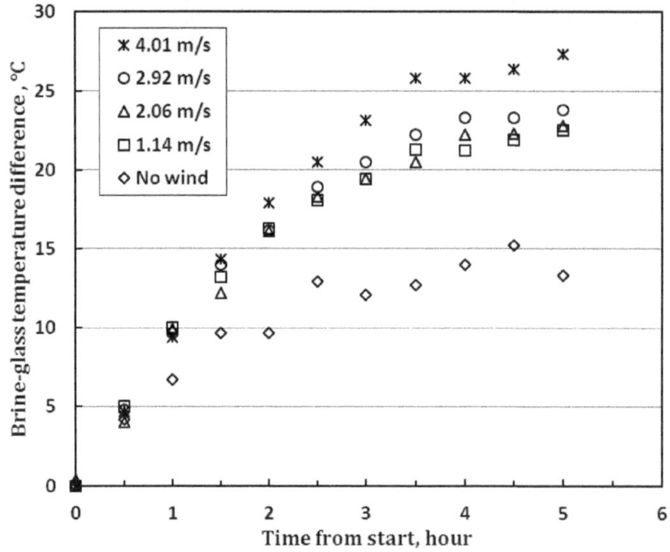

Figure 6. Variation of brine-glass temperature difference with time for different wind speeds

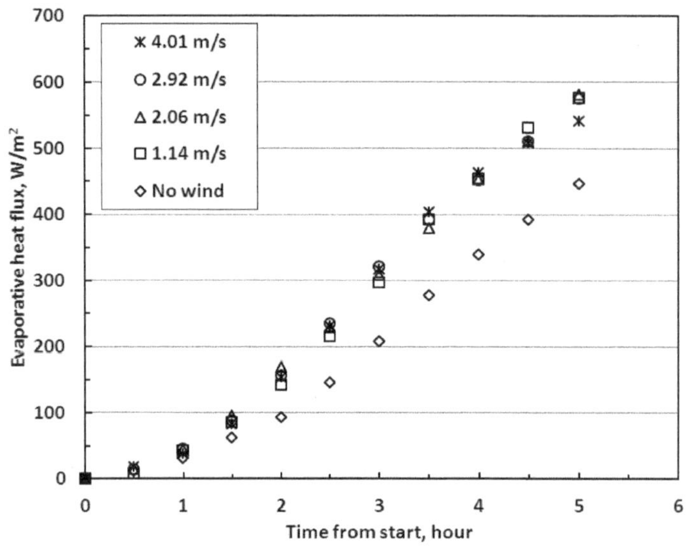

Figure 7. Variation of evaporative heat flux inside the still with time for different wind speeds

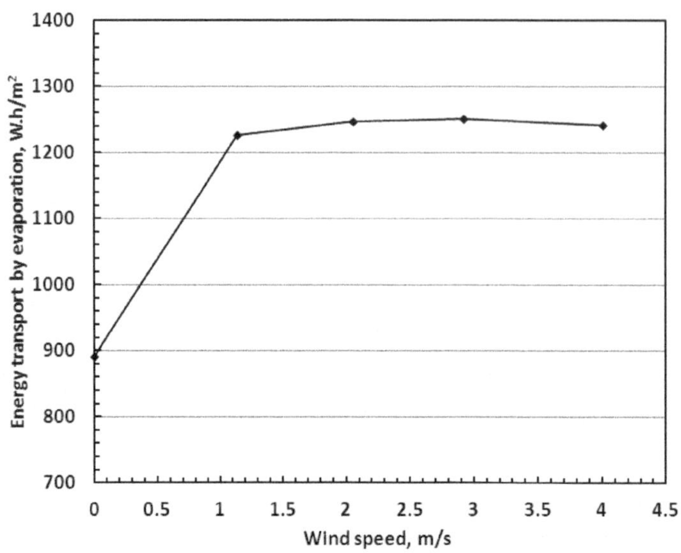

Figure 8. Variation of total energy transported by evaporation with wind speed

5. Conclusions

Indoor tests were conducted to investigate the effect of wind speed on the performance of basin type stills. This effect is examined by using a fan to generate airflow analogous to the outdoor wind, and heaters that provide uniform heat flux to the basin. The tests were conducted for four different wind speeds of 1.14, 2.06, 2.92 and 4.01 m/s in addition to tests with stationary air, all with an input power of 459 W and 4 cm brine depth. It can be concluded that increasing wind speed will definitely increase the yield of solar stills. However further increase in wind speed will not increase the yield much further. In the contrary, high wind speeds may give less improvement in productivity than moderate wind speed; in this study, this critical value was found to be around 4 m/s. The increase in productivity was found to cause an increase in the heat transport by evaporation inside the still due to the increase in the brine-glass temperature difference.

References

[1] Elsherbiny S.M., Fath H.E.S. Solar distillation under climatic conditions of Egypt. Renewable Energy1993, 3, 61-65.

[2] Elsafty A.F., Fath H.E.S., Amer A.M. Mathematical model development for a new solar desalination system (SDS). Energy Conversion and Management 2008, 49, 3331-3337.

[3] Madhlopa A., Johnstone C. Numerical study of a passive solar still with separate condenser. Renewable Energy 2009, 34, 1668-1677.

[4] Morse R.N., Read R.W. A rational basis for the engineering development of a solar still. Solar Energy 1968, 12, 5-17.

[5] Nafey A.S., Abdelkader M., Abdelmotalip A., Mabrouk A.A. Parameters affecting solar still productivity. Energy Conversion and Management 2000, 41, 1797-1809.

[6] Fath H.E.S., Hosny H.M. Thermal performance of a single-sloped basin still with an inherent built-in additional condenser. Desalination 2002, 142, 19-27.

[7] Fath H.E.S., Elsherbiny S., Ghazy A. A naturally circulated humidifying /dehumidifying solar still with a built-in passive condenser. Desalination 2004, 169, 129-149.

[8] Mamlook R., Badran O. Fuzzy sets implementation for the evaluation of factors affecting solar still production. Desalination 2007, 203, 394-402.

[9] Tiwari G.N., Dimri V., Chel A. Parametric study of an active and passive solar distillation system: Energy and exergy analysis. Desalination 2009, 242, 1-18.

[10] Toure S., Meukam P. A numerical model and experimental investigation for a solar still in climatic conditions in Abidjan (Cote d'ivoire). Renewable Energy1997, 11, 319-330.

[11] Abdenacer K., Nafila S. Impact of temperature difference (water-solar collector) on solar-still global efficiency. Desalination 2007, 209, 298-305.

[12] Dimri V., Sarkar B., Singh U., Tiwari G.N. Effect of condensing cover material on yield of an active solar still: an experimental validation. Desalination 2008, 227, 178-189.

[13] Zurigat Y.H., Abu-Arabi M.K. Modelling and performance analysis of a regenerative solar desalination unit. Applied Thermal Engineering 2004, 24, 1061-1072.

[14] El-Sebaii A.A. Effect of wind speed on some designs of solar stills. Energy Conversion and Management 2000, 41, 523-538.

[15] El-Sebaii A.A. Effect of wind speed on active and passive solar stills. Energy Conversion and Management 2004, 45, 1187-1204.

[16] El-Sebaii A.A. On effect of wind speed on passive solar still performance based on inner/outer surface temperatures of the glass cover. Energy 2011, 36, 4943-4949.

[17] Kumar S., Tiwari G.N. Analytical expression for instantaneous exergy efficiency of a shallow basin passive solar still. International Journal of Thermal Sciences 2011, 50, 2543-2549.

[18] Farid M., Hamad F. Performance of a single-basin solar still. Renewable Energy1993, 3, 75-83.

[19] Younis S.M., El-Shakweer M.H., Eldanasary M.M., Gharieb A.A., Mourad R.I., Effect of some factors on water distillation by solar energy. Misr J. Ag. Eng. 2010, 586-599.

[20] Badran O. Experimental study of the enhancement parameters on a single slope solar still productivity. Desalination 2007, 209,136-143.

[21] Maalej A.Y. Solar still performance. Desalination 1991, 82, 207-219.

[22] Khalifa A.J.N. On the effect of cover tilt angle of the simple solar still on its productivity in different seasons and latitudes. Energy Conversion and Management 2011, 52, 431-436.

[23] Dunkle R.V. Solar water distillation: The roof type still and a multiple effect diffusion still. Proceedings of the International Conference of Heat Transfer1961. University of Colorado, Publication No. 108, 895-902.

Updated equation to compute the soil voltage contour under fault condition

M. Nassereddine, J. Rizk, M. Nagrial, A. Hellany

School of Computing, Engineering & Mathematics, University of Western Sydney, Australia.

Abstract

The neighboring rates between high voltage substation and residential buildings are amplified due to populations increase. Under fault or system malfunction, earth potential rise could reach an unsafe condition. This earth potential rise extends to the surrounding soil and known as the soil voltage. To guarantee safety acquiescence of the new system, earthing system design is required. Earthing system offers a safe working atmosphere for employees and people transitory by during a fault or malfunction of a power system. The soil voltage due to high voltage EPR plays important roles when it comes to step and touch voltage computation. In this paper, a new equation is studied to compute the soil voltage at distance x from the fault location. Numerous field tests is included to verify the proposed method.

Keywords: Earth grid; High voltage substation; EPR Contour, Soil voltage.

1. Introduction

The neighbouring rates among the high voltage (HV) infrastructure and residential are rapidly inclining due to the growing population. This neighboring forces the obligation of a satisfactory earthing system to guarantee safety compliance of the HV infrastructure. Under fault situation, the earth potential rise (EPR) can grasp unsafe limits that may leads to human and property damage. The EPR is directly related to the substation earth grid resistance [1, 2]. The earth grid computation is directly related to the soil resistivity of the area [3-5]. The step and touch voltages are directly related to the soil voltage at the feet of the personnel [6].

When personal touching the pole, the person is standing at X distance away from the pole, as shown in Figure 1, the touch voltage is the different between the hand and feet potential, as the EPR dive toward the zero value with distance, the larger the X distance the higher is the touch voltage.

Many researchers address the soil voltage at distance X from the EPR [6-8]. The works is based on single electrode earth grid and on the estimation of the equivalent radius of the semi hemisphere which represent the single electrode. This method becomes complicated when it comes to complex earth grid.

The efforts in this paper work on establishing an equation to compute the soil voltage at distance X for single electrode and complex electrodes gird. Numerous field tests are included to verify the proposed method.

2. Theoretical study

Under fault condition, the earth potential rise extends to the surrounding soil area. The work in [7] shows the contour of the EPR around a transmission pole. Transmission poles are formed of conductive materials, concrete or steel poles. Therefore, touch voltage to the pole becomes an issue under fault condition. The touch voltage is the difference between the pole EPR and the soil voltage at the feet of the person standing 1 or 1.5meters away from the pole. According to [8], the relative value of the potential at a distance X from the center of the circular flat earthing electrode, with diameter D is given in equation (1).

$$V(x) = \frac{2}{\pi} EPR \times Arc\sin\left(\frac{D}{2x}\right) \tag{1}$$

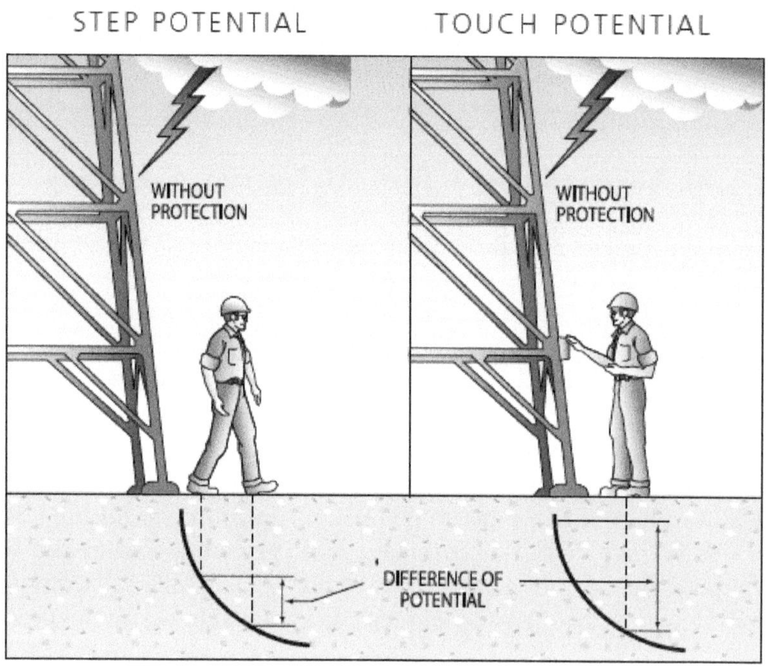

Figure 1. Step and touch voltage

Another approach to compute the soil voltage at distance X is presented in equation (2) [9].

$$V(x) = EPR\frac{r_e}{x + r_e} \tag{2}$$

where r_e is the radius of the equivalent hemispherical electrode in meters

Based on the author published works in [6], figure 2 shows the percentage of the maximum touch voltages based on equation (1) and (2). For person standing up to 1.5meters from the pole, the maximum possible touch voltage cannot exceed the 45% of the pole EPR.

2.1 New soil voltage equation

The following analysis is used to establish the soil voltage at distance X from the EPR as per this paper. The resistance of a cylindrical electrode in respect of a point at distance X can be found using equation (3) [10]:

$$R_{respect-to-x} = \int_r^x \frac{\rho}{S_{electrode}} dr \tag{3}$$

Equation (4) represents the surface of a cylindrical electrode with a semi hemisphere end:

$$S_{electrode} = 2\pi rl + 2\pi r^2 \tag{4}$$

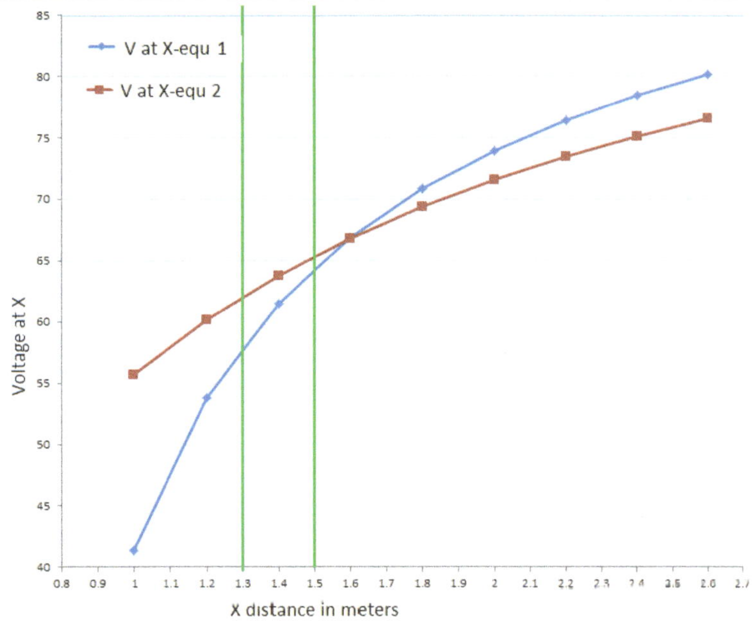

Figure 2. Voltages % at distance X against EPR as per equation 1 and 2

Equation (3) can be rearranged as shown in equation (5):

$$R_{respect-to-x} = \int_{r}^{x} \frac{\rho}{2\pi r l + 2\pi r^2} dr \tag{5}$$

Please note when X is infinite, equation (5) represents the true resistance of the electrode. Equation (6) represents the analysis of equation (5):

$$R_{respect-to-x} = \frac{\rho}{2\pi L}\left[Ln\left(\frac{x}{x+l}\right) + Ln\left(\frac{r+l}{r}\right)\right] \tag{6}$$

Using the EPR and the voltage at distance "X", equation (7) represents the resistance in respect to distance "X":

$$R_{respect-to-x} = \frac{EPR_E - V_x}{I} \tag{7}$$

where I is the current that generate EPR_E
Equation (8) is generated by analysing equations (6) and (7):

$$V_x = EPR_E\left|\left(\frac{Ln(A)}{Ln(B)}\right)\right| \tag{8}$$

where the terms A and B are defined in equations (9) and (10):

$$A = \frac{x}{x+l} \tag{9}$$

$$B = \frac{r+l}{r} \tag{10}$$

Equation 8 represents the new equation to compute the soil voltage at distance x under EPR condition. Figure 3 shows the simulation of equations (2) and (8) for a single electrode with EPR of 1 volt. The simulation is completed for a 4.8meter electrode with 14mm diameter. It should be noted, the results in Figure 3 show that equation (8) represents a more conservative approach when it comes to touch voltage.

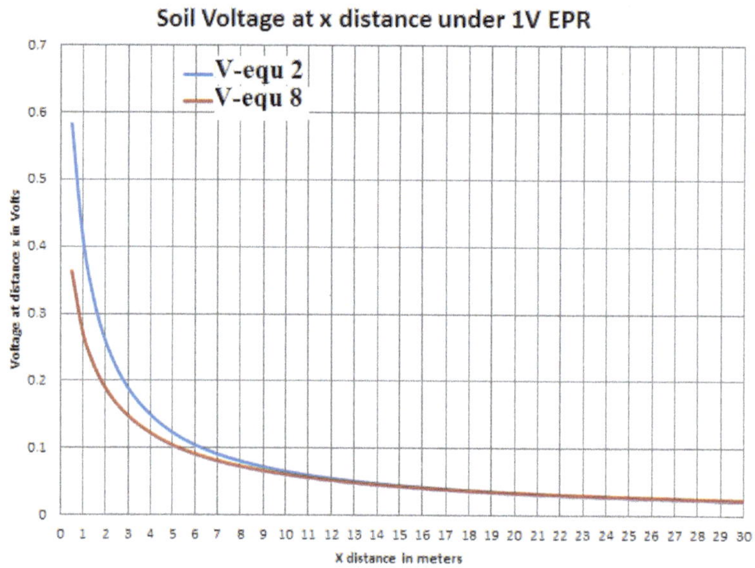

Figure 3. Soil voltage at distance x

3. Soil voltage field tests

The field tests were completed at UWS test ground as shown in Figure 3. The tests were completed on single electrode earth grid and on complex grid system as shown in Figures 4 and 5.

Figure 4. UWS field test area and the single electrode installation

As per this paper, there are two existing equations to compute the soil voltage due to existing EPR (equations (1) and (2)) and one proposed equation as per this paper analysis (equation (8)). The below works is to verify the validity of equation (8).

The first field test is completed on a single electrode. The installed electrode has the characteristics as shown in Figure 6, the equivalent radius of a full copper electrode is computed to be 1.6mm.

The field data is obtained by completing the tests at UWS ground and at another location which is located at Elizabeth Crescent in Kingswood NSW 2747. The test at Elizabeth is completed on an electrode length of 0.4meters. The computed and measured soil voltage is shown in Figure 7. The results show the accuracy of the proposed equation as per this paper. It should be noted that the soil resistivity at Elizabeth crescent for the depth of 0.5meters is homogenous

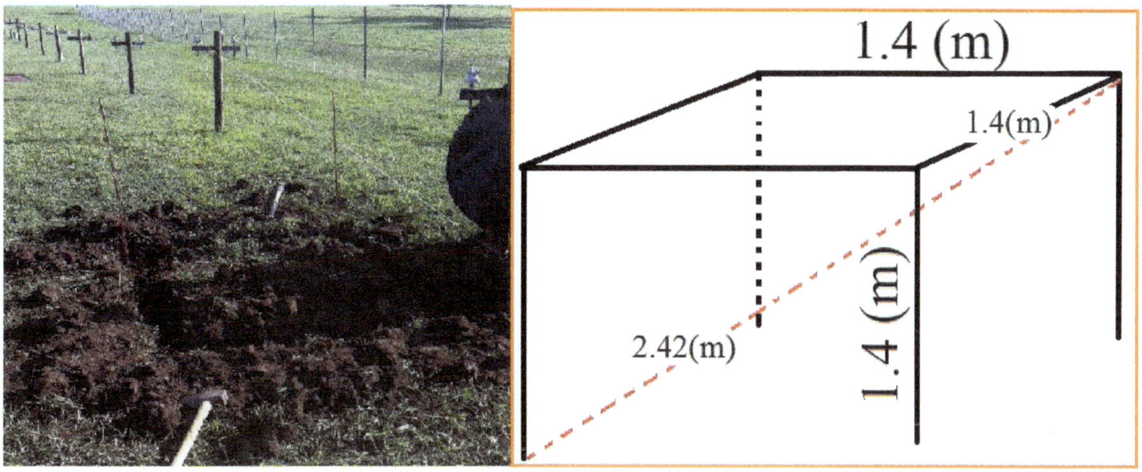

Figure 5. UWS complex earth grid field installation

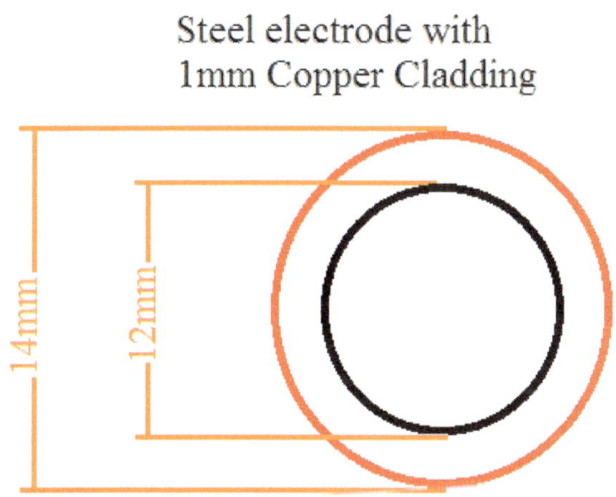

Figure 6. Steel electrode with 1mm copper cladding electrode

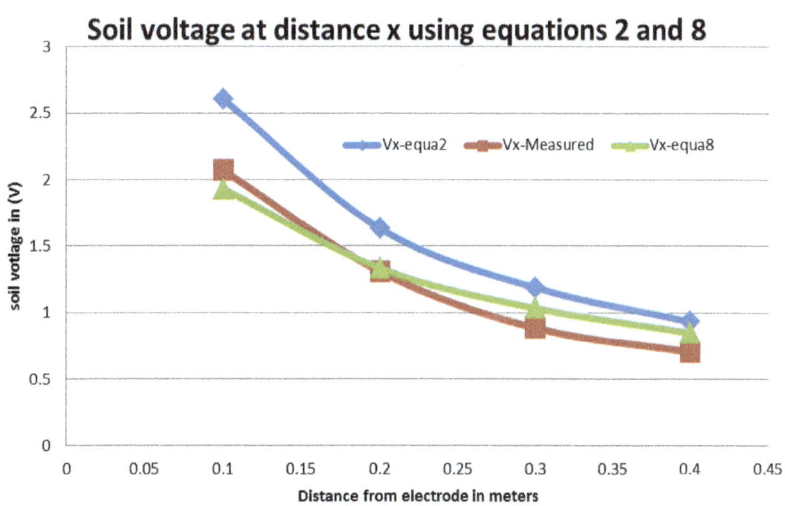

Figure7. Computation and measured data for Elizabeth street field test

Based on Figure 7, the use of equation 8 will yield lower error in comparison to equation 2 against the measured results. It should be noted that the soil resistivity structure at Elizabeth cresent was measured to be homogenous. The field works as shown below didn't find any deviation in the results when the soil resistivity structure is not homogenous. Figure 8 shows the layout of the installed electrode which covers different soil structure. When the surface soil resistivity is different to the first layer soil resistivity, the current in the electrode splits between both layers as shown in Figure 9. The field test shows that this current split will have no impact on the measurement and computation of the soil voltage.

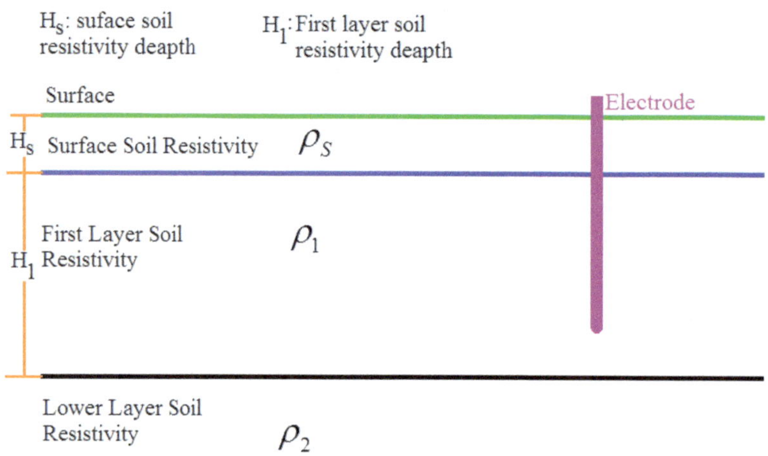

Figure 8. Electrode installed in non-homogenous soil

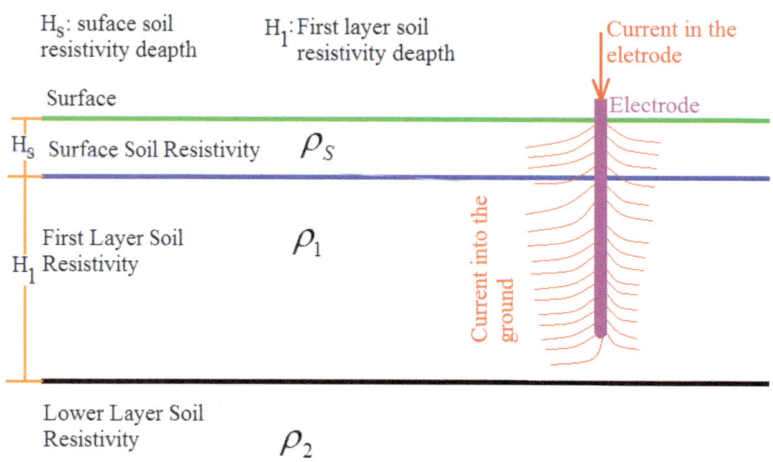

Figure 9. electrode current splits between both soil layers

At UWS field test, an electrode is installed in the middle of a track where crush rock and disturbed ground is installed. Figure 10 shows the computed soil resistivity of the area using the measured field data with the aid of engineering software. Figure 11 shows the computed results as per equations (2), (8) and the field tests results. The figure shows the advance accuracy of the developed equation over the existing one.

Further to the single electrode test works, the project completed the voltage soil measurements for the complex earth grid which is installed at UWS ground. The installed complex electrode grid is shown in Figure 5. Figure 12 represents the soil voltage simulation using equation 8 for an EPR of 1V. The figure shows that the soil voltage drops with distance which also supports the behavior of the EPR contour when assessing the step and touch voltages as illustrated in figure 1. The author works in [11] is used to compute the equivalent radius of the complex earth grid system.

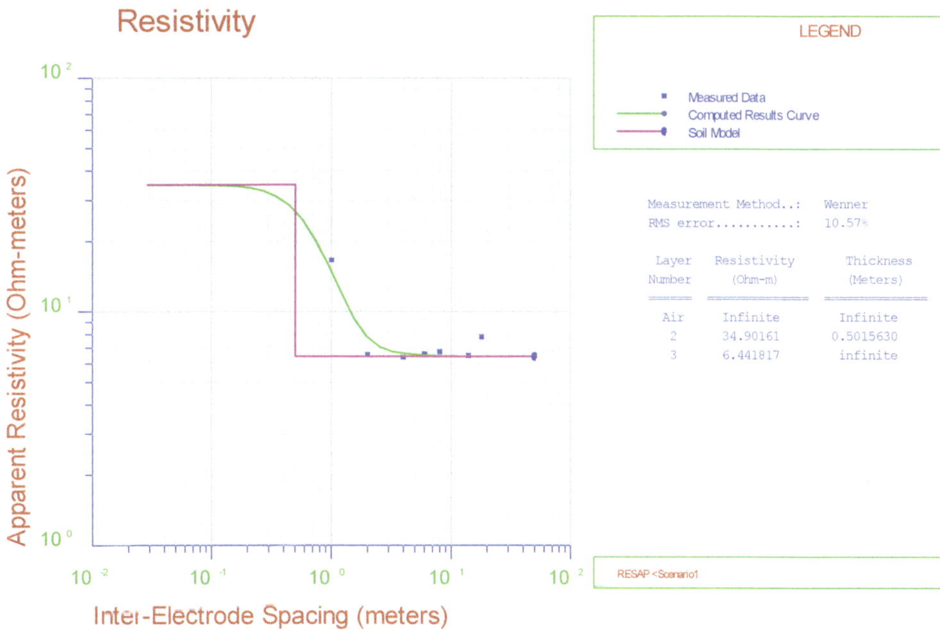

Figure 10. Soil Resisivity strucutre at the tested electrode at UWS

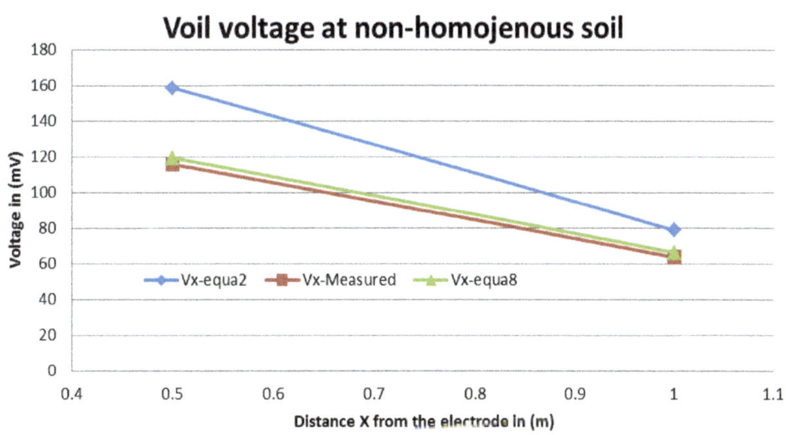

Figure 11. UWS field test to verify equation (8)

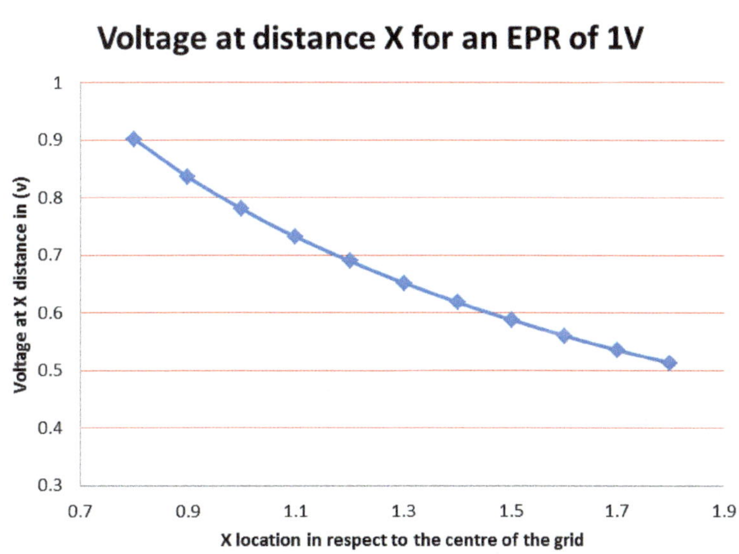

Figure 12. Voltage at distance X from the installed earth grid of EPR 1V

The field test followed the following steps to measure the voltage at distance X:

1. Locate the injection probe 24meters away from the grid. This represents 10 time the diagonal dimension of the grid
2. Measure the voltage using the potential probe
3. Measure the injected current for each measurements
4. Use the actual grid resistance which was previously measured to compute the EPR of the grid.
5. The voltage at distance X is computed by subtracting the voltage measured in step (2) from the grid EPR.

Figure 13 represents the test layout. The test results are shown in Figure 14.

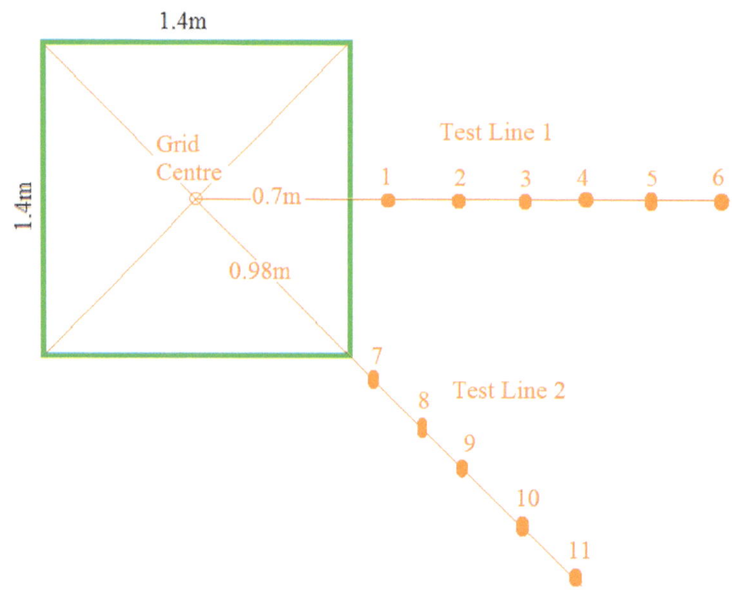

Figure13. Field test layout

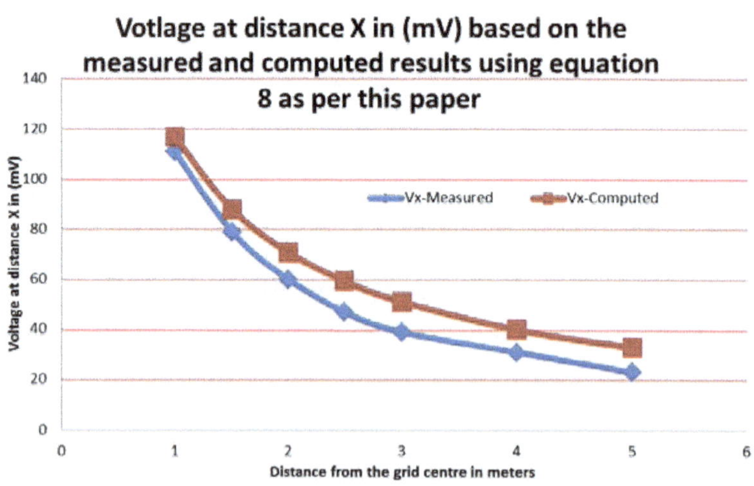

Figure 14. Voltage at distance X field test for line 1 as per Figure 12 layout and computed results

4. Soil voltage contour discussion

Under fault condition, the EPR of the faulted structure extends to the surrounding soil area. This EPR causes the step and touch voltages. The cited equations (1) and (2) do not yield accurate results as shown in the field test. The works in this paper establish the new equation (8) to compute the soil voltage contour. The field tests show the accuracy of the proposed methods over the existing ones. Furthermore, the work shows that the proposed equation is valid for homogenous and non-homogenous soil resistivity structure. In addition, the field's tests show that equation (8) along with the works in [11] yield accurate results when it comes to soil voltage computation for complex earth grid system.

The main benefits of this finding are:

- Accurate compute of the soil voltage contour
- Accurate computation of the touch voltage
- As shown in the next section, the soil voltage contour aid in accurate computation of the fault current distribution within the OHEW system

5. Conclusion

This paper highlights the relation between the soil voltage and the touch voltages. The work introduces the new equation to compute the soil voltage at distance x from the EPR. The field test shows the high accuracy reading of the new equation in regards to the existing ones. The novel findings of this paper add valuable information to the soil voltage computation which leads to higher accuracy computation when it comes to step and touch voltages.

Acknowledgment

This works was supported by the National Electrical Engineering Consultancy (NEEC PTY LTD)

References

[1] Nassereddine M. Hellany A, Nagrial M. and Rizk J."Soil resistivity structure and its implication on the earth grid of HV substation" 2011 World Academy of Science, engineering and Technology, Vol 60, pp 1322-1326

[2] IEEE 80 guide to safety in AC substation grounding, 2000' (IEEE, New York, 2000).

[3] Nassereddine M, Rizk J, Nasserddine G, Nagrial M and Hellany A. "Soil resistivity structure and its implication on the pole grid resistance for transmission lines" 2013 World Academy of Science, engineering and Technology, Dubai, UEA

[4] Nassereddine M, Hellany A and Rizk J "How to design an effective earthing system to ensure the safety of the people", 2009 International Conference on Advances in Computational Tools for Engineering Applications, pp 416-421

[5] Nassereddine M. Rizk J., Hellany A. and Nagrial M. "Estimation of apparent soil resistivity for two layers soil structure using the reflection coefficient "K""-International Journal of Energy and Environment, Vol.4, Issue4, 2013, PP. 573-580, 2013

[6] Nassereddine M., Rizk J., Hellany A. and Nagrial M. "Transmission Mains pole earth potential rise (EPR) under substation fault" International journal of emerging electric power systems-Volume 14, Issue 5, Pages 449-507. 2013

[7] Nassereddine M., Hellany A., Nagrial M. and Rizk J. "Safety Compliance of Substation Earthing Design" World Academy of Science, engineering and Technology, Vol 60, PP. 525-529. 2011

[8] Popovic L. "Proximity Effect Between an Earthing Grid and External Electrodes in an Earthing System of High-Voltage Insulations" IEE Proc. Gen., Trans. and Dist. Vol. 133 Issue 6. PP. 346-352. 1986.

[9] Datta M., Basu A., and Chowdhury M., "Determination of earth resistance of multiple driven-rod electrodes" Proceeding of the instribution of electrical engineers, Vol 114, Issue 7. PP.1001-1006. 1967

[10] Nassereddine M., Rizk J., Hellany A. and Nagrial M. "Estimation of Novel Position for the Current and Potential Probe for the Fall of Potential Method" IET Generation, Transmission & Distribution. Vol. 8. Issue 4. PP. 774-782. 2014

[11] Nassereddine M., Rizk J., Hellany A., and Nagrial M. "Substation Earth Grid Measurement using the Fall of Potential Method (FOP) for a Limited Test Area" Australasian University Power Engineering Conference. 2014

Exergy analyses of an endoreversible closed regenerative Brayton cycle CCHP plant

Bo Yang[1,2,3], Lingen Chen[1,2,3], Yanlin Ge[1,2,3], Fengrui Sun[1,2,3]

[1] Institute of Thermal Science and Power Engineering, Naval University of Engineering, Wuhan 430033, P. R. China.
[2] Military Key Laboratory for Naval Ship Power Engineering, Naval University of Engineering, Wuhan 430033, P. R. China.
[3] College of Power Engineering, Naval University of Engineering, Wuhan 430033, P. R. China.

Abstract
An endoreversible closed regenerative Brayton cycle CCHP (combined cooling, heating and power) plant coupled to constant-temperature heat reservoirs is presented using finite time thermodynamics (FTT). The CCHP plant includes an endoreversible closed regenerative Brayton cycle, an endoreversible four-heat-reservoir absorption refrigerator and a heat recovery device of thermal consumer. The heat-resistance losses in the hot-, cold-, thermal consumer-, generator-, condenser-, evaporator- and absorber-side heat exchangers and regenerator are considered. The performance of the CCHP plant is studied from the exergetic perspective, and the analytical formulae about exergy output rate and exergy efficiency are derived. Through numerical calculations, the pressure ratio of regenerative Brayton cycle is optimized, the effects of heat conductance of regenerator and ratio of heat demanded by the thermal consumer to power output on dimensionless exergy output rate and exergy efficiency are analyzed.

Keywords: Finite time thermodynamics; Endoreversible closed regenerative Brayton cycle CCHP plant; Endoreversible four-heat-reservoir absorption refrigerator; Exergy output rate; Exergy efficiency.

1. Introduction

To solve energy crisis and reduce environmental pollution in the world, in recent years, people have paid much attention to new thermodynamic systems which are energy saving and environment friendly. Cogeneration which obeys energy cascade utilization principle is the simultaneous production of several forms of energy from one energy source. In general, cogeneration has two forms: CHP (combined heating and power) and CCHP (combined cooling, heating and power). Compared to conventional centralized cooling, heating or power generated systems, cogeneration has an advantage of high energy utilization efficiency and low emission of harmful pollution. Some researchers have studied CHP and CCHP plants using classical thermodynamics. Ertesvag [1] introduced relative avoided irreversibility (RAI) to analyze and compare the exergetic consequences of various legislations for CHP systems. Ferdelji *et al.* [2] performed exergy analysis (exergy losses and exergy efficiency) of a steam turbine CHP plant, and provided detailed information about magnitudes of losses and their distribution throughout the systems. Sanaye *et al.* [3] investigated the optimal design of a gas turbine CHP plant, defined an objective function as the sum of the operating cost related to the fuel consumption and the capital investment for equipment purchase and maintenance costs. Khaliq and Dincer [4] investigated the energetic and exergetic performances of a CHP plant with absorption inlet cooling and evaporative aftercooling. Temir and Bilge [5] investigated the thermoeconomic performance of CCHP system taking investment and operation costs of the system into account. Mago and Chamra [6] evaluated and optimized the operation strategies of CCHP plant with considerations of primary energy consumption, operating costs and carbon dioxide emissions Khaliq [7] carried out the exergy analysis of a gas turbine CCHP system for combined production of power, heat and refrigeration. Kavvadias and Maroulis [8] developed a multi-objective optimization method for the design of CCHP plants considering technical, economical, energetic and environmental performance indicators.

Finite-time thermodynamics (FTT) [9-17] is a powerful tool for analyzing and optimizing performance of various thermodynamic cycles and devices. In recent years, some authors have carried out the performance analyses and optimization for various Brayton cycle CHP plants by using FTT. Yilmaz [18] optimized the exergy output rate and exergy efficiency of an endoreversible simple Brayton closed cycle CHP plant and found that the lower the consumer-side temperature, the better the exergy performance. Hao and Zhang [19, 20] optimized the total useful-energy rate (including power output and useful heat rate output) and the exergy output rate of an endoreversible Joule-Brayton CHP cycle by optimizing the pressure ratio. Ust *et al.* [21] introduced a new objective function called the exergetic performance coefficient (EPC), and optimized an irreversible regenerative Brayton closed cycle CHP plant with heat resistance and internal irreversibility. By using finite time exergoeconomic analysis [22-26], Tao *et al.* [27-29] performed the finite time exergoeconomic performance analyses and optimization for endoreversible simple [27] and regenerative [28] and irreversible simple [29] Brayton closed cycle CHP plants, and found that there existed an optimal heat consumer-side temperature through a new method of calculating thermal exergy output rate. Further, Chen *et al.* [30] and Yang *et al.* [31-34] investigated the finite time exergoeconomic performances of endoreversible constant-temperature heat reservoir [30, 31] and variable-temperature heat reservoir [32, 33] and irreversible constant-temperature heat reservoir [34] closed intercooled regenerative Brayton cycle CHP plants, respectively. Also Chen *et al.* [35] and Yang *et al.* [36] carried out performance analyses and optimization of exergy output rate and exergy efficiency for an endoreversible constant-temperature heat reservoir closed intercooled regenerative Brayton cycle CHP plant.

In the recent years, absorption refrigeration cycle which can be driven by 'low-grade' heat energy has attracted increasing attention, and some work on absorption refrigeration cycle using FTT has been developed. Chen [37] investigated the maximum specific cooling load of an irreversible four- heat-reservoir absorption refrigeration cycle with heat resistance and internal irreversibility by optimizing the distribution of the heat transfer areas of the heat exchangers. Chen *et al.* [38-40] and Zheng *et al.* [41-43] performed the cooling load and coefficient of performance (COP) performance analyses and optimization for endoreversible [38] and irreversible [39-43] four-heat-reservoir absorption refrigeration cycles with Newton's [39-41] and linear phenomenological [38, 42, 43] heat transfer laws. Qin *et al.* [44, 45] analyzed and optimized the thermoeconomic performance [44] and the cooling load and COP performance [45] of constant- temperature [44] and variable-temperature [45] four-heat-reservoir absorption refrigeration cycles. Tao *et al.* [46] studied the optimal ecological function performance of an endoreversible four-heat- reservoir absorption refrigeration cycle.

Using FTT, Chen *et al.* [47] and Feng *et al.* [48] established endoreversible [47] and irreversible [48] closed simple Brayton cycle CCHP plants which contain an endoreversible four- heat-reservoir absorption refrigerator, and performed finite time exergoeconomic performance optimization by optimizing the pressure ratio and the heat conductance distribution of the hot-, cold-, thermal consumer-, generator-, condenser-, evaporator- and absorber-side heat exchangers.

In the open literature, there is no work concerning FTT performance of regenerative Brayton cycle CCHP plant. Thus, in present study, an endoreversible regenerative Brayton cycle CCHP plant coupled to constant-temperature heat reservoirs is provided using FTT. The exergy output rate and exergy efficiency of the plant are investigated by optimizing pressure ratio of the regenerative Brayton cycle, and the effects of design parameters on the general and optimal performances are investigated by numerical calculation.

2. Thermodynamic model of the CCHP plant

Figure 1 shows the flow chart of an endoreversible closed regenerative Brayton cycle CCHP plant coupled to constant-temperature heat reservoirs. Figure 2 shows the *T-s* diagram. The whole cycle is finished through the state changes of the working fluid. Process 1-2 is an isentropic adiabatic compression process in the compressor. Process 2-3 is an isobaric absorbed heat process in the regenerator. Process 3-4 is an isobaric absorbed heat process in the hot-side heat exchanger. Process 4-5 is isentropic adiabatic expansion process in the turbine. Process 5-6 is an isobaric evolved heat process in the regenerator. Process 6-7 is an isobaric supplied heat process in the generator-side heat exchanger. Process 7-8 is an isobaric supplied heat process in the thermal consumer-side heat exchanger. Process 8-1 is an isobaric evolved heat process in the cold-side heat exchanger.

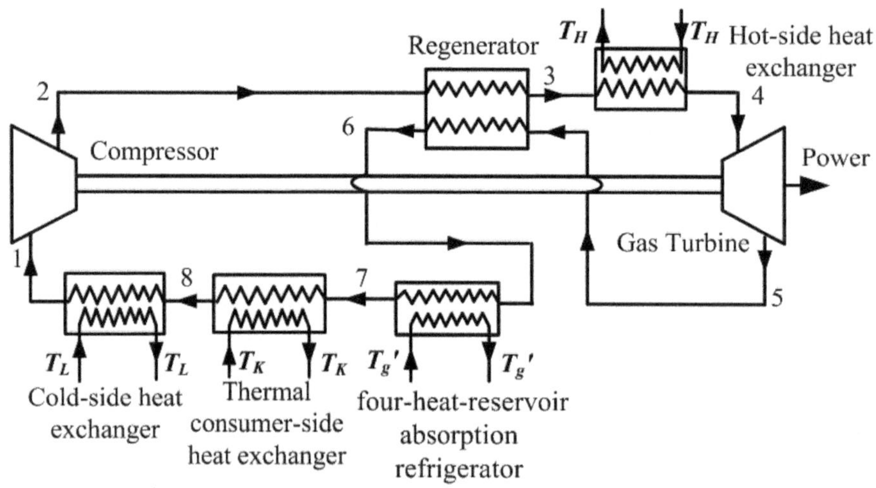

Figure 1. Schematic diagram of an endoreversible closed regenerative Brayton cycle CCHP plant

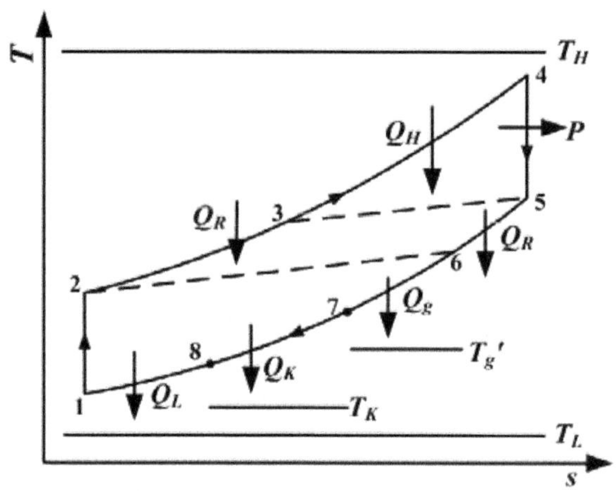

Figure 2. *T-s* diagram of an endoreversible closed regenerative Brayton cycle CCHP plant

Assuming that the working fluid used in the Brayton cycle is an ideal gas with constant thermal capacity rate C_{wf}. The hot-, cold- and thermal consumer-side heat reservoir temperatures are T_H, T_L and T_K respectively, and the temperature of working fluid in the generator is T_g'. The heat exchangers between the working fluid and the heat reservoir and the regenerator are counter-flow. The heat conductances (heat transfer surface area and heat transfer coefficient product) of the hot-, cold-, generator- and thermal consumer-side heat exchangers, and the regenerator are U_H, U_L, U_g U_K, U_R respectively. Assuming that the heat transfer obeys a linear law, according to the properties of working fluid and the theory of heat exchangers, the heat transfer rate (Q_H) from the hot-side heat reservoir to the working fluid, the heat transfer rate (Q_L) from the working fluid to the cold-side heat reservoir, the heat transfer rate (Q_R) regenerated in the regenerator, the heat transfer rate (Q_g) from the working fluid of Brayton cycle to the working fluid in the generator, and the heat transfer rate (Q_K) from the working fluid of Brayton cycle to the thermal consumer device can be expressed as:

$$Q_H = C_{wf}(T_4 - T_3) = C_{wf}E_H(T_H - T_3) \tag{1}$$

$$Q_L = C_{wf}(T_8 - T_1) = C_{wf}E_L(T_8 - T_L) \tag{2}$$

$$Q_R = C_{wf}(T_3 - T_2) = C_{wf}(T_5 - T_6) = C_{wf}E_R(T_5 - T_2) \tag{3}$$

$$Q_g = C_{wf}(T_6 - T_7) = C_{wf}E_g(T_6 - T_g') \tag{4}$$

$$Q_K = C_{wf}(T_7 - T_8) = C_{wf}E_K(T_7 - T_K) \tag{5}$$

where E_H, E_L, E_R, E_g and E_K are the effectivenesses of the hot- and cold-side heat exchangers, the regenerator, the generator- and the thermal consumer-side heat exchanger respectively, which are used to reflect the heat resistance losses, and are defined as:

$$\begin{gathered} E_H = 1-\exp(-N_H),\ E_L = 1-\exp(-N_L),\ E_R = 1-\exp(-N_R) \\ E_g = 1-\exp(-N_g),\ E_K = 1-\exp(-N_K) \end{gathered} \tag{6}$$

where $N_i(i = H, L, R, g, K)$ are the numbers of heat transfer units of the hot- and cold-side heat exchangers, the regenerator, the generator- and the thermal consumer-side heat exchanger respectively, and are defined as:

$$N_H = U_H / C_{wf},\ N_L = U_L / C_{wf},\ N_R = U_R / C_{wf},\ N_g = U_g / C_{wf},\ N_K = U_K / C_{wf} \tag{7}$$

Defining that the pressure ratio of the regenerative Brayton cycle is π and the working fluid isentropic temperature ratio for the compression process 1-2 is y, i.e. $T_2 = yT_1$. According to the thermodynamic knowledge, one has:

$$T_4 = yT_5,\ y = \pi^{(k-1)/k} \tag{8}$$

where k is the specific heat ratio of the working fluid.

Figure 3 shows a model of an endoreversible four-heat-reservoir absorption refrigerator, which is composed of a generator, a condenser, an evaporator and an absorber. And there are four corresponding heat reservoirs, the temperature of the generator heat reservoir is variable, from T_6 to T_7, while the temperatures of the condenser, evaporator and absorber heat reservoir are constant, which are T_c, T_e and T_a, respectively. Assuming that the working fluid used in the absorption refrigerator flows steadily, and the temperatures of the working fluid in the condenser, evaporator and absorber are T_c', T_e' and T_a' respectively. The heat conductances of the condenser-, evaporator- and absorber-side heat exchanger are U_c, U_e and U_a, respectively. According to the theory of heat exchangers, the heat transfer rates (Q_c, Q_e and Q_a) which go through the condenser, evaporator and absorber can be, respectively, expressed as:

$$Q_c = U_c(T_c' - T_c) \tag{9}$$

$$Q_e = R = U_e(T_e - T_e') \tag{10}$$

$$Q_a = U_a(T_a' - T_a) \tag{11}$$

where R is the cooling load of the absorption refrigerator.

In addition, the power input required by the solution pump and heat loss rate caused by the flow of the working fluid in the absorption refrigerator can be negligible compared with the energy input to the generator. They are often neglected in the analyses and optimization of absorption refrigeration cycle

[37-48]. Therefore, the distribution ratio (n) of the total heat rejection between the absorber and condenser can be defined as:

$$n = Q_a / Q_c \tag{12}$$

For the endoreversible absorption refrigeration cycle, according to the first and second law of the thermodynamics, one has:

$$Q_g + Q_e - Q_a - Q_c = 0 \tag{13}$$

$$Q_g / T_g^{'} + Q_e / T_e^{'} - Q_a / T_a^{'} - Q_c / T_c^{'} = 0 \tag{14}$$

When U_c, U_e, U_a, T_c, T_e and T_a are fixed, combining equations (9)-(14), one can obtain the function relation between the cooling load (R) of the absorption refrigerator and the heat transfer rate (Q_g) that goes through the generator [37]:

$$\frac{Q_g}{T_g^{'}} + \frac{RU_e}{T_e U_e - R} - \frac{nU_a(Q_g + R)}{(n+1)T_a U_a + n(Q_g + R)} - \frac{U_c(Q_g + R)}{(n+1)T_c U_c + Q_g + R} = 0 \tag{15}$$

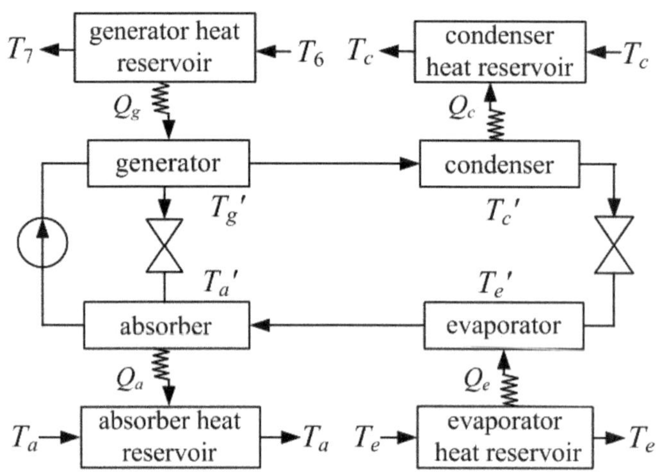

Figure 3. An endoreversible four-heat-reservoir absorption refrigerator model

3. Exergy performance analyses

According to the first law of thermodynamics, the power output (the exergy output rate of power) of the CCHP plant is:

$$P = Q_H - Q_L - Q_g - Q_K \tag{16}$$

The ratio of heat demanded by the thermal consumer to power output is defined as:

$$w = Q_K / P \tag{17}$$

Combining equations (1)-(5) with (8), (16) and (17) yields the temperatures (T_1, T_3, T_4, T_5, T_6, T_7, T_8, $T_g^{'}$) and the heat transfer rates (Q_H, Q_L, Q_R, Q_g and Q_K):

$$T_1 = \frac{wE_H T_H (y-1)(1-E_L)(1-E_K) - AE_K(E_L T_K - E_L T_L - T_K)}{AE_K - w(1-E_L)(1-E_K)[A - y^2 E_H - y(1-E_H)(yE_R + 1 - 2E_R)]} \tag{18}$$

$$T_3 = \frac{y^2 T_1 (1-E_R) + E_H E_R T_H}{A} \tag{19}$$

$$T_4 = \frac{y E_H T_H + y^2 T_1 (1-E_H)(1-E_R)}{A} \tag{20}$$

$$T_5 = \frac{E_H T_H + y T_1 (1-E_H)(1-E_R)}{A} \tag{21}$$

$$T_6 = \frac{(1-E_R) E_H T_H + y T_1 [y E_R + (1-E_H)(1-2E_R)]}{A} \tag{22}$$

$$T_7 = \frac{T_1 - E_L T_L - E_K T_K + E_L E_K T_K}{(1-E_L)(1-E_K)} \tag{23}$$

$$T_8 = \frac{T_1 - E_L T_L}{1-E_L} \tag{24}$$

$$T_g' = \frac{A(T_1 - E_L T_L - E_K T_K + E_L E_K T_K) - (1-E_L)(1-E_K)(1-E_g)\{(1-E_R)E_H T_H + y T_1[y E_R + (1-E_H)(1-2E_R)]\}}{A E_g (1-E_L)(1-E_K)} \tag{25}$$

$$Q_H = \frac{C_{wf} E_H [(y-E_R)T_H - y^2 T_1 (1-E_R)]}{A} \tag{26}$$

$$Q_L = \frac{C_{wf} E_L (T_1 - T_L)}{1-E_L} \tag{27}$$

$$Q_R = \frac{C_{wf} [y T_1 (y - y E_R - A) + E_H E_R T_H]}{A} \tag{28}$$

$$Q_g = \frac{C_{wf}(1-E_L)(1-E_K)\{(1-E_R)E_H T_H + y T_1[y E_R + (1-E_H)(1-2E_R)]\} - A C_{wf}(T_1 - E_L T_L - E_K T_K + E_L E_K T_K)}{A(1-E_L)(1-E_K)} \tag{29}$$

$$Q_K = \frac{C_{wf} E_K (T_1 + E_L T_K - T_K - E_L T_L)}{(1-E_L)(1-E_K)} \tag{30}$$

where $A = y - (1-E_H)E_R$.

Then the power output can be expressed as:

$$\begin{aligned} P &= Q_H - Q_L - Q_g - Q_K \\ &= \frac{\begin{array}{c} C_{wf} E_H (1-E_L)(1-E_K)[(y-E_R)T_H - y^2 T_1 (1-E_R)] - A C_{wf} E_L (T_1 - T_L)(1-E_K) - C_{wf}(1-E_L)(1-E_K)[y^2 E_R + (1-E_R)E_H T_H + y T_1 (1-E_H)(1-2E_R)] + \\ A C_{wf}(T_1 - E_L T_L - E_K T_K + E_L E_K T_K) - A C_{wf} E_K (T_1 + T_K E_L - T_K - E_L T_L) \end{array}}{A(1-E_L)(1-E_K)} \end{aligned} \tag{31}$$

Assuming that the ambient temperature is T_0, the net total exergy input rate of the CCHP plant is:

$$e_{in} = Q_H(1 - T_0 / T_H) - Q_L(1 - T_0 / T_L) \tag{32}$$

The thermal exergy output rate supplied for the thermal consumer is:

$$e_K = Q_K(1 - T_0 / T_K) = \frac{C_{wf} E_K (T_K - T_0)(T_1 + T_K E_L - T_K - E_L T_L)}{(1 - E_L)(1 - E_K) T_K} \tag{33}$$

The cooling exergy output rate of the absorption refrigeration cycle is:

$$e_e = R(T_0 / T_e - 1) \tag{34}$$

where the cooling load (R) is decided by equations (15), (18), (25) and (29).
Applying the exergy conservation principle to the CCHP plant, one has:

$$e_{in} = P + e_K + e_e + T_0 \sigma \tag{35}$$

Combining equations (26)-(35), the entropy generation rate (σ) of the CCHP plant can be yielded:

$$\begin{aligned}\sigma &= Q_L / T_L + Q_K / T_K + Q_g / T_0 + R(1/T_0 - 1/T_e) - Q_H / T_H \\ &= C_{wf} E_L (T_1 - T_L) / [(1 - E_L) T_L] + C_{wf} E_K (T_1 + T_K E_L - T_K - E_L T_L) / [(1 - E_L) \times \\ &\quad (1 - E_K) T_K] + \{C_{wf}(1 - E_L)(1 - E_K)\{(1 - E_R) E_H T_H + y T_1 [y E_R + (1 - E_H)(1 - \\ &\quad 2E_R)]\} - A C_{wf}(T_1 - E_L T_L - E_K T_K + E_L E_K T_K)\} / [A T_0 (1 - E_L)(1 - E_K)] + \\ &\quad R(1/T_0 - 1/T_e) - C_{wf} E_H [(y - E_R) T_H - y^2 T_1 (1 - E_R)] / (A T_H)\end{aligned} \tag{36}$$

The exergy output rate and exergy efficiency of the CCHP plant are defined as:

$$e_{out} = P + e_K + e_e \tag{37}$$

$$\eta_{ex} = e_{out} / e_{in} \tag{38}$$

Defining the nondimensionalized exergy output rate by using $C_{wf} T_0$:

$$\begin{aligned}\overline{e_{out}} &= (P + e_K + e_e) / (C_{wf} T_0) \\ &= \Big[T_K T_e \{C_{wf} E_H (1 - E_L)(1 - E_K)[(y - E_R) T_H - y^2 T_1 (1 - E_R)] - A C_{wf} E_L (T_1 - \\ &\quad T_L)(1 - E_K) - C_{wf}(1 - E_L)(1 - E_K)[y^2 E_R + (1 - E_R) E_H T_H + y T_1 (1 - E_H) \times \\ &\quad (1 - 2E_R)] + A C_{wf}(T_1 - E_L T_L - E_K T_K + E_L E_K T_K) - A C_{wf} E_K (T_1 + T_K E_L - \\ &\quad T_K - E_L T_L)\} + A C_{wf} E_K T_e (T_K - T_0)(T_1 + T_K E_L - T_K - E_L T_L) + A R T_K \times \\ &\quad (T_0 - T_e)(1 - E_L)(1 - E_K) \Big] \Big/ \Big[A C_{wf} T_K T_e T_0 (1 - E_L)(1 - E_K) \Big]\end{aligned} \tag{39}$$

The exergy efficiency can be expressed as:

$$\eta_{ex} = \frac{P + e_K + e_e}{P + e_K + e_e + T_0 \sigma} = 1 - \frac{\sigma}{\overline{e_{out}} C_{wf} + \sigma} \tag{40}$$

In addition, in order to make sure that the design state of the CCHP plant is meaningful, the power output (P), heat transfer rate (Q_g) supplied to the generator and heat transfer rate (Q_K) supplied to the thermal consumer device should be larger than zero, and the following equations about temperatures should be satisfied:

$$T_1 < T_2, T_3 < T_4, T_5 < T_4, T_7 < T_6, T_8 < T_7, T_1 < T_8 \tag{41}$$

4. Numerical examples

In order to see how the design parameters influence the dimensionless exergy output rate and exergy efficiency of the CCHP plant, detailed numerical examples are given. The following temperature ratios are defined: $\tau_H = T_H / T_0$, $\tau_L = T_L / T_0$, $\tau_K = T_K / T_0$, $\tau_e = T_e / T_0$. In the calculations, without special illustration, the numerical values of the parameters are set as follows: $k = 1.4$, $C_{wf} = 1.0kW/K$, $U_H = 2kW/K$, $U_L = 2kW/K$, $U_R = 2kW/K$, $U_K = 2kW/K$, $U_g = 2kW/K$, $U_c = 2kW/K$, $U_e = 2kW/K$, $U_a = 2kW/K$, $\tau_H = 5.0$, $\tau_L = 1$, $\tau_K = 1.2$, $T_c = 300K$, $T_e = 280K$, $T_a = 300K$, $T_0 = 300K$, $w = 0.4$, and $n = 1$.

4.1 Optimal pressure ratio

Figures 4 and 5 show the effects of U_R and w on the characteristics of $\overline{e_{out}}$ and η_{ex} versus π, respectively. It can be seen from Figure 4 that $\overline{e_{out}}$ and η_{ex} exist optimal values ($(\overline{e_{out}})_{opt}$ and $(\eta_{ex})_{opt}$) with respect to π respectively, and the corresponding optimal pressure ratios are notated as $\pi_{(\overline{e_{out}})_{opt}}$ and $\pi_{(\eta_{ex})_{opt}}$. A critical pressure ratio exists, and when π is smaller than the critical pressure ratio, the calculations indicate that $T_5 > T_2$ and $Q_R > 0$, thus $\overline{e_{out}}$ and η_{ex} increase with the increase of U_R; when π is larger than the critical pressure ratio, one has $T_5 < T_2$ and $Q_R < 0$, thus $\overline{e_{out}}$ and η_{ex} decrease with the increase of U_R, which is similar to the effect of regeneration on Brayton power cycle [49-51]. It can be seen from Figure 5 that $\overline{e_{out}}$ and η_{ex} increase equably with the increase of w. The broken lines in Figures 4 and 5 indicate that when U_R or w is large, and when π is smaller than certain values, one can obtain $T_6 < T_7$, $Q_g < 0$, which is shown in Figure 6, thus the CCHP plant becomes the CHP plant.

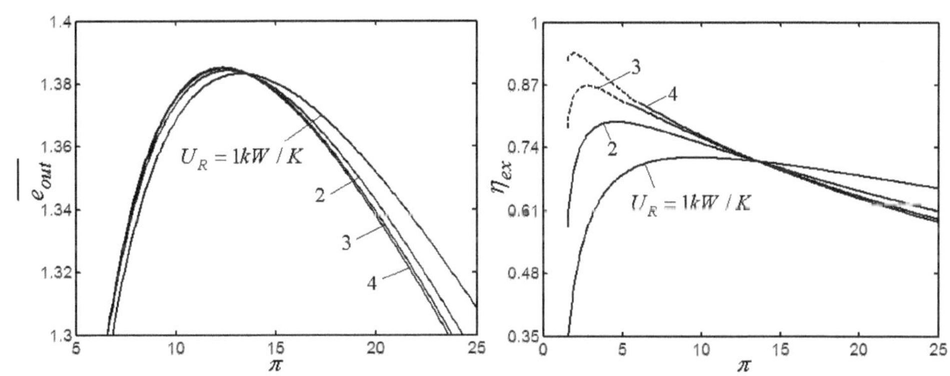

Figure 4. Effects of U_R on the characteristics of $\overline{e_{out}}$ and η_{ex} versus π

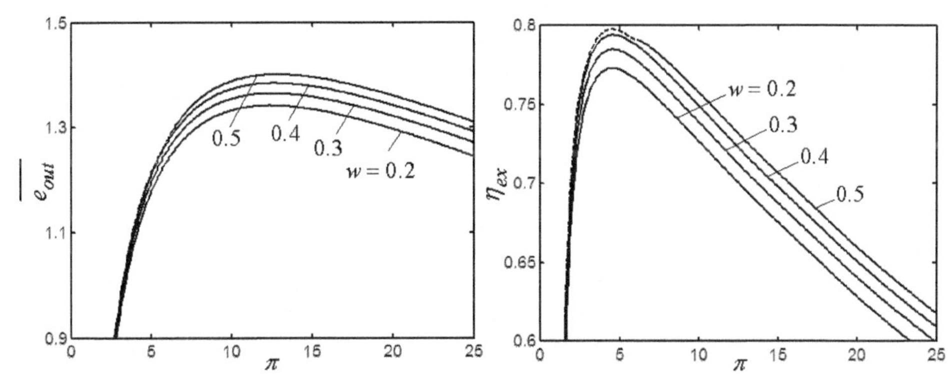

Figure 5. Effects of w on the characteristics of $\overline{e_{out}}$ and η_{ex} versus π

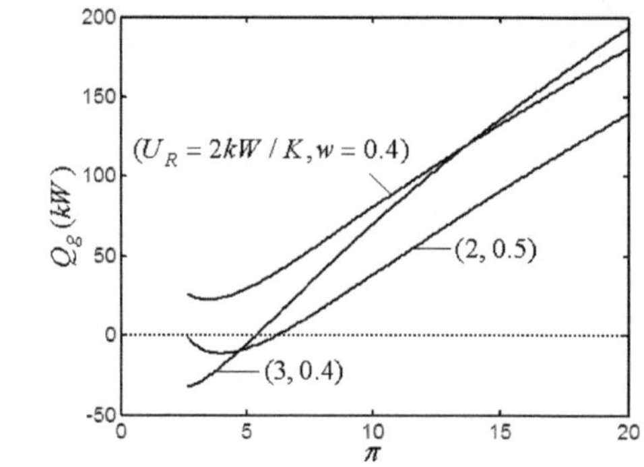

Figure 6. Relation of Q_g versus π for different U_R and w

4.2 Optimal dimensionless exergy output rate and optimal exergy efficiency

Figures 7-11 show the relations of the optimal dimensionless exergy output rate ($(\overline{e_{out}})_{opt}$) and corresponding exergy efficiency ($(\eta_{ex})_{(\overline{e_{out}})_{opt}}$), the optimal exergy efficiency ($(\eta_{ex})_{opt}$) and corresponding dimensionless exergy output rate ($(\overline{e_{out}})_{(\eta_{ex})_{opt}}$), and the two optimal pressure ratios ($\pi_{(\overline{e_{out}})_{opt}}$ and $\pi_{(\eta_{ex})_{opt}}$, which correspond to $(\overline{e_{out}})_{opt}$ and $(\eta_{ex})_{opt}$) versus τ_H, U_H, τ_e, $U_g = U_c = U_e = U_a$ and τ_K, respectively.

It can be seen from Figure 7 that $(\overline{e_{out}})_{opt}$, $(\overline{e_{out}})_{(\eta_{ex})_{opt}}$, $(\eta_{ex})_{opt}$, $(\eta_{ex})_{(\overline{e_{out}})_{opt}}$, $\pi_{(\overline{e_{out}})_{opt}}$ and $\pi_{(\eta_{ex})_{opt}}$ increase monotonically with the increase of τ_H, and one has $(\overline{e_{out}})_{opt} > (\overline{e_{out}})_{(\eta_{ex})_{opt}}$, $(\eta_{ex})_{opt} > (\eta_{ex})_{(\overline{e_{out}})_{opt}}$ and $\pi_{(\overline{e_{out}})_{opt}} > \pi_{(\eta_{ex})_{opt}}$ for the same τ_H, which indicates that the optimal design scope of the CCHP plant should be $(\overline{e_{out}})_{(\eta_{ex})_{opt}} < \overline{e_{out}} < (\overline{e_{out}})_{opt}$, $(\eta_{ex})_{(\overline{e_{out}})_{opt}} < \eta_{ex} < (\eta_{ex})_{opt}$. It also can be seen that with the increase of τ_H, the range $[\pi_{(\eta_{ex})_{opt}}, \pi_{(\overline{e_{out}})_{opt}}]$ expands.

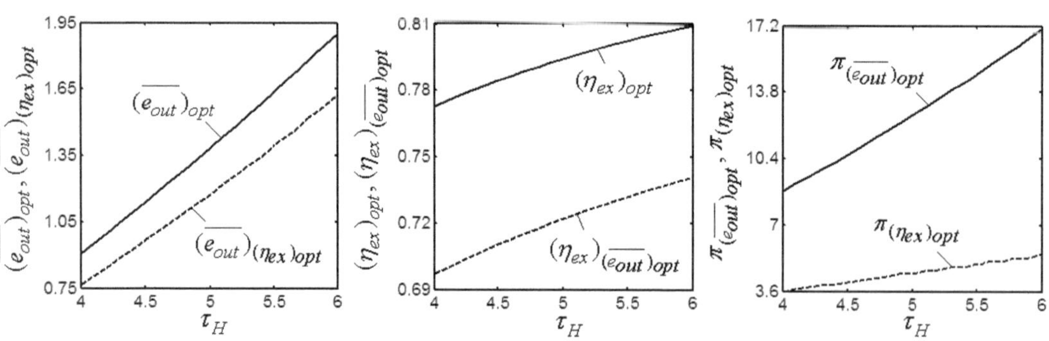

Figure 7. Relations of $(\overline{e_{out}})_{opt}$, $(\overline{e_{out}})_{(\eta_{ex})_{opt}}$, $(\eta_{ex})_{opt}$, $(\eta_{ex})_{(\overline{e_{out}})_{opt}}$, $\pi_{(\overline{e_{out}})_{opt}}$, and $\pi_{(\eta_{ex})_{opt}}$ versus τ_H

It can be seen from Figure 8 that $(\overline{e_{out}})_{opt}$, $(\overline{e_{out}})_{(\eta_{ex})_{opt}}$, $(\eta_{ex})_{opt}$, $(\eta_{ex})_{(\overline{e_{out}})_{opt}}$, $\pi_{(\overline{e_{out}})_{opt}}$ and $\pi_{(\eta_{ex})_{opt}}$ increase with the increase of U_H, and increase slowly when U_H is large. The calculations also indicate that the influences of U_L and U_K on the exergy performances of the CCHP plant are similar to U_H.

It can be seen from Figure 9 that $(\overline{e_{out}})_{opt}$, $(\overline{e_{out}})_{(\eta_{ex})_{opt}}$, $(\eta_{ex})_{opt}$, $(\eta_{ex})_{(\overline{e_{out}})_{opt}}$, $\pi_{(\overline{e_{out}})_{opt}}$ and $\pi_{(\eta_{ex})_{opt}}$ decrease nearly linearly with the increase of τ_e.

It can be seen from Figure 10 that $(\overline{e_{out}})_{opt}$, $(\overline{e_{out}})_{(\eta_{ex})_{opt}}$, $(\eta_{ex})_{opt}$, $(\eta_{ex})_{(\overline{e_{out}})_{opt}}$, $\pi_{(\overline{e_{out}})_{opt}}$ and $\pi_{(\eta_{ex})_{opt}}$ increase with the increase of $U_g = U_c = U_e = U_a$, but the variations of the numerical values are slight.

It can be seen from Figure 11 that with the increase of τ_K, $(\overline{e_{out}})_{opt}$ and $(\eta_{ex})_{(\overline{e_{out}})_{opt}}$ increase first and then decrease, i.e. there exists an optimal value of thermal consumer-side temperature which makes $(\overline{e_{out}})_{opt}$

reach a maximum dimensionless exergy output rate. For $(\eta_{ex})_{opt}$, the calculations indicate that when τ_K is increased to a certain value (about 1.3 in this example), the heat transfer rate is smaller than zero, i.e. $Q_g < 0$, thus in a meaningful design range of τ_K, $(\eta_{ex})_{opt}$ and $(\overline{e_{out}})_{(\eta_{ex})_{opt}}$ increase with the increase of τ_K. It also can be seen from Figure 11 that with the increase of τ_K, $\pi_{(\overline{e_{out}})_{opt}}$ decreases first and then increases, and the change of $\pi_{(\eta_{ex})_{opt}}$ is not obviously.

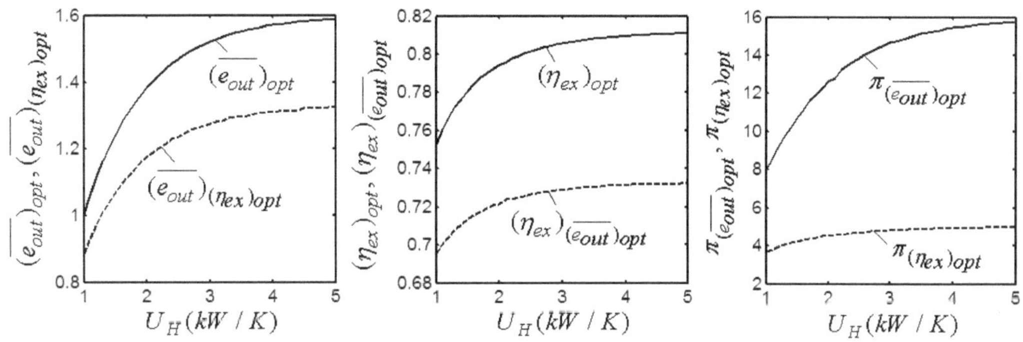

Figure 8. Relations of $(\overline{e_{out}})_{opt}$, $(\overline{e_{out}})_{(\eta_{ex})_{opt}}$, $(\eta_{ex})_{opt}$, $(\eta_{ex})_{(\overline{e_{out}})_{opt}}$, $\pi_{(\overline{e_{out}})_{opt}}$, and $\pi_{(\eta_{ex})_{opt}}$ versus U_H

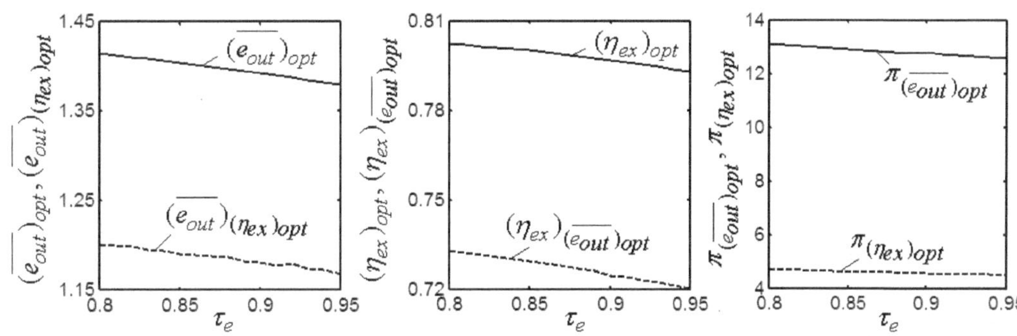

Figure 9. Relations of $(\overline{e_{out}})_{opt}$, $(\overline{e_{out}})_{(\eta_{ex})_{opt}}$, $(\eta_{ex})_{opt}$, $(\eta_{ex})_{(\overline{e_{out}})_{opt}}$, $\pi_{(\overline{e_{out}})_{opt}}$, and $\pi_{(\eta_{ex})_{opt}}$ versus τ_e

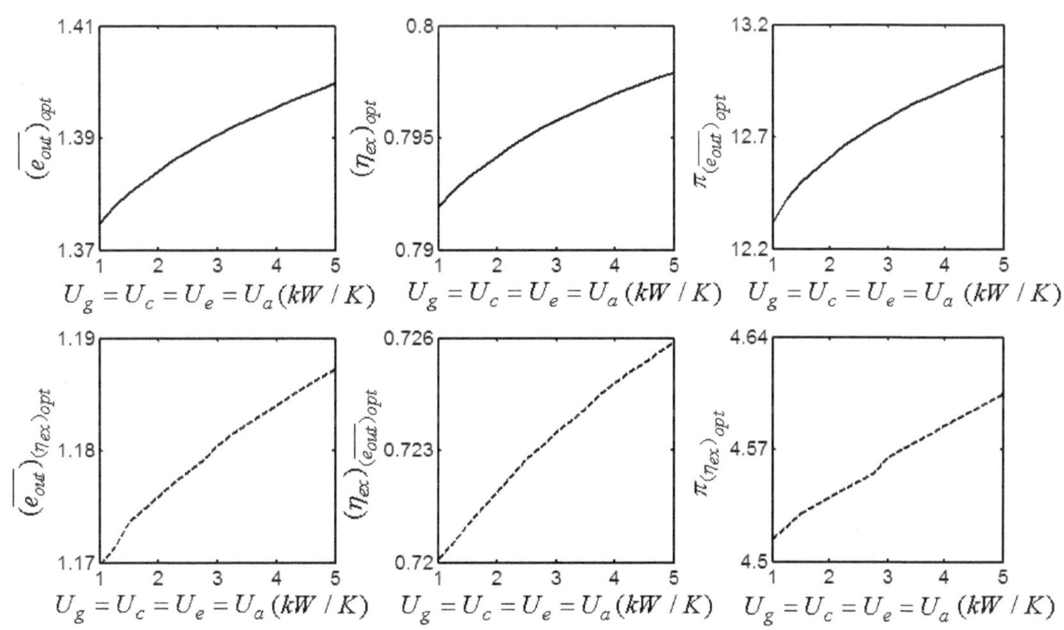

Figure 10. Relations of $(\overline{e_{out}})_{opt}$, $(\overline{e_{out}})_{(\eta_{ex})_{opt}}$, $(\eta_{ex})_{opt}$, $(\eta_{ex})_{(\overline{e_{out}})_{opt}}$, $\pi_{(\overline{e_{out}})_{opt}}$, and $\pi_{(\eta_{ex})_{opt}}$ versus $U_g = U_c = U_e = U_a$

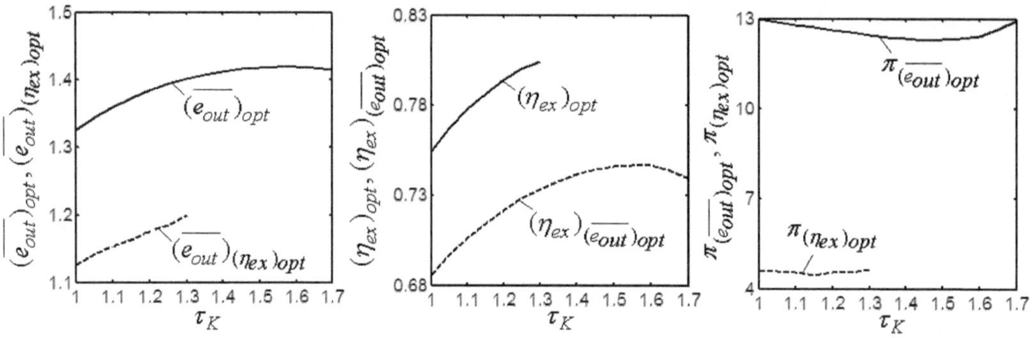

Figure 11. Relations of $(\overline{e_{out}})_{opt}$, $(\overline{e_{out}})_{(\eta_{ex})_{opt}}$, $(\eta_{ex})_{opt}$, $(\eta_{ex})_{(\overline{e_{out}})_{opt}}$, $\pi_{(\overline{e_{out}})_{opt}}$, and $\pi_{(\eta_{ex})_{opt}}$ versus τ_K

5. Conclusion

In present work, FTT is used to establish a CCHP plant model composed of an endoreversible closed regenerative Brayton cycle, an endoreversible four-heat-reservoir absorption refrigerator and a heat recovery device of thermal consumer. The exergy output rate and exergy efficiency of the plant are researched by theoretical analyses and numerical calculations, and the significant results are as follows:

(1) Both dimensionless exergy output rate and exergy efficiency have optimal values with respect to the pressure ratio of regenerative Brayton cycle. For regeneration, a critical pressure ratio exists, when pressure ratio is smaller than the critical pressure ratio, dimensionless exergy output rate and exergy efficiency increase with the increase of heat conductance of regenerator, and when pressure ratio is larger than the critical pressure ratio, dimensionless exergy output rate and exergy efficiency decrease with the increase of heat conductance of regenerator.

(2) The larger the ratio of heat demanded by the thermal consumer to power output of the plant, the better the exergy performances. But when the heat to power output ratio or the heat conductance of regenerator is too large and the pressure ratio is smaller than certain value, the CCHP plant will become a CHP plant.

(3) The appropriate design scope of the CCHP plant is determined by four paramters (optimal dimensionless exergy output rate and corresponding exergy efficiency, as well as optimal exergy efficiency and corresponding dimensionless exergy output rate). The optimal exergy performances can be further improved by increasing the ratio of hot-side heat reservoir temperature to environment temperature, the heat conductances of the hot-, cold- and thermal consumer-side heat exchangers, and decreasing the ratio of evaporator heat reservoir temperature to environment temperature. The influences of the heat conductances of the generator-, condenser-, evaporator- and absorber-side heat exchanger on the exergy performances are slight.

(4) The optimal dimensionless exergy output rate has a maximum with respect to the thermal consumer temperature, and in a meaningful design range, exergy efficiency increases with the increase of thermal consumer temperature.

The investigation in this paper may provide some guidelines for the optimal design and parameters selection of practical Brayton cycle CCHP plant.

Acknowledgments

This paper is supported by the National Key Basic Research and Development Program of China (973) (Project No. 2012CB720405) and The National Natural Science Foundation of P. R. China (Project No. 10905093).

Nomenclature

C	heat capacity rate (kW / K)	τ_L	ratio of cold-side heat reservoir temperature to environment temperature
E	effectiveness of the heat exchanger	τ_K	ratio of thermal consumer-side temperature to environment temperature
e	exergy flow rate (kW)	***Subscripts***	
k	ratio of the specific heats	a	absorber
N	number of heat transfer units	c	condenser
n	distribution ratio of heat rejection between absorber and condenser	e	evaporator

P	power output of the CCHP plant (kW)	ex	exergy
Q	rate of heat transfer (kW)	g	generator
R	cooling load of the absorption refrigerator (kW)	H	hot-side
T	temperature (K)	in	input
U	heat conductance (kW / K)	K	thermal consumer-side
w	ratio of heat demanded by the thermal consumer to power output	L	cold-side
y	isentropic temperature ratio for compression process or expansion process	opt	optimal
Greek symbols		out	output
η	efficiency	R	regenerator
π	pressure ratio	wf	working fluid
σ	entropy generation rate of the CCHP plant (kW / K)	0	ambient
τ_e	ratio of evaporator heat reservoir temperature to environment temperature	$1,2,3,4,5,6,7,8$	state points of the cycle
τ_H	ratio of hot-side heat reservoir temperature to environment temperature	$-$	dimensionless

References

[1] Ertesvag I S. Exergetic comparison of efficiency indicators for combined heat and power (CHP). Energy, 2007, 32(11): 2038-2050.

[2] Ferdelji N, Galovic A , Guzovic Z. Exergy analysis of a co-generation plant. Therm. Sci., 2008, 12(4): 75-88.

[3] Sanaye S, Ziabasharhagh M and Ghazinejad M. Optimal design of gas turbine CHP plant with preheater and HRSG. Int. J. Energy Res., 2009, 39(8): 766-777.

[4] Khaliq A, Dincer I. Energetic and exergetic performance analyses of a combined heat and power plant with absorption inlet cooling and evaporative aftercooling. Energy, 2011, 36(5): 2662-2670.

[5] Temir G, Bilge D. Thermoeconomic analysis of a trigeneration system. Appl. Therm. Eng., 2004, 24(17/18): 2689-2699.

[6] Mago P J, Chamra L M. Analysis and optimization of CCHP systems based on energy, economical, and environmental considerations. Energy & Buildings, 2009, 41(10): 1099-1106.

[7] Khaliq A. Exergy analysis of gas turbine trigeneration system for combined production of power heat and refrigeration. Int. J. Refri., 2009, 32(3): 534-545.

[8] Kavvadias K C, Maroulis Z B. Multi-objective optimization of a trigeneration plant. Energy Policy, 2010, 38(2): 945-954.

[9] Bejan A. Entropy generation minimization: The new thermodynamics of finite-size devices and finite-time process. J. Appl. Phys., 1996, 79(3): 1191-1218.

[10] Hoffmann K H, Burzler J M, Schubert S. Endoreversible thermodynamics. J. Non-Equilib. Thermodyn., 1997, 22(4): 311-355.

[11] Berry R S, Kazakov V A, Sieniutycz S, Szwast Z, Tsirlin A M. Thermodynamic Optimization of Finite Time Processes. Chichester: Wiley, 1999.

[12] Chen L, Wu C, Sun F. Finite time thermodynamic optimization of entropy generation minimization of energy systems. J. Non-Equilib. Thermodyn., 1999, 24(4): 327-359.

[13] Chen L, Sun F. Advances in Finite Time Thermodynamics: Analysis and Optimization. New York: Nova Science Publishers, 2004.

[14] De Vos A. Thermodynamics of Solar Energy Conversion. Berlin: Wiley-Vch, 2008.

[15] Sieniutycz S, Jezowski J. Energy Optimization in Process Systems. Elsevier, Oxford, UK, 2009.

[16] Andresen B. Current trends in finite-time thermodynamics. Angew. Chem. Int. Ed., 2011, 50(12): 2690-2704.

[17] Feidt M. Thermodynamics of energy systems and processes: A review and perspectives. J. Appl. Fluid Mechanics, 2012, 5(2): 85-98.

[18] Yilmaz T. Performance optimization of a gas turbine-based cogeneration system. J. Phys. D: Appl. Phys., 2006, 39(11): 2454-2458.

[19] Hao X, Zhang G. Maximum useful energy-rate analysis of an endoreversible Joule-Brayton cogeneration cycle. Appl. Energy, 2007, 84(11): 1092-1101.

[20] Hao X, Zhang G. Exergy optimisation of a Brayton cycle-based cogeneration plant. Int. J. Exergy, 2009, 6(1): 34-48.

[21] Ust Y, Sahin B, Yilmaz T. Optimization of a regenerative gas-turbine cogeneration system based on a new exergetic performance criterion: exergetic performance coefficient. Proc. IMechE, Part A: J. Power Energy, 2007, 221(4): 447-458.

[22] Chen L, Sun F, Chen W. Finite time exergoeconomic performance bound and optimization criteria for two-heat-reservoir refrigerators. Chin. Sci. Bull., 1991, 36(2): 156-157 (in Chinese).

[23] Wu C, Chen L, Sun F. Effect of heat transfer law on finite time exergoeconomic performance of heat engines. Energy, 1996, 21(12): 1127-1134.

[24] Chen L, Sun F, Wu C. Exergoeconomic performance bound and optimization criteria for heat engines. Int. J. Ambient Energy, 1997, 18(4): 216-218.

[25] Wu C, Chen L, Sun F. Effect of heat transfer law on finite time exergoeconomic performance of a Carnot heat pump. Energy Convers. Mgmt., 1998, 39(7): 579-588.

[26] Chen L, Wu C, Sun F. Effect of heat transfer law on finite time exergoeconomic performance of a Carnot refrigerator. Int. J. Exergy, 2001, 1(4): 295-302.

[27] Tao G, Chen L, Sun F, Wu C. Exergoeconomic performance optimization for an endoreversible simple gas turbine closed-cycle cogeneration plant. Int. J. Ambient Energy, 2009, 30(3): 115-124.

[28] Tao G, Chen L, Sun F. Exergoeconomic performance optimization for an endoreversible regenerative gas turbine closed-cycle cogeneration plant. Rev. Mex. Fis., 2009, 55(3): 192-200.

[29] Chen L, Tao G, Sun F. Finite time exergoeconomic optimal performance for an irreversible gas turbine closed-cycle cogeneration plant. Int. J. Sustainable Energy, 2012, 31(1): 43-58.

[30] Chen L, Yang B, Sun F. Exergoeconomic performance optimization of an endoreversible intercooled regenerated Brayton cogeneration plant. Part 1: thermodynamic model and parameter analyses. Int. J. Energy & Environment, 2011, 2(2): 199-210.

[31] Yang B, Chen L, Sun F. Exergoeconomic performance optimization of an endoreversible intercooled regenerated Brayton cogeneration plant. Part 2: heat conductance allocation and pressure ratio optimization. Int. J. Energy & Environment, 2011, 2(2): 211-218.

[32] Yang B, Chen L, Sun F. Exergoeconomic performance analyses of an endreversible intercooled regenerative Brayton cogeneration type model. Int. J. Sustainable Energy, 2011, 30(2): 65-81.

[33] Yang B, Chen L, Sun F. Exergoeconomic performance optimization of an endoreversible intercooled regenerative Brayton combined heat and power plant coupled to variable- temperature heat reservoirs. Int. J. Energy & Environment, 2012, 3(4): 505-520.

[34] Yang B, Chen L, Sun F. Finite time exergoeconomic performance of an irreversible intercooled regenerative Brayton cogeneration plant. J. Energy Inst., 2011, 84(1): 5-12.

[35] Chen L, Yang B, Sun F, Wu C. Exergetic performance optimisation of an endoreversible intercooled regenerated Brayton cogeneration plant. Part 1: thermodynamic model and parametric analysis. Int. J. Ambient Energy, 2011, 32(3): 116-123.

[36] Yang B, Chen L, Sun F, Wu C. Exergetic performance optimisation of an endoreversible intercooled regenerated Brayton cogeneration plant. Part 2: exergy output rate and exergy efficiency optimisation. Int. J. Ambient Energy, 2012, 33(2): 98-104.

[37] Chen J. The optimum performance characteristics of a four-temperature-lever irreversible absorption refrigerator at maximum specific cooling load. J. Phys. D: Appl. Phys., 1999, 32(24): 3085-3091.

[38] Chen L, Zheng T, Sun F, Wu C. Optimal cooling load and COP relationship of a four-heat-reservoir endoreversible absorption refrigerator cycle. Entropy, 2004, 6(3): 316-326.

[39] Chen L, Zheng T, Sun F, Wu C. Performance limits of real absorption refrigerators. J. Energy Inst., 2005, 78(3): 139-144.

[40] Chen L, Zheng T, Sun F, Wu C. Irreversible four-temperature-level absorption refrigerator. Sol. Energy, 2006, 80(3): 347-360.

[41] Zheng T, Chen L, Sun F, Wu C. Performance of a four-heat-reservoir absorption refrigerator with heat resistance and heat leak. Int. J. Ambient Energy, 2003, 24(3): 157-168.

[42] Zheng T, Chen L, Sun F, Wu C. Performance optimization of an irreversible four-heat- reservoir absorption refrigerator. Appl. Energy, 2003, 76(4): 391-414.

[43] Zheng T, Chen L, Sun F, Wu C. The influence of heat resistance and heat leak on the performance of a four-heat-reservoir absorption refrigerator with heat transfer law of $Q \propto \Delta(T^{-1})$, Int. J. Therm. Sci., 2004, 43(12): 1187-1195.

[44] Qin X, Chen L, Sun F Wu C. Thermoeconomic optimization of an endoreversible four-heat-reservoir absorption-refrigerator. Appl. Energy, 2005, 81(4): 420-433.

[45] Qin X, Chen L, Sun F. Thermodynamic modeling and performance of variable-temperature heat reservoir absorption refrigeration cycle. Int. J. Exergy, 2010, 7(4): 521-534.

[46] Tao G, Chen L, Sun F, Wu C. Optimization between cooling load and entropy-generation rate of an endoreversible four-heat-reservoir absorption refrigerator. Int. J. Ambient Energy, 2009, 30(1): 27-32.

[47] Chen L, Feng H, Sun F. Exergoeconomic performance optimization for a combined cooling, heating and power generation plant with an endoreversible closed Brayton cycle. Math. Compu. Model., 2011, 54(11-12): 2785-2801.

[48] Feng H, Chen L, Sun F. Exergoeconomic optimal performance of an irreversible closed Brayton cycle combined cooling, heating and power plant. Appl. Math. Model., 2011, 35(9): 4661-4673.

[49] Wu C, Chen L, Sun F. Performance of a regenerating Brayton heat engines. Energy, 1996, 21(2): 71-76.

[50] Chen L, Sun F, Wu C. Theoretical analysis of the performance of a regenerated closed Brayton cycle with internal irreversibilities. Energy Convers. Manage., 1997, 38(9): 871-877.

[51] Chen L, Sun F, Wu C. Optimum heat conductance distribution for power optimization of a regenerated closed Brayton cycle. Int. J. Green Energy, 2005, 2(3): 243-258.

Costing energy efficiency improvements in buildings Case study: Braşov, Romania

Elena Eftimie

Department of Product Design, Mechatronics and Environment, Transilvania University of Braşov, Eroilor 29, 500036, Romania.

Abstract

One of the methods of buildings' energetic streamline consists of reducing the thermal energy needs (i.e. the building heating/cooling demand) at the level of building. In this regard, this study provides the opportunity of performing a comparative analysis between the values of energy demand for space heating/cooling, based on a case study in which for a building have been modified, at a time, the insulation material of exterior walls, the thermopane windows and the roof insulation. To evaluate the energy consumption in buildings, it is proposed an advanced hourly calculation method using simulations with TRNSYS program, in order to obtain values as close to reality of the energy demand for their space heating and cooling. It is envisaged that the use of building performance simulation programs allow the modelling and computer simulation of building performance in order to obtain a solution that to approximate to a large extent an actual case. Also it should be noted that the estimation and the analysis of the building energy behaviour – still from the design phase or prior to its rehabilitation – is more efficient and economical than solving problems in the use phase of the building.

Keywords: Building performance; Indoor thermal comfort; Low-energy buildings; Space heating/cooling demand; TRNSYS.

1. Introduction

Globally, about 40% of the total energy consumption is represented by the thermal energy demand of buildings. In addition, the construction sector is growing, which will lead to the increase of energy consumption [1, 2]. The residential buildings and the trade ones (offices, commercial areas, hotels, restaurants, schools, hospitals, gyms, indoor swimming pools) are the largest final consumers of energy, particularly for heating, lighting, home appliances and equipment [1].

At present, the care for global energy depletion makes from the increasing of building energy efficiency a necessary economical standard; so besides the aesthetic factors underlying the construction of a building, it needs to be also designed from the point of view of energy efficiency.

A major objective of low energy buildings is to minimize the amount of external energy purchased – providing indoor thermal comfort of occupants – regardless of the season and outdoor climatic conditions [3]. Low energy buildings usually use a high level of insulation and energy efficient windows to reduce heating and cooling demand, and obtaining of high energy efficiency.

The energy efficiency increase of buildings consists of a set of methods and techniques that consider both the buildings as a whole, as well as that centres of energy exchange with the environment.

Increasing the energy performance of a building can be achieved by different methods such as:

- at the building level by creating the indoor comfort conditions, respectively a good insulation of walls and the use of windows with a high degree of thermal protection;
- increasing the performance of heating systems;
- increasing the performance of air conditioning systems and those concerning the electrical installations.

This paper aims to analyze some of the most effective methods to improve the energy performance of a building at construction level, namely:

- the thermal insulation of exterior walls, a method by which, once with the building envelopment it is also provided an increase of their lifetime;
- the use of thermopane windows, preferably modern windows with triple-pane insulation glass (glazed windows with double effect), that to assure the maintaining of heat indoor during the winter, but also to prevent its excessive influx during summer;
- the thermal insulation of roofs with lightweight materials that do not load the building but ensure its higher lifetime.

The objective of the study is to quantify by energy simulation, which are the values of the energy demand for space heating and cooling, thus highlighting the differences between the energy performances of a building in various embodiments; in this regard it will be achieved a comparative analysis of energy demand values obtained by the use of three different types of insulation materials, by replacing windows and by replacing the roof insulation. The exemplification will be performed by a case study for a multi-zone building located in urban area of Braşov.

2. Materials and methods used

2.1 Possibilities to increase the energy efficiency of building

The best solution and the one with the best efficiency of heat gain and of heat carrier saving is the insulation of whole building, both of the roof as well as of the basement, by which large amounts of heat are lost.

The highest losses of a building are found in the field of thermal energy. For this reason, there are required a number of additional measures that take into account the following aspects:

- the building thermal envelope must ensure the comfortable indoor climate with low energy consumption, regardless of the season (both in warm seasons as well in the cold ones) [4-6];
- the windows must have a coefficient of thermal loss as low as possible and the highest solar gain, for saving more energy;
- the proper insulation of the roof especially for buildings with a few floors.

For a detailed study of the energetic behaviour of buildings, the constructive properties and also the materials for walls, ceilings, floors, windows and roof must be known.

2.1.1 Exterior walls insulation

As first method of rehabilitation of a building, the exterior walls insulation is considered. It is envisaged that the thickness and quality of envelope have a significant influence on the amount of energy that is lost due to excessive transfer between the inside and the outside thereof.

The insulation defects have as effects the heat losses during the winter (these causing the condensation on inside walls) and the excessive power consumption of air-conditioning equipments in the summer.

One of the easiest methods to maintain the indoor thermal comfort of a building consists of the thermal insulation that will reduce the costs for thermal energy [4].

The most commonly used insulation materials are the expanded polystyrene (EPS) and extruded polystyrene (XPS).

To determine the most effective choice of polystyrene type, the following aspects must be considered:

- Expanded polystyrene (according to EN 13163 [7])
  - ✓ it can be used successfully in buildings located in areas with high humidity; the fungi, bacteria or mould do not affect it.
  - ✓ the vapour permeability of the material, if it is mounted on the outside, does not favour the "lock" of moisture between the polystyrene plate and the wall, thus the mould does not appears.
  - ✓ in the last years, while the price for the most construction materials has fluctuated much, the cost of expanded polystyrene remained constant.

- Extruded Polystyrene (in accordance with EN 13164[8])
  - ✓ it does not allow the vapour crossing, the humidity remains between the wall and the insulation material and thus the condensation occurs;
  - ✓ reliability and high resistance to the destructive effects of nature;
  - ✓ high resistance to heat transfer when the temperature drops;
  - ✓ higher strength compared to of expanded polystyrene to chemical agents such as acids, alkalis, alcohol and alcohol-based dyes, salt water, cement, asphalt etc.;
  - ✓ an extruded polystyrene plate exposed to sunlight, even if it changes its colour, it will not change significantly its thermal insulation values.

Regardless of the used polystyrene type, its thickness influences the heat loss. The recommended minimum thickness for insulation of facades is by 10cm; the thickness increase of polystyrene makes that the investment to be more profitable in the long term [9].

Still the polystyrene – that does not allow the air crossing from the outside to the inside to save energy – can represent a significant disadvantage; in a hermetically sealed building, energy is saved, but over time the construction is not protected due to the mould occurrence and condensation.

In a building must exist transfer between the air from the inside – that has already been used – and the air from the outside, but the polystyrene has not the property of being a good air conductive that to let building "to breathe".

In these situations, there can be used new alternative materials, more efficient and even cheaper. In this category are included, facade systems that includes insulation made of polyurethane.

- Polyurethane thermal insulation (in accordance with EN 13165 [10])
  - ✓ its heat transfer coefficient has a value of about 0.020 W/mK, compared to expanded polystyrene that has a value of about 0.036 W/mK; (these values for both materials may vary depending on the density of the material and the manufacturer);
  - ✓ it is resistant to damages caused by chemical substances; expanded polystyrene is sensitive to petroleum-based solvents such as gasoline, several insect sprays and ordinary adhesives;
  - ✓ it has fireproof properties, it does not burn and does not sustain combustion;
  - ✓ it can be applied without interruption, eliminating the thermal bridges between the panels, on the entire surface of building, regardless its size and form;
  - ✓ the properties of polyurethane foam in terms of soundproofing are far superior to those of expanded polystyrene;
  - ✓ it has waterproofing properties that makes it from this point of view to be preferable compared to polystyrene that can absorb and retain water, which can lead to increasing of the structure weight on what this was mounted (in these situations, the detachment of insulation material may result).

2.1.2 New modern windows with triple-pane insulation glass (low-E)

The exterior windows are part of the building envelope so that in a rehabilitation process the characteristics of windows are important.

The energy efficient glazed windows reduce the thermal energy consumption.

In the field of high quality windows, currently there were developed a number of modern technologies that allow reducing costs for space heating. The current trend in this field is directed to windows with triple-pane insulation glass [11]; these windows have double effect, respectively of maintaining the heat indoors in winter but do not allow excessive influx from outside in summer.

2.1.3 Roof insulation

The thermal protection improvement of the roof represents an effective measure that can be applied to existing buildings, in view of the rehabilitation and their thermal energy modernization.

The proper insulation of roof prevents heat loss, respectively energy, that occur at the roof level; therefore its efficient thermal insulation is essential to ensure indoor thermal comfort of the building.

Providing an additional insulation layer for this construction element does not require major investments, it is relatively simple to perform, and the investment recovery time is reduced.

A substantial increase in the thermal resistance of the roof is much more effective and appropriate if the number of floors is more reduced.

2.2 Computational methods

The energy performance simulation of building proposed to study was performed using transient analysis software, TRNSYS (TRansient System Simulation) [12]. This software is dynamic simulation program that benefits by a modular structure that makes possible its supplement with mathematical models.

The energy simulation was performed using the weather data recorded by a local weather station (Braşov urban area) by implementing them in TRNSYS subroutines; it was considered the fact that to achieve an energy calculation as accurately is important to have accurate weather data (solar radiation data, ambient temperature, relative humidity, wind speed and direction).

Implementing the building model consisted of the following steps:

- the defining of thermal zones and their characteristics;
- the detailed specification of envelope elements for building, the optical properties of windows, the working programme of the equipment;
- defining of the orientation for building and for glazed surfaces;
- the specification of infiltration due to leaks and the type of air conditioning;
- the specification of heating and cooling regimes (temperatures during the day and the night, supplied heating power);
- specifying the internal gains distributed in the three components (persons, artificial lighting, electrical devices);
- the detailed description of shading type.

3. Results and discussions

3.1 Case study

Energy calculation is applied for an office building of Transilvania University of Braşov; the building has two floors with a built area of 260m^2. The North and South oriented exterior walls of the second floor are formed mostly from windows (Figure 1).

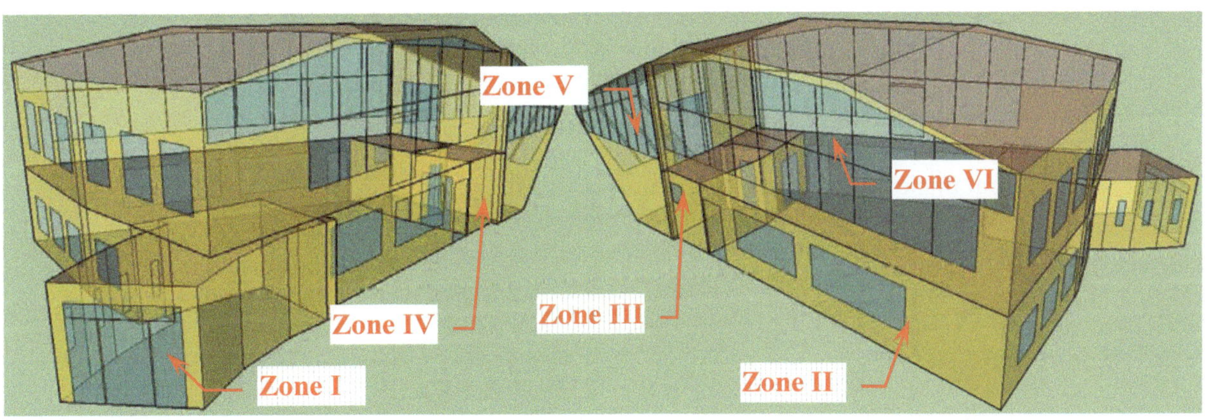

Figure 1. Sample of figure for the international journal of energy and environment

To define in detail the characteristics and the thermal behaviour of proposed building this was divided into 6 thermal zones (Zone I: Entrance Hall, Zone II: Office (First Floor), Zone III: Bathroom, Zone IV: Small Lobby, Zone V: Staircase; Zone VI: Second Floor), for each zone being possible to define a different thermal regime.

The two offices are located each on one floor having each 10 occupants.

The occupants schedule was considered from 8.00 until 20.00 during the weekdays, from Monday to Friday.

The control strategy for electronics apparatus from offices (PC, laptop, printer, and photocopier) was defined according to the same schedule.

The artificial lighting scenario is the same as that for occupants activity, lighting being available from 8.00 to 20.00, during the days from Monday to Friday.

For setting of the heating requirements, the air temperature of the zone was set to 21°C during workday and for the periods of night, Saturday and Sunday, the air temperature was set to 19°C.

Regarding the cooling control, for the set-point temperature above which the cooling is active, it was considered a constant value of 25°C.

The air flow from outside into the building zones was specified by defining the type of infiltrations [12, 13]. The air change rate of the infiltration was considered as having a value of 0.5 1/h for the following thermal zones: Entrance Hall, Bathroom, Small Lobby, First Floor, respectively 0.3 1/h for the Second Floor and Staircase (it is considered that airing takes place only during work schedule).
TRNSYS simulations were performed for specific climatic conditions of Braşov urban area (Romania); thus the meteorological database implemented in TRNSYS subroutines contains the data monitored using a local Weather Station (Delta T) placed near the building subjected to analysis.
The geographic coordinates of Braşov are 23.1° East longitude and 45.5° North latitude, this area being characterized by a climate profile of continental temperate type.
The proposed simulations were performed for the same configuration of building, during the same time period, but this being affected by a series of successive modifications. Therefore, the simulated model is a transformation of an existing reference building into a low energy building.
Thus, Table 1 shows the building variants for which the simulations were made respectively using three types of insulation materials (and different thicknesses) for exterior walls, different types of windows (double-pane insulation glass (low-E) and triple-pane insulation glass (low-E)) and two different types of thermal insulation of the roof.

Table 1. Building variants for which the energy simulations were carried out

Variant	Characteristics	Energy demand for space heating (kWh/m^2 / year)
Reference building v_0	- building without insulation of the exterior walls; u-value for exterior walls, 0.908 W/(m^{2}K); - exterior windows with a standard spacer and a standard double pane low-e glazing (u=1.27 W/(m^{2}K); g=0.591); - roof insulated with mineral wool of 15cm (according to EN 13162 [14]), u-value for roof, 0.299 W/(m^{2}K); [12]	87.87
v_1	- reference building + extruded polystyrene insulation (XPS), 10 cm thickness, u-value for exterior walls, 0.247 W/(m^{2}K);	65.09
v_2	- reference building + expanded polystyrene insulation (EPS), 10 cm thickness, u-value for exterior walls, 0.278 W/(m^{2}K);	66.29
v_3	- reference building + expanded polystyrene insulation (EPS), 20 cm thickness, u-value for exterior walls, 0.164 W/(m^{2}K);	62.23
v_4	- reference building + polyurethane insulation, 10cm thickness, u-value for exterior walls, 0.213 W/(m^{2}K);	63.98
v_5	- reference building + polyurethane insulation, 15cm thickness, u-value for exterior walls, 0.154 W/(m^{2}K);	61.79
v_6	- reference building + polyurethane insulation, 15cm thickness, u-value for exterior walls, 0.154 W/(m^{2}K); - new exterior windows with triple-pane low-e glazing (u=0.4 W/(m^{2}K); g=0.408);	48.42
v_7	- reference building + polyurethane insulation, 15cm thickness, u-value for exterior walls, 0.154 W/(m^{2}K); - new exterior windows with triple-pane low-e glazing (u=0.4 W/(m^{2}K); g=0.408); - new roof (polyurethane insulation of 20cm), u-value for roof, 0.132 W/(m^{2}K);	42.79

3.2 Energetic simulation of building - Space heating demand

3.2.1 The exterior walls insulation

The correct insulation of a building leads to reducing of energy consumption for heating space, in the same time ensuring a constant temperature indoor and improving also its sustainability.
Considering the building without thermal insulation of exterior walls and with their thermal insulation with extruded polystyrene of 10cm (v_1 versus v_0, Figure 2) it can be noticed a considerable decrease of

heating demand (annually this decrease is about 30%) and an increasing of cooling demand (annually with about16%).

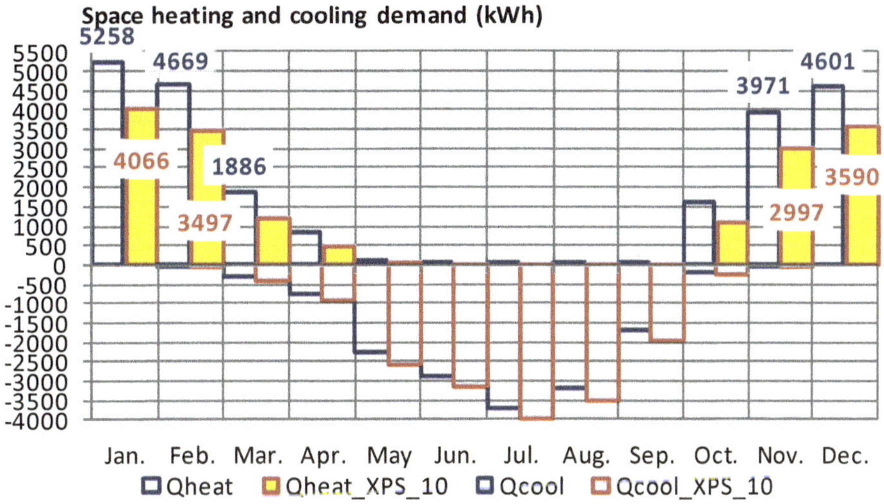

(a)

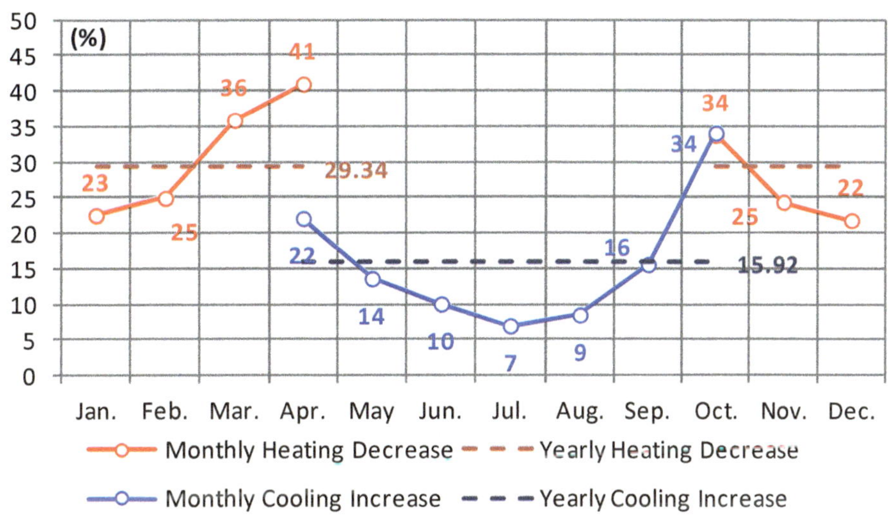

(b)

Figure 2. Exterior walls insulation (v_0 versus v_1): (a) Monthly demand of space heating and cooling; (b) Monthly percentage differences between variants

However it should be taken into account that in order to ensure a certain degree of indoor thermal comfort, there are needed both the avoidance of condensation on the inside surfaces of the walls as well the discomfort avoidance; the lower the difference between the indoor air temperature and the inside surface temperature of the wall is, the lower the discomfort of cold radiation is.

Considering this aspect in Figure 3 is represented the monthly variation of the differences between indoor air temperature and the inside surface temperature of the North wall for the Second Floor.

The exterior walls insulation with extruded polystyrene leads to a significant decrease of the difference between the indoor air temperature and the North wall temperature, during the months, January to March and October to December (a monthly average of 1.8°C); however, during the months of April to September there is an increase of this difference, respectively the inside surface temperature of wall is higher than indoor air temperature (an average increase of 0.9°C).

A thermal insulation material is characterized by the thermal conductivity that measures the ability of a material to transmit thermal energy; therefore, the lower this coefficient is the better insulation is obtained.

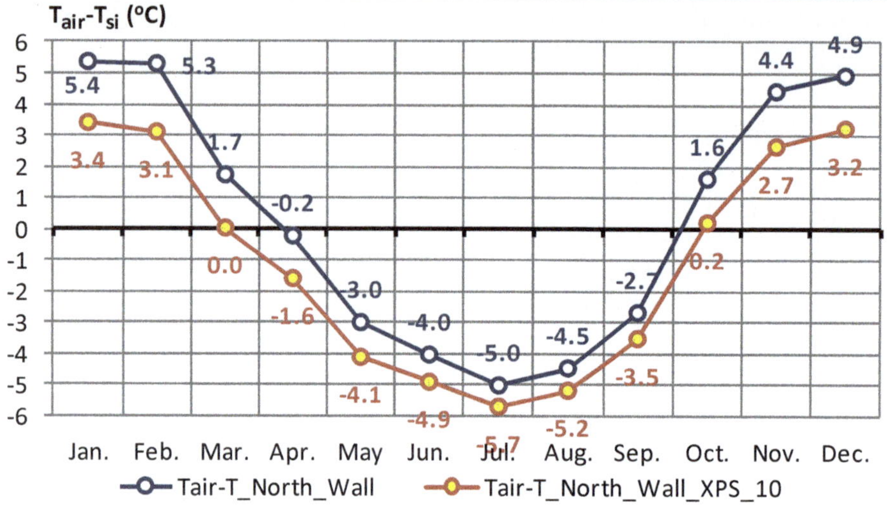

Figure 3. Differences between the monthly averages of indoor air temperature and monthly averages of inside surface temperature of the North wall (Second Floor)

The comparison of different insulation materials is achieved based on their thermal conductivity values, but there are not available concrete comparative analyzes in terms of energy demand for space heating / cooling. Although it can be said that: the polyurethane foam is better in terms of thermal conductivity, with 20.7% compared to extruded polystyrene and with 41.5% compared to expanded polystyrene (Table 2, [4]), these percentages do not reflect themselves in a similarly manner on the differences between thermal demands for building heating / cooling.

Therefore at the insulation material selection is recommended the comparative analysis of thermal demand for space heating / cooling, the values obtained for different types and thicknesses of insulation materials. In this way it can also takes account of factors such as the complete structure of the building and climatic conditions of the geographical area.

Thus for the considered building, the energy simulations were achieved for three types of insulation materials (all of 10cm thick), Figure 4.

Table 2. Insulation materials used in the building energy simulation [12]

Insulation	Thermal conductivity (W/mK)	Density (kg/m^3)
Extruded Polystyrene Average Density (XPS)	0.032	35
Expanded Polystyrene Average Density (EPS)	0.0375	25
Polyurethane Average Density	0.0265	40

The carried analysis led to the conclusion that the polyurethane foam provides the best insulation during the cold period but its use leads to an increased demand for the space cooling. The expanded polystyrene (EPS) provides the worst insulation during the cold period and the extruded polystyrene (XPS) leads to the lowest values of the space cooling demand.

However, among the annual obtained values (both for heating and cooling demand) small differences were recorded; thus,

- the percentage increase of heating demand compared to polyurethane foam,
  - when using the expanded polystyrene is 3.6% and,
  - when using the extruded polystyrene 1.73%,
- the percentage increase of cooling demand compared to extruded polystyrene,
  - when using the expanded polystyrene is 1.78%, and,
  - when using polyurethane foam of 3.0%.

One of the factors that make the difference between a good insulation and an inefficient one is the thickness of insulation material. In this respect Figure 5 shows the influence of insulation material thickness, respectively for expanded polystyrene (10cm and 20cm thick) and polyurethane foam (10cm and 15cm).

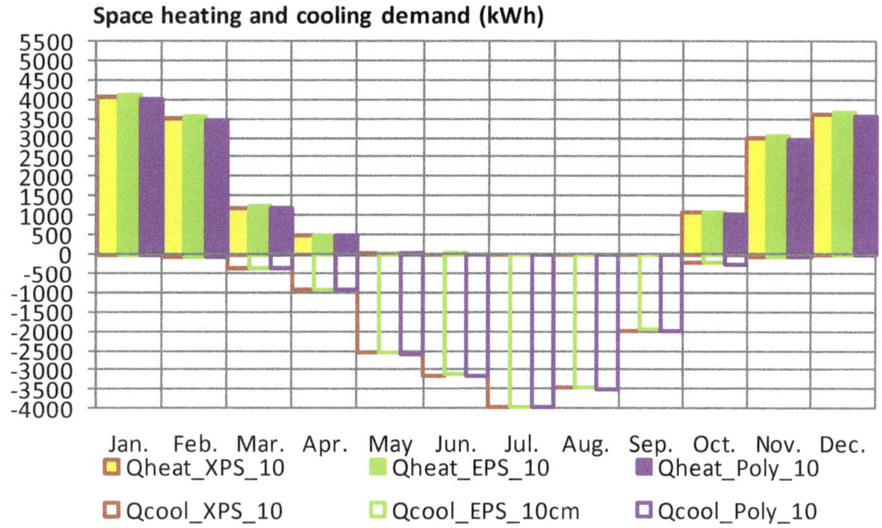

Figure 4. Insulation of exterior walls with different insulation materials with thickness of 10cm (the comparison of variants v_1, v_2, v_4)

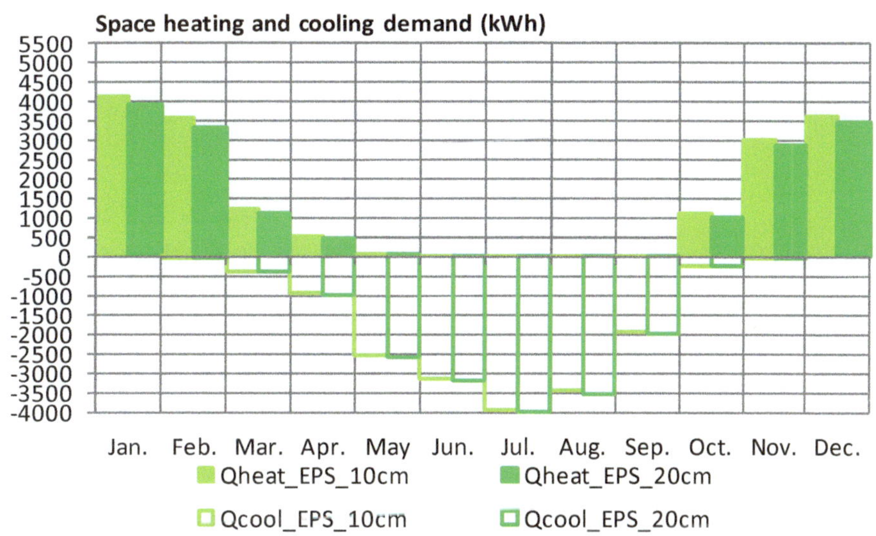

(a)

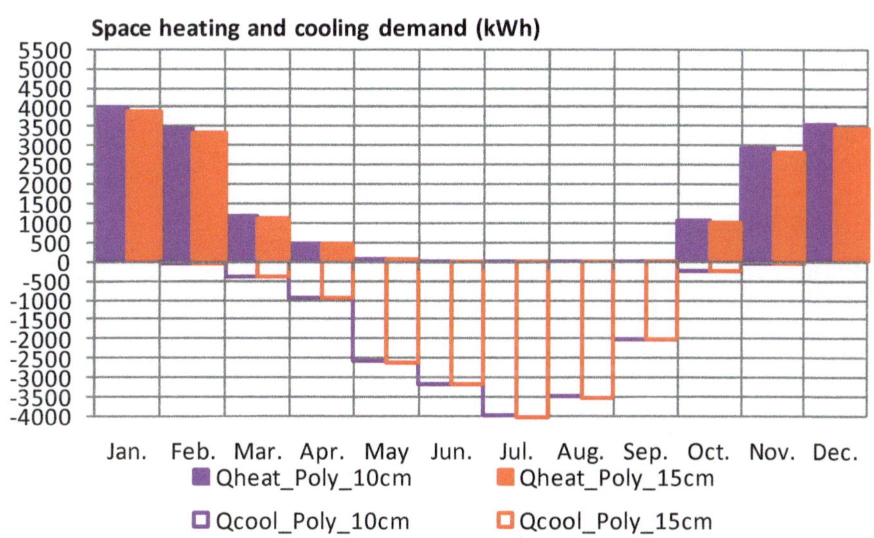

(b)

Figure 5. Insulation of exterior walls using different thicknesses of insulation material (v_2 versus v_3 and v_4 versus v_5): (a) expanded polystyrene; (b) polyurethane foam

In the case of expanded polystyrene use, doubling the thickness of the insulation layer leads to a decrease in the space heating demand with 6.1%, but to an increase in the cooling demand with 2.1% (Figure 5,(a)). However it is noted that an increasing of the polyurethane foam layer from 10cm to 15cm leads to a decrease of space heating demand with 3.4% and to a ***decrease*** of cooling demand with 1.3% (Figure 5,(b)).

In view of the above conclusion, for the building being analyzed it is recommended the use of polyurethane foam with 15cm thick as insulation material (the following simulations were made for variants of the building for what the thermal insulation of exterior walls was made with polyurethane foam of 15cm).

3.2.2 New modern windows with triple-pane insulation glass (low-E)

The thermal losses for windows are characterized by the value of heat transfer coefficient U; this value is inversely proportional to the thermal resistance, a low value for U leading to a better energy efficiency of the window [11, 13].

As a measure for the possible solar heat gain, the glazing g-value must be as high as possible.

Although the reference building variant is provided with double-pane insulation glass (low-E) for exterior windows (u=1.27 W/(m^{2}K); g=0.591) their change with triple-pane insulation glass windows (u=0.4 W/(m^{2}K); g=0.408) has positive effects both on the heating /cooling demand as well on the indoor thermal comfort (Figures 6 and 7).

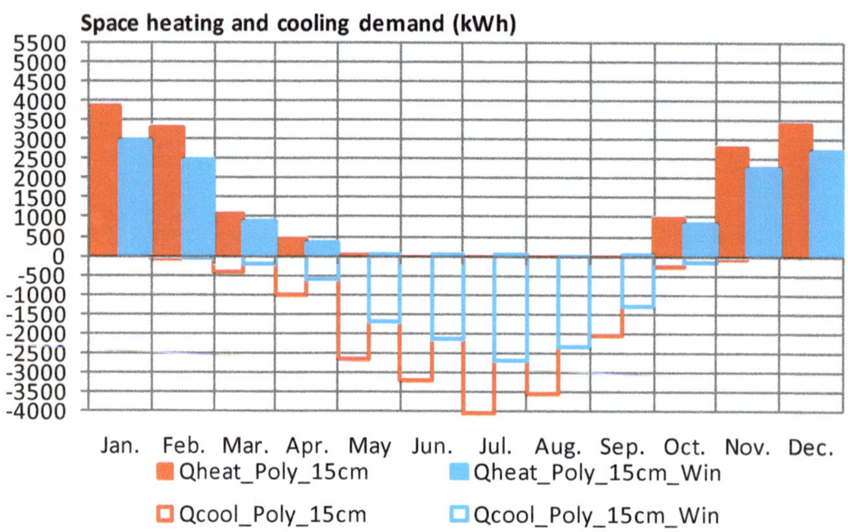

(a)

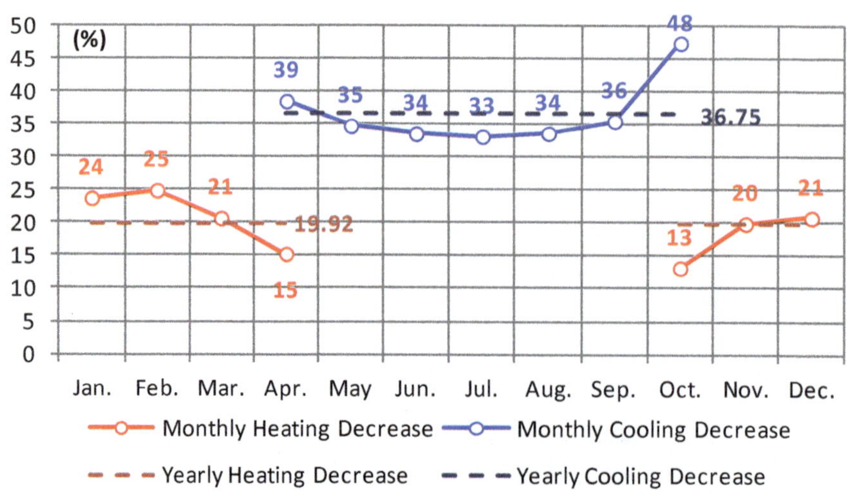

(b)

Figure 6. The replacement of the double-pane insulation glass windows by triple-pane insulation glass windows (v_5 versus v_6): (a) monthly demand for space heating and cooling; (b) monthly percentage differences between variants

Thus, the use of modern windows leads to a decrease of the annual value of energy demand for space heating with about 20%; the decrease for annual space cooling demand is about 37%.
The analysis of monthly temperature differences between the indoor air temperature and the inside surface temperature of the North wall for the Second Floor is presented in Figure 7,(a).

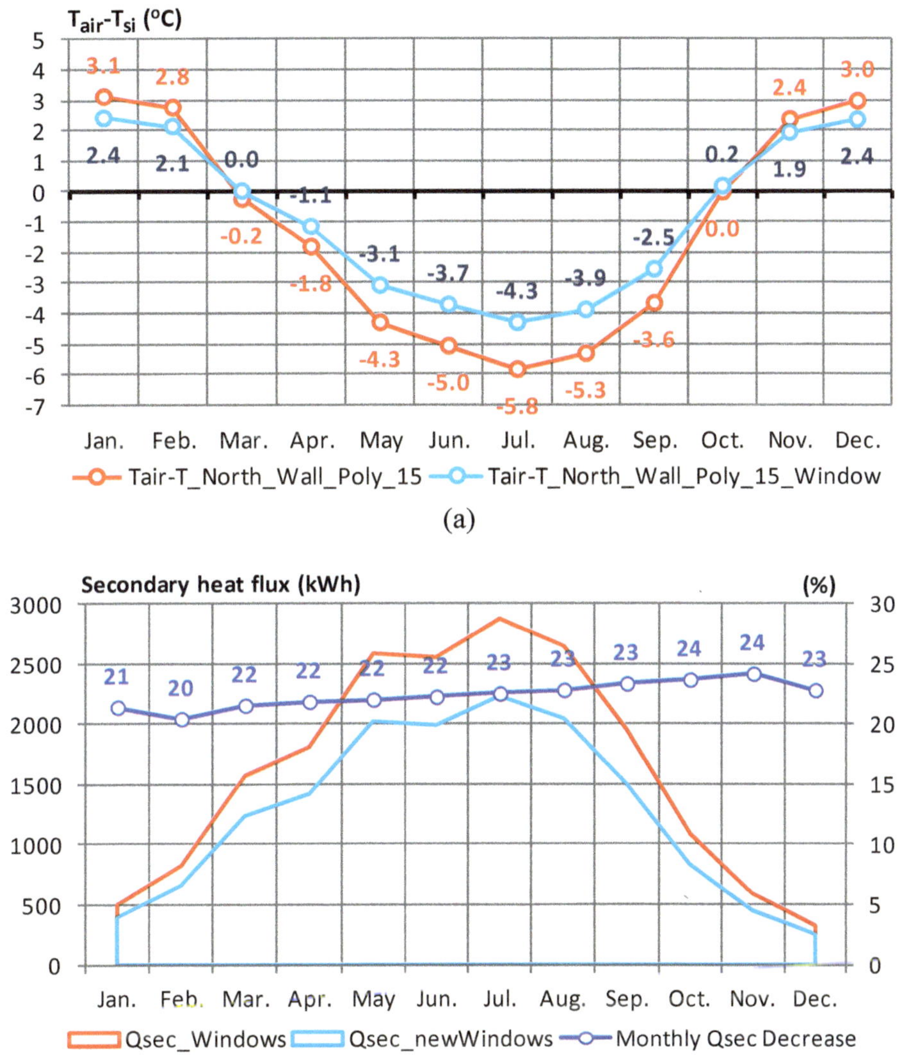

(a)

(b)

Figure 7. The use of triple-pane insulation glass windows (T_{air} – air temperature; T_{si} - inside surface temperature): (a) T_{air}-T_{si} – the North wall of the Second Floor; (b) secondary heat flux of all windows

It is noted that changing the windows has a positive influence on indoor comfort both in the cold periods January to March and October to December as well during the warm period April to October. Thus for the two cold periods mentioned above it is recorded a decrease of temperature difference on average with 0.4°C, a significant decrease of the temperature difference being obtained during the warm period, respectively of 1.2°C.
This fact is due to the decrease of secondary heat flux that is transmitted through the windows throughout the year (Figure 7,(b)); as it can be seen, the use of triple-pane insulation glass windows leads to a decrease of the monthly values of the secondary heat flux with 20-24%.

3.2.3 The roof insulation

For the thermal insulation of the roof, this paper proposes to replace the existent insulation of mineral wool with the polyurethane foam.
Considering the building (provided with exterior walls insulated with polyurethane foam of 15cm and triple-pane insulation glass windows) the replacement of roof insulation with polyurethane foam of 15cm

thickness, leads to a decrease in the annual demand of space heating with about 11%, but to an increase in cooling demand of about 7% (Figure 8).

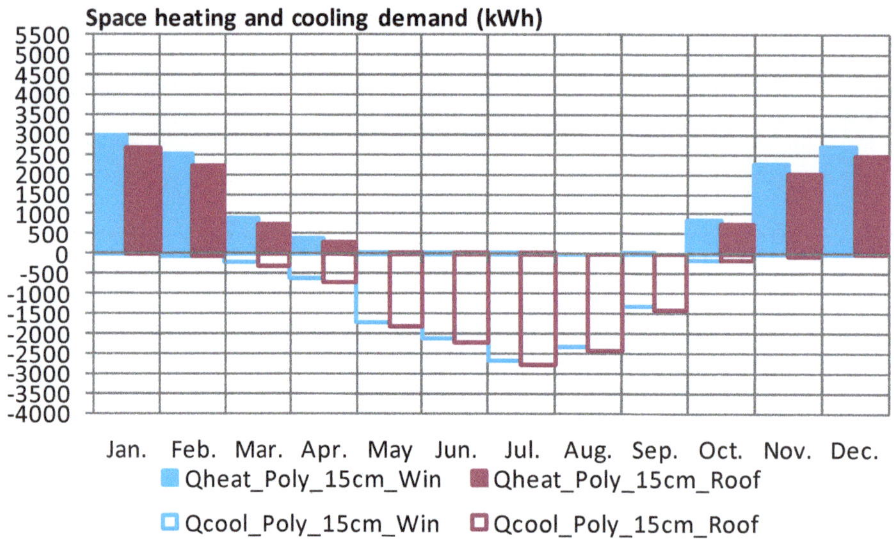

Figure 8. Monthly values of heating and cooling demand at the replacement of the roof insulation (v_6 versus v_7)

The last stage of the study proposes a comparative analysis between the building reference variant (v_0) and the latest variant proposed (v_7).

From the comparative analysis of results obtained following the energy simulations (Figures 9 and 10 and Table 3) it can be said that the rehabilitation of the entire building can lead to:

- the decrease of space heating demand with about 53% compared to the reference variant;
- the decrease of space cooling demand with about 17% compared to the reference variant;
- the temperature difference between indoor air temperature and North wall surface temperature for the Second Floor, decreases on average with about 2.6°C during the cold periods: January to March, October to December, and with about 0.3°C during period of May to September.

Table 3. The annual values of the space heating / cooling demand

Variant	Yearly Qheat [kWh]	Yearly Qcool [kWh]	Yearly decrease of heating demand (%) compared to v_0	Yearly increase of cooling demand (%) compared to v_0
v_0	22847	14663	-	-
v_1	16923	16295	25.93	11.13
v_2	17235	16587	24.56	13.12
v_3	16181	16940	29.18	15.53
v_4	16634	16785	27.19	14.47
v_5	16066	16570	29.68	13.01
v_6	12589	11060	44.90	-24.57
v_7	11127	11851	51.30	-19.18

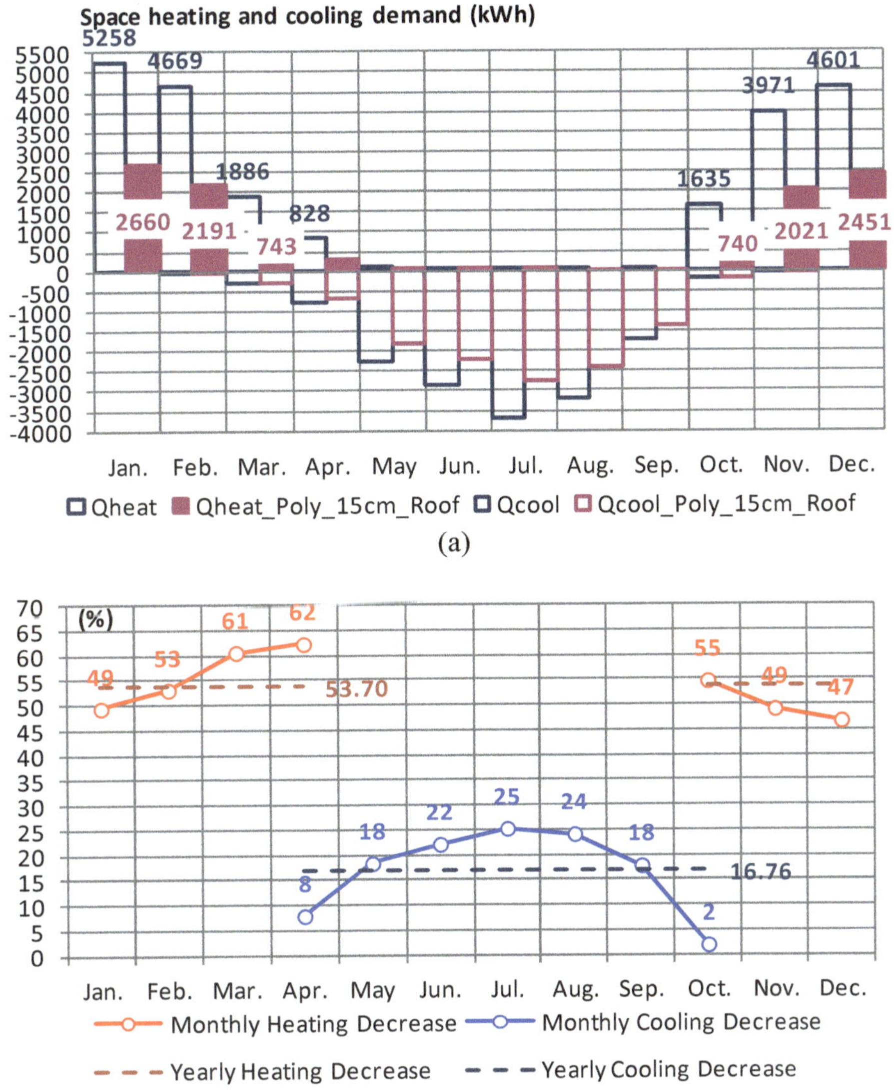

Figure 9. The reference variant versus the final one (v0 versus v7): (a) monthly demand of space heating / cooling; (b) monthly percentage differences between variants

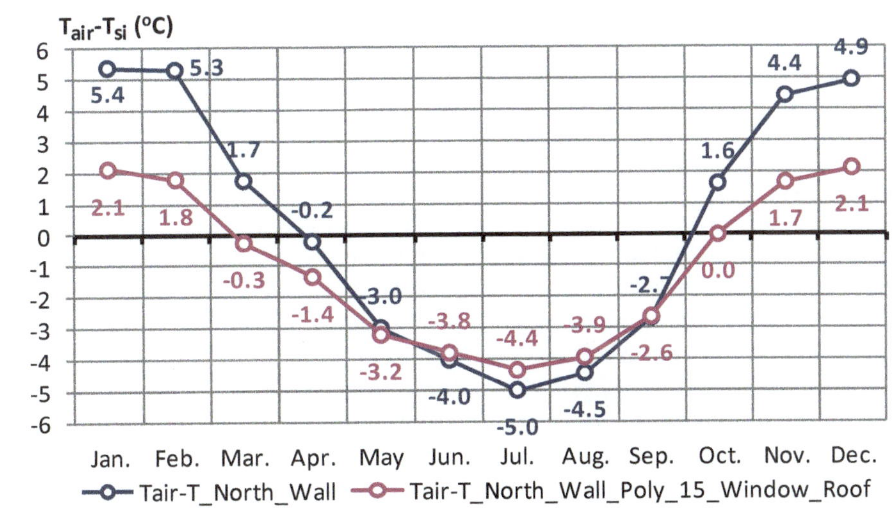

Figure 10. Differences between the monthly averages of indoor air temperature and the monthly averages of the North wall temperature (Second Floor)

4. Conclusions

For Braşov area, the energy demand for building heating in cold season represents an important share in energy consumption. Thus, as early as the construction phase of a building there can be considered a series of methods to reduce the energy demand for heating. To highlight the advantages of such methods, this paper proposed the energy analysis of an office building and the comparative study of the influence of various factors, such: the insulation material type of the exterior walls, the type of thermopane windows and the thermal insulation type of the roof. Following the achieved study, the following conclusions can be formulated:

- The climatic conditions of area have a significant influence both on the indoor comfort as well on the energy consumption in buildings. In view of this aspect, the energy demand simulation was performed using weather data recorded by a local weather station (data specific to Braşov area), data that were then implemented for simulations with TRNSYS software.
- The building envelope must be analyzed for its features of heat transmission, for its ability to control heat gain and losses considering the use of different insulation materials.
- A good insulation of exterior walls leads to a decrease in heating demands, through a good preservation of indoor temperature. Also, an increase in indoor thermal comfort is recorded due to eliminating the effect of "cold wall" on exterior walls (the difference between the wall surface temperature and indoor air temperature decreases). In addition, the risk of condensation is reduced and sudden temperature changes are prevented.
- High efficiency windows (which include multiple layers of insulation and low emissivity coatings) lead to energy savings by reducing the energy losses during the cold period, but also to maintaining of some reduced temperatures and reduced cooling demands during the warm period.
- The roof thermal insulation leads to a decrease of heating energy demand (for the considered case study the annual reduction of heating demand for the Second Floor is of 20%).

Finally it must be mentioned the important role of computer simulation in the design based on energy performance of buildings, in order to obtain some results based on which the final solution will decided. Using libraries of building materials, windows, weather data and standards for determining the buildings performance, through computer simulations there can be evaluated the parameters that ensure the energy efficiency of a building (calculation of building loads and of energy consumption; evaluation of thermal comfort conditions, thermal behaviour, etc.).

References

[1] International Energy Agency. SHC Task 40, Net Zero Energy Solar Buildings: International Projects of Carbon Neutrality in Buildings, 2011.

[2] Directive 2010/31/EU of the European Parliament and of the Council of 19 May 2010 on the energy performance of buildings. Official Journal of the European Union, 2010.

[3] Hui S.C.M. Low energy building design in high density urban cities. Renewable Energy 2001, 24, 627-640.

[4] Clarke J.A. Energy Simulation in Building Design, The second Edition. Butterworth-Heinemann A division of Reed Educational and Professional Publishing Ltd., 2001.

[5] Zalejska-Jonsson A. Evaluation of low-energy and conventional residential buildings from occupants' perspective. Building and Environment 2012, 58, 135-144.

[6] Laustsen J. Energy efficiency requirements in building codes, energy efficiency policies for new buildings. International Energy Agency, OECD/IEA France, 2008.

[7] EN 13163. Thermal insulation products for buildings - Factory made products of expanded polystyrene (EPS) - Specification, November 2008.

[8] EN 13164. Thermal insulation products for buildings - Factory made products of extruded polystyrene foam (XPS) - Specification, November 2008.

[9] EN 7345. Thermal insulation - Physical quantities and definitions, March 1996.

[10] EN 13165. Thermal insulation products for buildings - Factory made rigid polyurethane foam (PUR) products - Specification, November 2008.

[11] VFF/BF - Save more energy with new windows, Update of the study "In a new light: Energetic modernization of old windows". Verband Fenster + Fassade (VFF) and Bundesverband Flachglas (BF), Frankfurt am Main / Troisdorf, March 2014.

[12] Klein S.A., et al. TRNSYS 16, A transient System Simulation Program. University of Wisconsin Solar Energy Laboratory, 2006.

[13] Malkawi A.M., Augenbroe G. Advanced Building Simulation. Spon Press, Taylor & Francis Group, 2004.

[14] EN 13162. Thermal insulation products for buildings - Factory made mineral wool (MW) products - Specification, November 2008.

Exergoeconomic optimization of an irreversible regenerated air refrigerator with constant-temperature heat reservoirs

Yi Zhang[1], Lingeng Chen[2,3,4], Guozhong Chai[1]

[1] College of Mechanical Engineering, Zhejiang University of Technology, Hangzhou, 310014, China.
[2] Institute of Thermal Science and Power Engineering, Naval University of Engineering, Wuhan 430033, China.
[3] Military Key Laboratory for Naval Ship Power Engineering, Naval University of Engineering, Wuhan 430033, China.
[4] College of Power Engineering, Naval University of Engineering, Wuhan 430033, China.

Abstract
Based on the finite time exergoeconomic method, the performance analysis and optimization of an irreversible regenerated air refrigerator cycle are carried out by taking the profit rate as the optimization objective. The profit rate is defined as the difference between the revenue rate of output exergy and the cost rate of input exergy. The analytical expression for profit rate is derived, taking into account several irreversibilities, such as heat resistance, losses due to the pressure drop and the effects of non-isentropic expansion as well as compression. The influences of several parameters such as the temperature ratio of reservoirs, the efficiencies of both compressor and expander, the pressure recovery coefficient and so on are discussed by numerical examples. According to the simulation results, the double-maximum profit rate can be achieved when the pressure ratio and the distributions of heat conductance reach their optimal values respectively. By varying the price ratio, the relationship between the profit rate objective and other objectives can be established and the implementation of profit rate as objective can achieve higher COP compared to the cases using ecological function and cooling load as objectives.

Keywords: Finite-time thermodynamics; Exergoeconomic performance; Irreversible regenerated air.

1. Introduction

The world's energy reserves are decreasing as the intensive consumption and exhaustion of resources, leading to the rising costs of energy. Hence, from the economic perspective, optimizations of the performance of thermodynamic cycles are urgently required. In order to obtain the result closer to the real device, finite-time thermodynamics [1-8], as a powerful tool, is often used to optimize thermodynamic performances. For the Finite-time thermodynamic analyses of refrigeration cycles, the cooling load [9-14], the coefficient of performance (COP) [15-17] and exergy efficiency [18, 19] are often selected as optimization objectives. As conventional refrigerants contain chlorofluorocarbons (CFCs) which are implicated in ozone depletion, the environment friendly air refrigerator is becoming a popular topic of research with its application spreading to aviation industry, food storage and other cooling processes in modern industries [20-22]. Based on the theory of finite-time thermodynamics, the performance of air refrigeration cycle (inverse Brayton cycle) is analyzed and optimized with the consideration of external irreversibility (heat resistance between the reservoir and internal cycle) and internal irreversibilities (losses due to pressure-drop in the piping and nonisentropic expansion as well as compression). For simple Brayton refrigeration cycles, Chen *et al.* [11] derived the analytical relations between cooling load and pressure ratio and between COP and pressure ratio with the consideration of non-isentropic expansion as well as compression and heat resistance losses. Also the results show the cooling load has a parabolic dependence on COP. Luo *et al.* [23] optimized the allocation of heat exchanger inventory for maximizing the cooling load and the COP of the irreversible air refrigeration cycles. Zhou *et al.* [24-27] analyzed and optimized simple endoreversible and irreversible Brayton refrigeration cycles coupled to both constant- and variable-temperature heat reservoirs by taking the cooling load density, i.e., the ratio of cooling load to the maximum specific volume, as the optimization objective. Tu *et al.* [28] optimized simple endoreversible Brayton refrigeration cycles coupled to constant- temperature heat reservoirs by taking cooling load, ecological function and exergy efficiency as optimization objectives. The performance analyses and optimizations considering these objectives were also compared. Ust [29] optimized an simple irreversible air refrigeration cycle based on ecological coefficient of performance (ECOP) criterion which is defined as the ratio of cooling load to the loss rate of availability. Compared with simple Brayton refrigeration cycles, regenerated cycles are more common in industrial applications. Chen *et al.* [30] optimized the performance of an externally and internally irreversible regenerated Brayton refrigerator by taking the cooling load as an objective. Zhou *et al.* [31, 32] carried out the performance analyses and optimizations for regenerated air refrigeration cycles coupled to constant- and variable-temperature heat-reservoirs by taking cooling load density as an optimization objective. Tu *et al.* [33, 34] optimized cooling load, COP and exergy efficiency for real regenerated air refrigerator. Tyagi *et al.* [35] optimized the performance of an irreversible regenerative Brayton refrigerator cycle by taking the cooling load per unit cost as an optimization objective. Ust [36] compared the performance analyses and optimizations by taking ecological coefficient of performance, exergetic efficiency and COP as optimization objectives for an irreversible regenerative air refrigerator cycle.

Nowadays, systems are analyzed and designed based on the consideration of both thermodynamic parameters and cost accounting requirements after the research of Salamon and Nitzan [37, 38] which is to maximize the profit rate of an endoreversible heat engine. In order to distinguish this method from the analysis optimizing pure thermodynamic objectives, Chen *et al.* [39-45] analyzed the profit rate of thermal systems by attributing costs to input and output exergy and termed this method as finite-time exergoeconomic analysis, and its performance bound at maximum profit rate as finite-time exergoeconomic performance bound. Other researches seeking for best economic performance of thermal systems were carried out on endoreversible heat engines, refrigerators and heat pumps by Ibrahim *et al.* [46], De Vos [47, 48] and Bejan [49], with the only irreversibility restricted to the heat transfer between the working fluid and the heat reservoirs. De Vos [47, 48] applied the Newtonian (linear) heat transfer law to derive the relation between the optimal efficiency and economic returns when carrying out thermoeconomics analysis for heat engine. Chen *et al.* [50] investigated the endoreversible thermoeconomic performance of heat engine with the heat transfer between the working fluid and the heat reservoirs obeying linear phenomenological law. Sahin *et al.* [51-53] proposed an optimization criterion considering thermodynamic parameters per unit total cost.

Based on the exergoeconomic analysis of Carnot cycle [39] and the study for a regenerated air refrigeration cycle with cooling load as objective [30], the profit rate optimization for an irreversible regenerated refrigerator is investigated in this paper.

2. Irreversible regenerated air refrigeration cycle

The model of an irreversible regenerated air (Brayton) refrigerator with constant-temperature (hot reservoir temperature T_H and cold reservoir temperature T_L) heat reservoirs to be considered in this paper is shown in Figures 1 and 2. The heat reservoirs are assumed to have infinite thermal capacitance rates and the working fluid is considered as ideal gas with constant thermal capacitance rate C_{wf}. Process 5-2 is a heat addition process with air flowing through the regenerator. In process 2-3, air is compressed non-isentropically considering the irreversibility effect of the compressor. In process 3-6, heat is rejected to the heat sink in hot-side heat exchanger when air flowing through the latter cooler. Process 6-4 is a heat rejection process with air flowing through the regenerator. In process 4-1 air is expanded non-isentropically considering the irreversibility effect of the expander. Process1-5 is a heat addition process with air flowing through the regenerator. Processes 2-3$_s$ and 4-1$_s$ are the isentropic compression and expansion processes in ideal Brayton cycle corresponding to the processes 2-3 and 4-1, respectively. In order to analytically express the compression and expansion irreversibilities, the efficiencies of the compressor and expander are introduced and defined as:

$$\eta_c = (T_{3s} - T_2)/(T_3 - T_2), \eta_t = (T_4 - T_1)/(T_4 - T_{1s}) \tag{1}$$

Counter-flow heat-exchanger model is applied to all the heat-exchangers including the hot- and cold-side heat-exchangers as well as the regenerator. Their heat conductance rates (the product of heat-transfer coefficient k and the heat-exchange surface area A) can be expressed as $U_H = k_H A_H$, $U_L = k_L A_L$ and $U_R = k_R A_R$, respectively. The pressure drop in the piping for low pressure part and for high pressure part can be expressed by the pressure recovery coefficients $D_1 = P_2 / P_1$ and $D_2 = P_4 / P_3$ respectively.

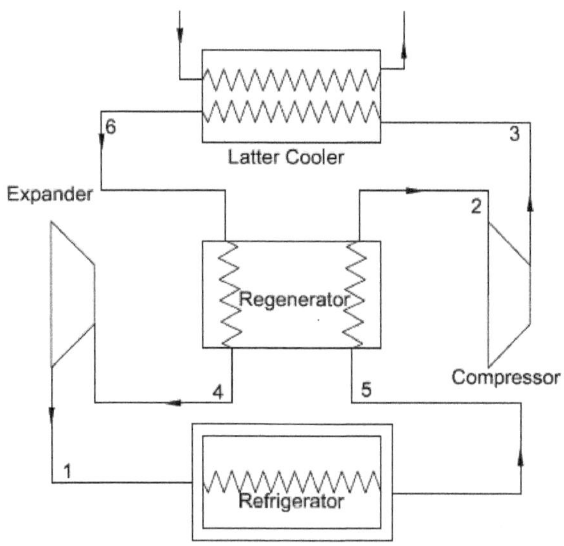

Figure 1. The schematic of a regenerated air refrigerator

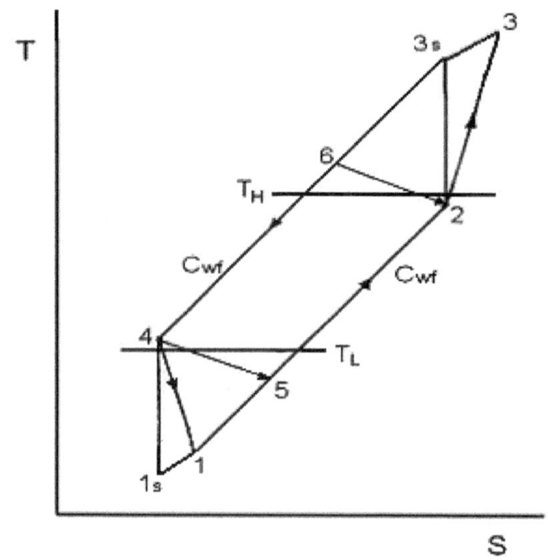

Figure 2. The temperature versus entropy diagram of a regenerated irreversible Brayton refrigeration cycle

3. Analytical expression for the profit rate

According to the properties of heat reservoir and working fluid, the heat transfer law (linear heat transfer law is applied between the heat reservoir and the working fluid) and the theory of heat exchangers, the rate of heat transfer (Q_H) released to the heat sink, the rate of heat transfer (Q_L) supplied by the heat source (the cooling rate R), and the rate of heat transfer happened in the regenerator (Q_R) can be expressed as, respectively,

$$\begin{aligned} Q_H &= U_H[(T_3 - T_H) - (T_6 - T_H)] / \ln[(T_3 - T_H)/(T_6 - T_H)] \\ &= C_{wf}(T_3 - T_6) = C_{wf} E_H (T_3 - T_H) \end{aligned} \tag{2}$$

$$\begin{aligned} R = Q_L &= U_L[(T_L - T_5) - (T_L - T_1)] / \ln[(T_L - T_5)/(T_L - T_1)] \\ &= C_{wf}(T_5 - T_1) = C_{wf} E_L (T_L - T_1) \end{aligned} \tag{3}$$

$$Q_R = C_{wf}(T_6 - T_4) = C_{wf}(T_2 - T_5) = C_{wf} E_R (T_6 - T_5) \tag{4}$$

where E_H, E_L and E_R are the effectivenesses of the hot- as well as cold-side heat exchangers and the regenerator, respectively, and are defined as:

$$E_H = 1 - \exp(-N_H), E_L = 1 - \exp(-N_L), E_R = N_R/(1 + N_R) \tag{5}$$

where N_H, N_L and N_R are the heat transfer units for the hot- as well as cold-side heat exchangers and the regenerator, respectively, and are defined as:

$$N_H = U_H / C_{wf}, N_L = U_L / C_{wf}, N_R = U_R / C_{wf} \tag{6}$$

The temperature ratio of the compressor operating isentropically is,

$$x = T_{3s}/T_2 = (P_3/P_2)^m = \pi^m, x \geq 1 \tag{7}$$

where $m = (k-1)/k$ with k being the adiabatic index, π is the pressure ratio of the compressor and P is the pressure. By defining the total pressure recovery coefficient as $D = D_1 D_2$, the temperature ratio of the expandor operating isentropically is given as,

$$T_4 / T_{1s} = (P_4 / P_1)^m = D^m x \tag{8}$$

The rate of exergy input to the system is equal to the net power input, and the first law of thermodynamics gives,

$$E_{in} = Q_H - Q_L \tag{9}$$

As the refrigerator is utilized to absorb heat from the cold space, the rate of output exergy is given as

$$E_{out} = Q_L (T_0/T_L - 1) - Q_H (T_0/T_H - 1) \tag{10}$$

The profit rate, defined as the difference between the revenue rate of output exergy and the cost rate of input exergy, is selected as the objective function of this refrigeration system, and is expressed as

$$M = \psi_1 E_{out} - \psi_2 E_{in} \tag{11}$$

where ψ_1 and ψ_2 are the prices of exergy output rate and power input, respectively.
Combining Eqs. (1)-(11) gives,

$$M = \frac{\begin{aligned} & C_{wf} E_L [\psi_1 (T_0 / T_L - 1) + \psi_2] \{[\eta_c - (x + \eta_c - 1)(1 - E_H) E_R] T_L - (D^{-m} x^{-1} \eta_t - \eta_t \\ & + 1)\{[E_R \eta_c + (x + \eta_c - 1)(1 - E_H)(1 - 2E_R)] T_L + (1 - E_R) E_H T_H \eta_c\}\} \\ & - C_{wf} E_H [\psi_1 (T_0 / T_H - 1) + \psi_2] \{(x + \eta_c - 1)(1 - E_R) E_L T_L + [(x + \eta_c - 1) E_R - \eta_c] T_H \\ & + (D^{-m} x^{-1} \eta_t - \eta_t + 1)(1 - E_L)[(x + \eta_c - 1)(1 - 2E_R) + E_R \eta_c] T_H\} \end{aligned}}{\begin{aligned} & \eta_c - (x + \eta_c - 1)(1 - E_H) E_R - (D^{-m} x^{-1} \eta_t - \eta_t + 1)(1 \\ & - E_L)[E_R \eta_c + (x + \eta_c - 1)(1 - E_H)(1 - 2E_R)] \end{aligned}} \tag{12}$$

By defining the heat reservoir temperature ratio as $\tau_1 = T_H / T_L$ and the ratio of hot-side heat reservoir temperature to the ambient temperature as $\tau_2 = T_H / T_0$, yields the expression of the dimensionless profit rate as,

$\overline{M} = M / (C_{wf} T_L \psi_1)$

$$= \frac{\begin{array}{l} E_L[\tau_1/\tau_2 - (1-\psi_2/\psi_1)]\{\eta_c - (x+\eta_c-1)(1-E_H)E_R - (D^{-m}x^{-1}\eta_t - \eta_t + 1)\{[E_R\eta_c \\ +(x+\eta_c-1)(1-E_H)(1-2E_R)] + (1-E_R)E_H\tau_1\eta_c\}\} \\ -E_H[1/\tau_2 - (1-\psi_2/\psi_1)]\{(x+\eta_c-1)(1-E_R)E_L + [(x+\eta_c-1)E_R - \eta_c]\tau_1 \\ +(D^{-m}x^{-1}\eta_t - \eta_t + 1)(1-E_L)[(x+\eta_c-1)(1-2E_R) + E_R\eta_c]\tau_1\} \end{array}}{\begin{array}{l} \eta_c - (x+\eta_c-1)(1-E_H)E_R - (D^{-m}x^{-1}\eta_t - \eta_t + 1)(1 \\ -E_L)[E_R\eta_c + (x+\eta_c-1)(1-E_H)(1-2E_R)] \end{array}} \quad (13)$$

4. Results and discussion

Eq. (13) indicates that the dimensionless profit rate for an irreversible regenerated air refrigerator is influenced by the pressure ratio (π), the heat transfer irreversibilities (E_H, E_L and E_R), the price ratio (ψ_1/ψ_2), the internal irreversibilities (η_c, η_R and D), the heat reservoir temperature ratio (τ_1) and the ratio of hot-side heat reservoir temperature to the ambient temperature(τ_2). As the air refrigeration cycle model is a steady flow model, the total heat transfer surface area of the heat exchangers is a constant, and the distribution of heat exchange surface area can be optimized. Also pressure ratio as a fundamental design parameter can be optimized. Hence, numerical simulation is carried out using Matlab to optimize the pressure ratio and the distribution of heat conductances of the heat exchangers. The effects of other parameters on the relationship between profit rate and the pressure ratio will be investigated. The relationship between the profit rate and other objectives will be discussed by varying the price ratio.

4.1 Optimal pressure ratio

The effects of the heat transfer irreversibilities (E_H, E_L and E_R), the internal irreversibilities (η_c, η_t and D), and two temperature ratios (τ_1 and τ_2) on the characteristic of profit rate versus the pressure ratio are shown in Figures 3-8 with the price ratio ψ_1/ψ_2=20. The shape of these curves is parabolic-like with one maximum profit rate ($\overline{M}_{\max,\pi}$) at the optimum pressure ratio (π_{opt}). The profit rate increases with the increasings of the effectivenesses of the hot-and cold-side heat exchangers (Figure 3), the efficiencies of the compressor and expander (Figure 5), the pressure recovery coefficient (Figure 6), the heat reservoir temperature ratio (Figure 7) and the ratio of hot-side heat reservoir temperature to the ambient temperature (Figure 8). In Figure 4, with the increasing of the effectiveness of the regenerator, the profit rate increases when the pressure ratio is relatively small, and decreases when the pressure ratio is large. Also, the optimal pressure ratio (π_{opt}) increases with the increasings of the effectiveness of the hot- and cold- side heat exchangers, the efficiencies of the compressor and the expander and two temperature ratios, while decreases with the increasings of the effectiveness of the regenerator and the pressure recovery coefficient.

4.2 Optimal distribution of heat conductance

For the fixed heat-exchanger inventory ($U_T = U_H + U_L + U_R$), the distribution of heat conductance will influence the performance of the air refrigerator with the profit rate being the objective. By defining the hot-side distribution of heat conductance (u_H) and cold-side distribution of heat conductance (u_L) as

$$u_H = U_H / U_T, u_L = U_L / U_T \quad (14)$$

The heat conductances of the hot- and cold-side heat exchanger as well as the regenerator can be expressed as,

$$U_H = u_H U_T, U_L = u_L U_T, U_R = (1 - u_H - u_L)U_T \quad (15)$$

When $k = 1.4$, $C_{wf} = 0.8\,kW/K$, $U_T = 5\,kW/K$, $\psi_1/\psi_2 = 20$, $\eta_c = \eta_t = 0.95$, $\tau_1 = 1.25$, $\tau_2 = 1$, $D = 0.96$, $\pi = 5$ are set, a 3-D plot of the dimensionless profit rate versus the hot-side distribution of heat conductance (u_H) and cold-side distribution of heat conductance (u_L) is shown in Figure 9. Eq.(14) indicates that both u_H and u_L should be smaller than or equal to 1 and so as their summation. Hence, an angle-bisecting plane vertical to the u_H and u_L plane is created in Figure 9 as a limit for u_H and u_L, with points in this plane indicating the limiting case of air refrigerator without regenerator. It can be observed from Figure 9 that the surface is similar to a paraboloid, with one maximum point ($\overline{M}_{\max,u}$) at optimal distributions of heat conductance (u_{Hopt} and u_{Lopt}). As mentioned previously, there is a maximum profit rate for the curve of the dimensionless profit rate versus pressure ratio. Hence, there is a double maximum value for the dimensionless profit rate with the distributions of heat conductance (u_H and u_L) and pressure ratio (π) as variables. The double maximum dimensionless profit rate can be searched out using the Matlab optimization toolbox. When $k = 1.4$, $C_{wf} = 0.8\,kW/K$, $U_T = 5\,kW/K$, $\psi_1/\psi_2 = 20$, $\eta_c = \eta_t = 0.95$, $\tau_1 = 1.25$, $\tau_2 = 1$ and $D = 0.96$ are set, the double maximum dimensionless profit rate ($\overline{M}_{\max,\max}$) is 0.0217 with the optimal distributions of heat conductance (u_{Hopt} and u_{Lopt}) equaling to 0.4737 and 0.3339, respectively, and the optimal pressure ratio (π_{opt}) being 7.8316. In Figure 10, the influences of the total heat exchanger inventory (U_T) on the double maximum dimensionless profit rate ($\overline{M}_{\max,\max}$) and its corresponding optimal distributions of heat conductance (u_{Hopt} and u_{Lopt}) can be observed. When increasing the total heat exchanger inventory (U_T), the double maximum dimensionless profit rate ($\overline{M}_{\max,\max}$) increases, the hot-side distribution of heat conductance (u_{Hopt}) decreases and the cold-side distribution of heat conductance (u_{Lopt}) first increases and then decreases.

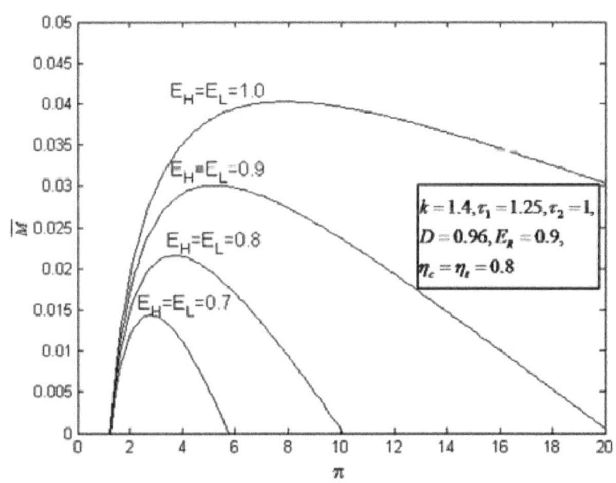

Figure 3. The effects of the effectivenesses of the hot- and cold- side heat exchangers on the profit rate versus the pressure ratio characteristic

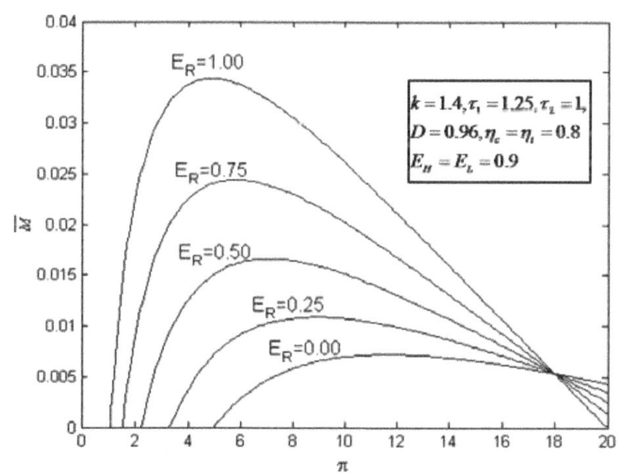

Figure 4. The effect of the effectiveness of the regenerator on the profit rate versus the pressure ratio characteristic

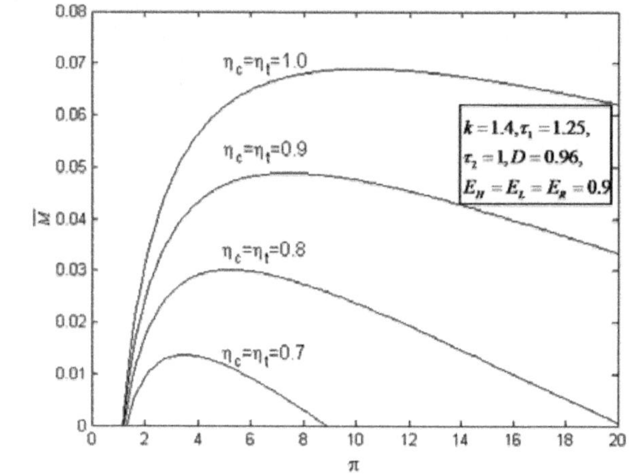

Figure 5. The effects of the efficiencies of the compressor and expander on the profit rate versus the pressure ratio characteristic

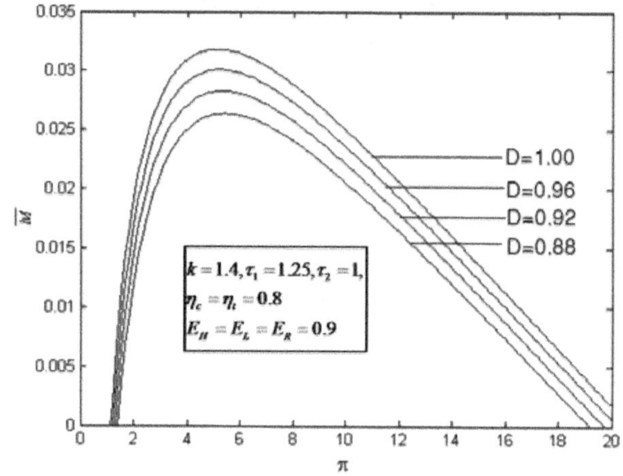

Figure 6. The effect of the pressure recovery coefficient on the profit rate versus the pressure ratio characteristic

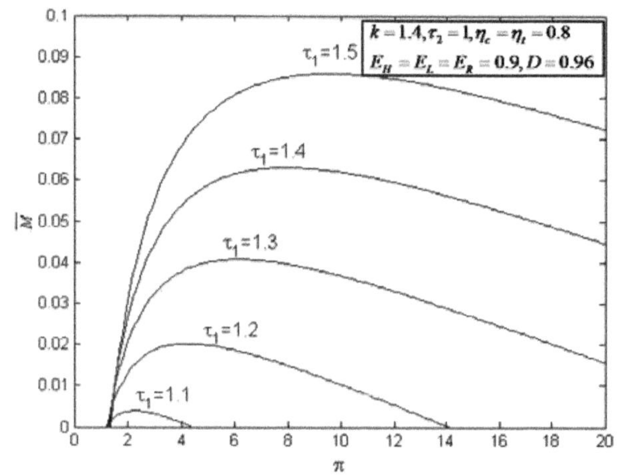

Figure 7. The effect of the heat reservoir temperature ratio on the profit rate versus the pressure ratio characteristic

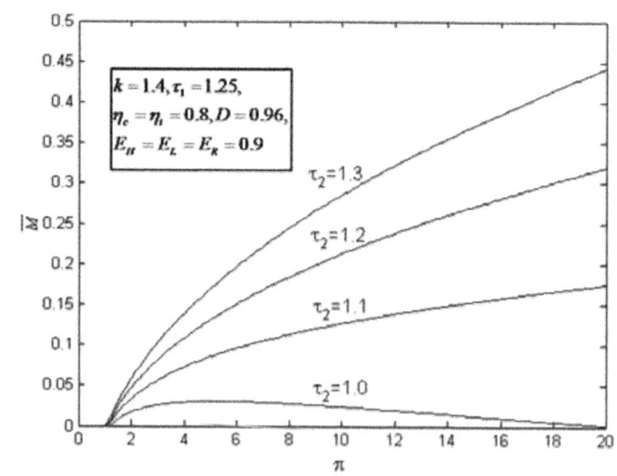

Figure 8. The effect of the ratio of hot-side heat reservoir temperature to the ambient temperature on the profit rate versus the pressure ratio characteristic

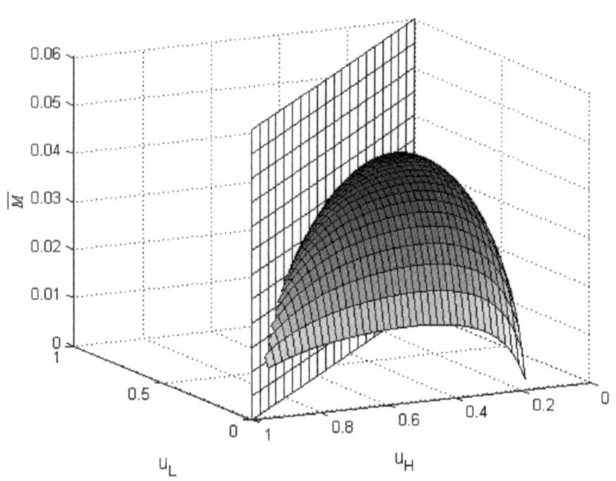

Figure 9. The profit rate versus the hot-side heat conductance distribution and the cold-side heat conductance distribution

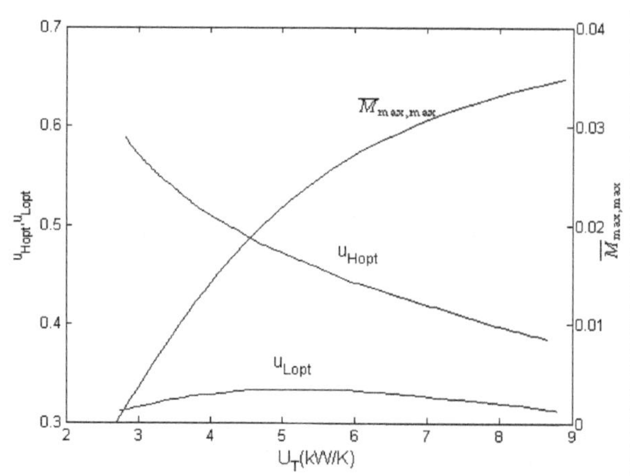

Figure 10. The double maximum profit rate and the corresponding optimum heat conductance distributions versus the total heat exchanger inventory

4.3 Influence of the price ratio

As the price of exergy output rate becomes very large compared with that of power input, i.e., $\psi_1/\psi_2 \to \infty$, the function of the dimensionless profit rate becomes

$$\overline{M} = M/(C_{wf}T_L\psi_1)$$
$$= \frac{\begin{array}{l} E_L(\tau_1/\tau_2-1)\{\eta_c-(x+\eta_c-1)(1-E_H)E_R-(D^{-m}x^{-1}\eta_t-\eta_t+1)\{[E_R\eta_c \\ +(x+\eta_c-1)(1-E_H)(1-2E_R)]+(1-E_R)E_H\tau_1\eta_c\}\} \\ -E_H(1/\tau_2-1)\{(x+\eta_c-1)(1-E_R)E_L+[(x+\eta_c-1)E_R-\eta_c]\tau_1 \\ +(D^{-m}x^{-1}\eta_t-\eta_t+1)(1-E_L)[(x+\eta_c-1)(1-2E_R)+E_R\eta_c]\tau_1\} \end{array}}{\begin{array}{l} \eta_c-(x+\eta_c-1)(1-E_H)E_R-(D^{-m}x^{-1}\eta_t-\eta_t+1)(1 \\ -E_L)[E_R\eta_c+(x+\eta_c-1)(1-E_H)(1-2E_R)] \end{array}} \tag{16}$$

If the hot reservoir temperature equals the ambient temperature ($\tau_2 \to 1$), the function of the dimensionless profit rate becomes

$$\overline{M} = M/(C_{wf}T_L\psi_1)$$
$$= \frac{\begin{array}{l} E_L(\tau_1/\tau_2-1)\{\eta_c-(x+\eta_c-1)(1-E_H)E_R-(D^{-m}x^{-1}\eta_t-\eta_t+1)\{[E_R\eta_c \\ +(x+\eta_c-1)(1-E_H)(1-2E_R)]+(1-E_R)E_H\tau_1\eta_c\}\} \end{array}}{\begin{array}{l} \eta_c-(x+\eta_c-1)(1-E_H)E_R-(D^{-m}x^{-1}\eta_t-\eta_t+1)(1 \\ -E_L)[E_R\eta_c+(x+\eta_c-1)(1-E_H)(1-2E_R)] \end{array}} = \frac{(T_0/T_L-1)R}{C_{wf}T_L} \tag{17}$$

The optimization of the profit rate also leads to the maximization of the cooling load (R).

On the other hand, with the price of power input approaching the price of the exergy output rate, i.e. $\psi_2/\psi_1 \to 1$, the function of the profit rate becomes

$$\overline{M} = M/(C_{wf}T_L\psi_1)$$
$$= \frac{\begin{array}{l} E_L\tau_1/\tau_2\{\eta_c-(x+\eta_c-1)(1-E_H)E_R-(D^{-m}x^{-1}\eta_t-\eta_t+1)\{[E_R\eta_c \\ +(x+\eta_c-1)(1-E_H)(1-2E_R)]+(1-E_R)E_H\tau_1\eta_c\}\} \\ -E_H/\tau_2\{(x+\eta_c-1)(1-E_R)E_L+[(x+\eta_c-1)E_R-\eta_c]\tau_1 \\ +(D^{-m}x^{-1}\eta_t-\eta_t+1)(1-E_L)[(x+\eta_c-1)(1-2E_R)+E_R\eta_c]\tau_1\} \end{array}}{\begin{array}{l} \eta_c-(x+\eta_c-1)(1-E_H)E_R-(D^{-m}x^{-1}\eta_t-\eta_t+1)(1 \\ -E_L)[E_R\eta_c+(x+\eta_c-1)(1-E_H)(1-2E_R)] \end{array}} = -\frac{T_0\sigma}{C_{wf}T_L} \tag{18}$$

where σ is the rate of entropy production of the regenerative air refrigeration cycle. When maximizing the profit rate under this condition, minimization of the entropy generation (σ) can be achieved.

For the case when $\psi_1/\psi_2 \to 2$ is satisfied, the function of the profit rate becomes

$$\overline{M} = M/(C_{wf}T_L\psi_1)$$
$$= \frac{\begin{array}{l} E_L[\tau_1/\tau_2-1/2]\{\eta_c-(x+\eta_c-1)(1-E_H)E_R-(D^{-m}x^{-1}\eta_t-\eta_t+1)\{[E_R\eta_c \\ +(x+\eta_c-1)(1-E_H)(1-2E_R)]+(1-E_R)E_H\tau_1\eta_c\}\} \\ -E_H[1/\tau_2-1/2]\{(x+\eta_c-1)(1-E_R)E_L+[(x+\eta_c-1)E_R-\eta_c]\tau_1 \\ +(D^{-m}x^{-1}\eta_t-\eta_t+1)(1-E_L)[(x+\eta_c-1)(1-2E_R)+E_R\eta_c]\tau_1\} \end{array}}{\begin{array}{l} \eta_c-(x+\eta_c-1)(1-E_H)E_R-(D^{-m}x^{-1}\eta_t-\eta_t+1)(1 \\ -E_L)[E_R\eta_c+(x+\eta_c-1)(1-E_H)(1-2E_R)] \end{array}} = \frac{E}{2C_{wf}T_L} \tag{19}$$

where E is the ecological objective of the regenerative air refrigeration cycle. The optimization of the profit rate also leads to the maximization of the ecological function (E) objective.

Figure 11 shows the effect of the price ratio on the profit rate versus COP characteristic, with the COP function derived by Chen *et al.* [33]. From Figure 11, it can be observed that the optimal COP is larger when $\psi_1/\psi_2 = 20$ is satisfied, compared to the other two cases which corresponding to the cases of ecological objective ($\psi_1/\psi_2 \to 2$) and cooling load objective ($\psi_1/\psi_2 \to \infty$).

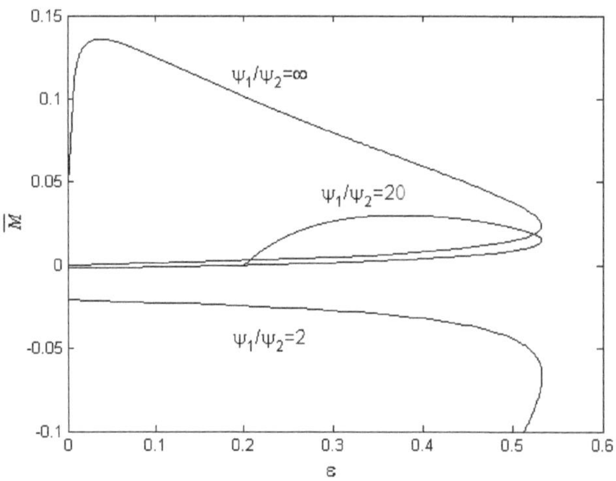

Figure 11. The profit rate versus COP with different price ratios

5. Conclusion

The analytical expression of the profit rate for the irreversible regenerated air refrigeration cycle coupled to constant heat reservoirs is derived based on the theoretical model. Numerical calculations are carried out to optimize the pressure ratio and the distributions of heat conductance between heat exchangers and regenerator with considering the influences of the internal irreversibilities (η_c, η_t and D), the heat reservoir temperature ratio (τ_1) and the ratio of hot-side heat reservoir temperature to the ambient temperature(τ_2). The relationship between the profit rate objective and other objectives is investigated. The comparison of these objectives has also been discussed.

There exists an optimal pressure ratio and a pair of optimal distributions of heat conductance corresponding to the double-maximum profit rate. Optimizing the air refrigeration cycle based on the profit rate objective can achieve higher COP compared to the cases using ecological function and cooling load as objectives.

Acknowledgments

This paper is supported by National Natural Science Foundation of China (Project No. 10905093).

References

[1] Andresen, B., Finite-time thermodynamics. 1983: University of Copenhagen Copenhagen.

[2] Bejan, A., Entropy generation minimization: The new thermodynamics of finite-size devices and finite?time processes. Journal of Applied Physics, 1996. 79(3): p. 1191-1218.

[3] Chen, L., C. Wu, and F. Sun, Finite time thermodynamic optimization or entropy generation minimization of energy systems. Journal of Non-Equilibrium Thermodynamics, 1999. 24(4): p. 327-359.

[4] Berry, R.S., et al., Thermodynamic optimization of finite-time processes. 2000: Wiley Chichester.

[5] Chen, L. and F. Sun, Advances in finite time thermodynamics: analysis and optimization. 2004: Nova Publishers.

[6] Durmayaz, A., et al., Optimization of thermal systems based on finite-time thermodynamics and thermoeconomics. Progress in Energy and Combustion Science, 2004. 30(2): p. 175-217.

[7] Andresen, B., Current Trends in Finite-Time Thermodynamics. Angewandte Chemie International Edition, 2011. 50(12): p. 2690-2704.

[8] Le Roux, W., T. Bello-Ochende, and J. Meyer, A review on the thermodynamic optimisation and modelling of the solar thermal Brayton cycle. Renewable and Sustainable Energy Reviews, 2013. 28: p. 677-690.

[9] Wu, C., Maximum obtainable specific cooling load of a refrigerator. Energy conversion and management, 1995. 36(1): p. 7-10.

[10] Chen, L., C. Wu, and F. Sun, Heat transfer effect on the specific cooling load of refrigerators. Applied thermal engineering, 1996. 16(12): p. 989-997.

[11] Chen, L., C. Wu, and F. Sun, Cooling load versus COP characteristics for an irreversible air refrigeration cycle. Energy Conversion and Management, 1998. 39(1): p. 117-125.

[12] Luo, J., et al., Optimum allocation of heat transfer surface area for cooling load and COP optimization of a thermoelectric refrigerator. Energy conversion and management, 2003. 44(20): p. 3197-3206.

[13] Chen, L., et al., Optimal cooling load and COP relationship of a four-heat-reservoir endoreversible absorption refrigeration cycle. Entropy, 2004. 6(3): p. 316-326.

[14] Assad, M.E.H., Cooling load optimization of an irreversible refrigerator with combined heat transfer. International Journal of Energy and Environment, 2013. 4(3): p. 377-386.

[15] Liu, X., et al., Cooling load and COP optimization of an irreversible Carnot refrigerator with spin-1/2 systems. International Journal of Energy and Environment, 2011. 2(5): p. 797-812.

[16] Chen, L., F. Meng, and F. Sun, Effect of heat transfer on the performance of thermoelectric generator-driven thermoelectric refrigerator system. Cryogenics, 2012. 52(1): p. 58-65.

[17] Hu, Y., et al., Coefficient of performance for a low-dissipation Carnot-like refrigerator with nonadiabatic dissipation. Physical Review E, 2013. 88(6): p. 062115.

[18] Chen, C.o.-K. and Su Y.-F., Exergetic efficiency optimization for an irreversible Brayton refrigeration cycle. International journal of thermal sciences, 2005. 44(3): p. 303-310.

[19] Su, Y. and C.o.-K. Chen, Exergetic efficiency optimization of a refrigeration system with multi-irreversibilities. Proceedings of the Institution of Mechanical Engineers, Part C: Journal of Mechanical Engineering Science, 2006. 220(8): p. 1179-1187.

[20] Elsayed, S., et al., Analysis of an air cycle refrigerator driving air conditioning system integrated desiccant system. International journal of refrigeration, 2006. 29(2): p. 219-228.

[21] Park, S.K., J.H. Ahn, and T.S. Kim, Off-design operating characteristics of an open-cycle air refrigeration system. International Journal of Refrigeration, 2012.

[22] Lu, Y., et al., Performance study on compressed air refrigeration system based on single screw expander. Energy, 2013.

[23] Luo, J., et al., Optimum allocation of heat exchanger inventory of irreversible air refrigeration cycles. Physica Scripta, 2002. 65(5): 410-415.

[24] Chen, L., et al., Performance optimisation for an irreversible variable-temperature heat reservoir air refrigerator. International journal of Ambient Energy, 2005. 26(4): p. 180-190.

[25] Zhou, S., et al., Cooling load density analysis and optimization for an endoreversible air refrigerator. Open Systems & Information Dynamics, 2001. 8(02): p. 147-155.

[26] Zhou, S., et al., Cooling load density characteristics of an endoreversible variable-temperature heat reservoir air refrigerator. International journal of energy research, 2002. 26(10): p. 881-892.

[27] Zhou, S., et al., Cooling load density optimization of an irreversible simple Brayton refrigerator. Open systems & information dynamics, 2002. 9(04): p. 325-337.

[28] Tu, Y., et al., Comparative performance analysis for endoreversible simple air refrigeration cycles considering ecological, exergetic efficiency and cooling load objectives. International journal of Ambient Energy, 2006. 27(3): p. 160-168.

[29] Ust, Y., Performance analysis and optimization of irreversible air refrigeration cycles based on ecological coefficient of performance criterion. Applied Thermal Engineering, 2009. 29(1): p. 47-55.

[30] Chen, L., et al., Performance of heat-transfer irreversible regenerated Brayton refrigerators. Journal of Physics D: Applied Physics, 2001. 34(5): 830-837.

[31] Zhou, S., et al., Theoretical optimization of a regenerated air refrigerator. Journal of Physics D: Applied Physics, 2003. 36(18): p. 2304-2311.

[32] Zhou, S., et al., Cooling-load density optimization for a regenerated air refrigerator. Applied energy, 2004. 78(3): p. 315-328.

[33] Chen, L., Y. Tu, and F. Sun, Exergetic efficiency optimization for real regenerated air refrigerators. Applied Thermal Engineering, 2011. 31(16): p. 3161-3167.
[34] Tu, Y., et al., Optimization of cooling load and coefficient of performance for real regenerated air refrigerator. Proceedings of the Institution of Mechanical Engineers, Part E: Journal of Process Mechanical Engineering, 2006. 220(4): p. 207-215.
[35] Tyagi, S., et al., A new thermoeconomic approach and parametric study of an irreversible regenerative Brayton refrigeration cycle. International journal of refrigeration, 2006. 29(7): p. 1167-1174.
[36] Ust, Y., Effect of regeneration on the thermo-ecological performance analysis and optimization of irreversible air refrigerators. Heat and Mass Transfer, 2010. 46(4): p. 469-478.
[37] Berry, R.S., P. Salamon, and G. Heal, On a relation between economic and thermodynamic optima. Resources and Energy, 1978. 1(2): p. 125-137.
[38] Salamon, P. and A. Nitzan, Finite time optimizations of a Newton's law Carnot cycle. The Journal of Chemical Physics, 1981. 74: p. 3546.
[39] Wu, C., L. Chen, and F. Sun, Effect of the heat transfer law on the finite-time, exergoeconomic performance of heat engines. Energy, 1996. 21(12): p. 1127-1134.
[40] Chen, L., F. Sun, and C. Wu, Effect of heat transfer law on the performance of a generalized irreversible Carnot engine. Journal of Physics D: Applied Physics, 1999. 32(2): p. 99-105.
[41] Chen, L., et al., Maximum profit performance of a three-heat-reservoir heat pump. International Journal of Energy Research, 1999. 23(9): p. 773-777.
[42] Wu, F., et al., Finite-time exergoeconomic performance bound for a quantum Stirling engine. International Journal of Engineering Science, 2000. 38(2): p. 239-247.
[43] Chen, L., C. Wu, and F. Sun, Effect of heat transfer law on the finite-time exergoeconomic performance of a Carnot refrigerator. Exergy, An International Journal, 2001. 1(4): p. 295-302.
[44] Chen, L., F. Sun, and C. Wu, Maximum-profit performance for generalized irreversible Carnot-engines. Applied Energy, 2004. 79(1): p. 15-25.
[45] Qin, X., et al., Thermo-economic optimization of an endoreversible four-heat-reservoir absorption-refrigerator. Applied energy, 2005. 81(4): p. 420-433.
[46] Ibrahim, O., S. Klein, and J. Mitchell, Effects of irreversibility and economics on the performance of a heat engine. Journal of Solar Energy Engineering;, 1992. 114(4).
[47] De Vos, A., Endoreversible thermoeconomics. Energy conversion and management, 1995. 36(1): p. 1-5.
[48] De Vos, A., Endoreversible economics. Energy conversion and management, 1997. 38(4): p. 311-317.
[49] Bejan, A., Power and refrigeration plants for minimum heat exchanger inventory. Journal of Energy Resources Technology, 1993. 115(2): p. 148-150.
[50] Chen, L., F. Sun, and C. Wu, Endoreversible thermoeconomics for heat engines. Applied Energy, 2005. 81(4): p. 388-396.
[51] Sahin, B. and A. Kodal, Finite time thermoeconomic optimization for endoreversible refrigerators and heat pumps. Energy Conversion and Management, 1999. 40(9): p. 951-960.
[52] Sahin, B., A. Kodal, and A. Koyun, Optimal performance characteristics of a two-stage irreversible combined refrigeration system under maximum cooling load per unit total cost conditions. Energy Conversion and Management, 2001. 42(4): p. 451-465.
[53] Sahin, B. and A. Kodal, Thermoeconomic optimization of a two stage combined refrigeration system: a finite-time approach. International Journal of Refrigeration, 2002. 25(7): p. 872-877.

Effect of the number and area of channels on the performance of rectangular, triangular and elliptical channel geometry PEM fuel cells

I. Khazaee

Department of Mechanical Engineering, Torbat-e-jam branch, Islamic Azad University, Torbat-e-jam, Iran.

Abstract
The performance of a PEM fuel cell is an important parameter that can be expressed with the polarization curve of it. In this study a complete three-dimensional and single phase model is used to investigate the effect of increasing the number of serpentine channels in the bipolar plates and also increasing the area (depth) of channels of a PEM fuel cell with rectangular, triangular and elliptical cross-section geometry. The proposed model is a full cell model, which includes all the parts of the PEM fuel cell, flow channels, gas diffusion electrodes, catalyst layers and the membrane. The results show that there are good agreement with the numerical results and experimental results of the previous work of authors. Also the results show that by increasing the number of channels from one to four and eight, the performance improved about 18% and by decreasing the area of channels from $2mm^2$ to $1mm^2$ the performance improved about 13%.

Keywords: PEM fuel cell; Elliptical channel; Rectangular channel; Numerical modeling; Current density.

1. Introduction

The polymer exchange membrane fuel cell (PEMFC) is considered to be the most promising candidate for electric vehicles by virtue of its high power density, zero pollution, low operating temperature, quick start-up capability and long lifetime. PEMFC can also be used in distributed power systems, submarines, and aerospace applications [1].

The single-cell PEMFC consists of a carbon plate, a gas diffusion layer (GDL), a catalyst layer, for each of the anode and the cathode sides, as well as a PEM membrane at the center that is shown in Figure 1.

Flow channel geometry is of critical importance for the performance of fuel cells containing proton exchange membranes (PEM) but is of less concern for solid oxide fuel cells (SOFC). The reactants, as well as the products, are transported to and from the cell through flow channels. Flow field configurations, including parallel, serpentine, interdigitated, and other combined versions, have been developed.

The performance of the fuel cell system is characterized by current-voltage curve *(i.e. polarization curve).* The difference between the open circuit potential of the electrochemical reaction and cell voltage occurs from the losses associated with the operation. The corresponding voltage drop is generally classified in three parts: (i) activation over-potential caused by the electrochemical reactions, (ii) ohmic drop across the polymer electrolyte, (iii) mass transfer limitations of reactants.

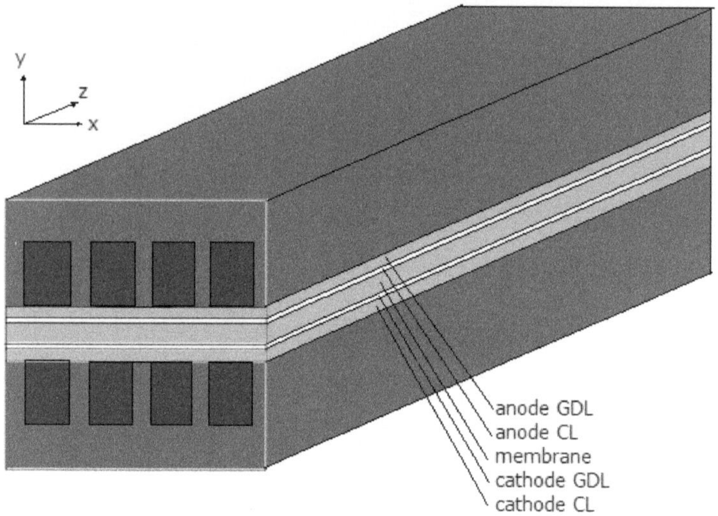

Figure 1. Schematic of a PEMFC

These associated losses dominate over different current density ranges. For low current densities; the activation over-potential is dominant. For high current densities, which are of particular interest for vehicle applications because of higher power density; the mass transfer limitations dominates the losses. For moderate current densities, the ohmic drop across the polymer membrane dominates.

Khazaee [2] investigated numerically the effect of using different obstacles on the performances, current density and gas concentration for different Aspect Ratios and compared the results with experimental results of a triangular channel geometry PEM fuel cell. He found that by increasing the hydrogen flow rate from 0.3 L/min to 0.7 L/min and the oxygen flow rate from 0.5 L/min to 0.9 L/min the cell performance enhances. Also he found that by increasing the number of the blocks in the channels and installation of rectangular block, the overall cell performance increases.

West and Fuller [3] proposed a two-dimensional numerical analysis of the rib spacing in PEM electrode assemblies on current and water distribution within the cell. The results indicated that increasing the rib width strongly affected the membrane water content before the catalyst utilization is reduced. Therefore, the two-dimensional effect has a significant influence on water management. Chiang and Chu [4] investigated the effects of transport components on the transport phenomena and performance of PEM fuel cells by using a three-dimensional model. The impacts of channel aspect ratio (AR) and GDL thickness were examined. It was found that a flat channel with a small AR or a thin GDL generates more current at low cell voltage due to the merits of better reactant gas transport and liquid water delivery. Wang et al. [5] developed a two-dimensional numerical model to study the two-phase flow transport in the air cathode of a PEMFC. In this paper, the model encompassed both single- and two-phase regimes corresponding to low and high current densities and was capable of predicting the transition between the two regimes. Jenn-Kun Kuo et al. [6] performed numerical simulations to evaluate the convective heat transfer performance and velocity flow characteristics of the gas flow channel design to enhance the performance of proton exchange membrane fuel cells (PEMFCs). Their study has simulated low Reynolds number laminar flow in the gas flow channel of a PEMFC. The heat transfer performance and enhanced gas flow velocity characteristics of four different channel geometries have been considered, namely a conventional straight gas flow channel and a gas flow channel with the three novel periodic patterns geometries. The results indicated that, compared to the conventional gas flow channel, the novel gas flow channels proposed in this study provide a significantly improved convective heat transfer performance and a higher gas flow velocity and, hence, an improved catalysis reaction performance in the catalyst layer. Khazaee et al. [7-9] proposed A complete three-dimensional and single phase model for annular-shaped and duct-shaped proton exchange membrane (PEM) fuel cell to investigate the effect of using different connections between bipolar plate and gas diffusion layer on the performances, current density, and temperature and gas concentration. They found that that the cell performance is increased as the number of connections between GDL and bipolar plate increases but for one connection conditions the effect of changing the location of connection on cell performance is negligibly small. Also they found that the water mole fraction gradually increases along the cell and the maximum of it is near the connections between GDL and bipolar plate. Yan et al. [10] developed a two-dimensional mass transport model to investigate the anode gas flow channel cross section and GDL porosity effects. They found that an increase in either the GDL porosity, channel width fraction, or the number of channels could lead to better cell performance. Chu et al. [11] also used a half-cell model to investigate the effect of the change of the GDL on the performance of a PEMFC. They found that a fuel cell embedded with a GDL with a larger averaged porosity will consume a greater amount of oxygen, so that a higher current density is generated and a better fuel cell performance is obtained.

This study numerically investigated the effect of changing the number of channels and the area of channels on cell performance of a PEM fuel cell with different cross-section geometry of channels. The objective of the current work is to show the effect of changing channel geometry that it may be of interest to engineers attempting to develop the optimization of a PEMFC and to researchers interested in the flow modification aspects of the PEMFC performance enhancement.

2. Numerical model

The numerical model for the fuel cell used here includes the anode flow channels, anode gas diffusion layer, anode catalyst layer, proton exchange membrane, cathode catalyst layer, cathode gas diffusion layer, and cathode flow channels. The thickness of the gas diffusion layer is 0.33mm thick, the catalyst layer is 0.01mm thick, and the proton exchange membrane is 0.051mm thick. Miniature fuel cells with dimensions of $45\times 95\times 101\ mm^3$ with an active area of $25\ cm^2$ and serpentine and rectangular, triangular and elliptical flow field geometries of channels are considered in this investigation. For one serpentine channel the width and depth were selected to be 4 and 1.27 mm for elliptical channel, 4 and 1 mm for rectangular channel and 4 and 2 mm for triangular channel and for four serpentine channels the width and depth were selected to be 1 and 1.27 mm for elliptical channel, 1 and 1 mm for rectangular channel and 1 and 2 mm for triangular channel and for eight serpentine channels the width and depth were selected to be 0.5 and 1.27 mm for elliptical channel, 0.5 and 1 mm for rectangular channel and 0.5 and 2 mm for triangular channel respectively.

Also when the cross section area of one channel changes from $1mm^2$ to $1.5mm^2$ and $2mm^2$ only for four serpentine channels, the width and depth were selected to be 1 and 1.93 mm for elliptical channel, 1 and 1.5 mm for rectangular channel and 1 and 3 mm for triangular channel geometry and 1 and 2.55 mm for elliptical channel, 1 and 2 mm for rectangular channel and 1 and 4 mm for triangular channel respectively. The geometry of cells and channels are shown in Figure 2.

The proposed model does not require any internal boundary conditions between the components of PEM Fuel Cell system. The different physical properties and transport parameters are incorporated into a single set of governing equations using a single domain formulation. The model aims to study the electrochemical kinetics, current distribution, reactant flow fields and multi-component transport of oxidizer and fuel streams in a multi-dimensional domain. The assumptions made in developing the model are as follows:

- Ideal gas mixtures
- Incompressible and Laminar flow because low flow velocities and low fuel utilization
- Isotropic and homogeneous porous electrodes, catalyst layers and membrane
- Negligible ohmic resistance at porous electrodes and current collectors

The model assumes that the system is steady; the inlet reactants are ideal gases; the flow is laminar; and the porous layers such as the diffusion layer, catalyst layer and PEM are isotropic. The model includes continuity, momentum and species equations for gaseous species, liquid water transport equations in the channels, gas diffusion layers, and catalyst layers, water transport equation in the membrane, electron and proton transport equations. The Bulter–Volumer equation was used to describe electrochemical reactions in the catalyst layers.

The conservation equations of mass, momentum, energy, species and charge are as follows:

$$\frac{\partial(\rho\varepsilon)}{\partial t}+\nabla.(\varepsilon\rho\vec{u})=0 \tag{1}$$

$$\frac{\partial(\rho\varepsilon\vec{u})}{\partial t} + \nabla.(\varepsilon\rho\vec{u}\vec{u}) = -\varepsilon\nabla p + \nabla.(\varepsilon\mu\nabla\vec{u}) + S_u \quad (2)$$

$$\frac{\partial(\rho\vec{u}(E+p))}{\partial t} + \nabla.((E+p)\rho\vec{u}\vec{u}) = \nabla.\left(\lambda_{eff}\nabla T - \sum_k h_k J_k + (\tau_{eff}.\vec{u})\right) + S_h \quad (3)$$

$$\frac{\partial(\varepsilon X_k)}{\partial t} + \nabla.(\varepsilon\vec{u}X_k) = \nabla.(D_k^{eff}\nabla X_k) + S_k \quad (4)$$

$$\nabla.(\sigma_g^{eff}\nabla\phi_e) + S_\phi = 0 \quad (5)$$

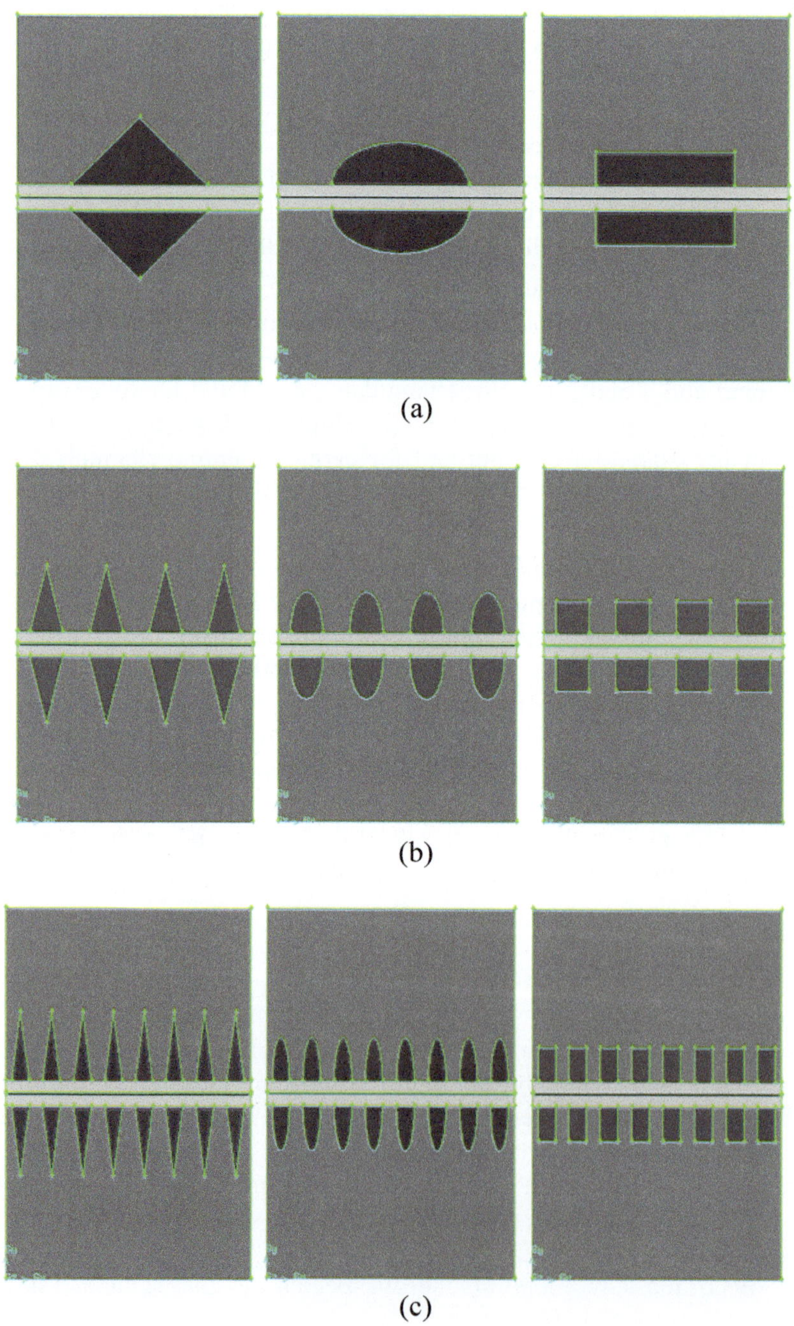

Figure 2. (a) Schematic of the one serpentine channel, (b) Schematic of the four serpentine channels, (c) Schematic of the eight serpentine channels

The corresponding source terms treating the electrochemical reactions and porous media are presented in Table 1.

Table 1. Source terms

	S_v (Momentum)	S_k (Species)	SØ (Phase potential)	S_h (Energy)
Gas channels	0	0	0	0
Backing layers	$-\frac{\mu}{k}\varepsilon^2\vec{u}$	0	0	0
Catalyst layers	$-\frac{\mu}{k_p}\varepsilon_m\varepsilon_{mc}\vec{u}-\frac{k_\phi}{k_p}c_f F\nabla\phi$	$-\frac{j_a}{2F}\frac{\rho}{\varepsilon_m\varepsilon_{mc}}$: anode H_2 $\frac{j_c}{2F}\frac{\rho}{\varepsilon_m\varepsilon_{mc}}$: cathode O_2 $-\frac{j_c}{F}\left(\frac{1}{2}+n_d\right)\frac{\rho}{\varepsilon_m\varepsilon_{mc}}$: cathode H_2O $-\frac{n_d}{F}j_a\frac{\rho}{\varepsilon_m\varepsilon_{mc}}$: anode H_2O	j	$\frac{\lvert j_c\rvert}{2F}T\lvert\Delta s\rvert+\lvert j_c\eta_c\rvert$ for cathode 0 for anode
Membrane	$-\frac{\mu}{k_p}\varepsilon_m\vec{u}-\frac{k_\phi}{k_p}c_f F\nabla\phi$	0	0	$\frac{I^2}{\sigma_{mem}}$

$$\lambda_{eff}=\varepsilon\lambda_f+(1-\varepsilon)\lambda_s \tag{6}$$

where λ_s is the thermal conductivity of the electrode solid matrix and λ_f is the thermal conductivity of the gas, which can be expressed as a function of temperature:

$$\lambda_f=a_0+a_1T+a_2T^2+a_3T^3 \tag{7}$$

It is of benefit to further explain the corresponding diffusivities of the governing equations. The diffusivities for species concentration equations and ionic conductivity for membrane phase potential equation are modified using Bruggman correlation to account for porous electrodes, which can be expressed as:

$$D_k^{eff}=\varepsilon_m^{1.5}D_k \tag{8}$$

$$\sigma_e^{eff}=\varepsilon_m^{1.5}\sigma_e \tag{9}$$

It is worth further explaining the mole fraction of oxygen appearing in Eq. 4 because oxygen is a gaseous species in the cathode flow channel and gas-diffusion electrode but becomes a species dissolved in the electrolyte in the catalyst layer and membrane regions. Our definition is given by

$$X_k=\begin{cases}C_k^g/C_{tot} & in\ gas\\ C_k^e/C_{tot} & in\ electrode\end{cases} \tag{10}$$

where C_k is the molar concentration of species k and superscripts g and e denote the gas and the electrolyte phases, respectively. Thus, X_k is a true mole fraction in the gas phase but is a pseudo mole fraction when species k is in the dissolved form. In addition, there is a discontinuity in the value of X_k at the interface between the gas-diffusion electrode and the catalyst layer due to the following thermodynamic relation

$$C_k^{e,sat} = \frac{RT}{H} C_k^g \tag{11}$$

where H is the Henry's law constant equal to $2 \times 10^5 \ atm\ cm^3 / mol$ for oxygen in the membrane. Either generation or consumption of chemical species k and the creation of electric current occurs only in the active catalyst layers where electrochemical reactions take place. The S_k and S_ϕ terms are therefore related to the transfer current between the solid matrix and the membrane phase inside each of the catalyst layers. These transfer currents at anode and cathode can be expressed as follows:

$$j_a = aj_{0,a}^{ref} \left(\frac{X_{H_2}}{X_{H_2,ref}} \right)^{1/2} \left(\exp\left(\frac{\alpha_a^a F}{RT} \eta_a \right) - \exp\left(-\frac{\alpha_c^a F}{RT} \eta_a \right) \right) \tag{12}$$

$$j_c = aj_{0,c}^{ref} \left(\frac{X_{O_2}}{X_{O_2,ref}} \right) \left(\exp\left(\frac{\alpha_a^c F}{RT} \eta_c \right) - \exp\left(-\frac{\alpha_c^c F}{RT} \eta_c \right) \right) \tag{13}$$

The above kinetics expressions are derived from the general Butler- Volmer equation based on the facts that the anode exhibits fast electrokinetics and hence a low surface overpotential to justify a linear kinetic rate equation, and that the cathode has relatively slow kinetics to be adequately described by the Tafel equation. In Eq. 12 and 13, the surface overpotential, $\eta(x, y)$, is defined as:

$$\eta(x, y) = \phi_s - \phi_e - V_{oc} \tag{14}$$

where ϕ_s and ϕ_e stand for the potentials of the electronically conductive solid matrix and electrolyte, respectively, at the electrode electrolyte interface. V_{oc} is equal to zero on the anode but is a function of temperature on the cathode namely

$$V_{oc} = 0.0025T + 0.2329 \tag{15}$$

where T is in Kelvin and V_{oc} is in volts. Notice that V_{oc} is not the true open-circuit potential of an electrode, which would then depend upon reactant concentrations according to the Nernst equation. The proton conductivity in the membrane phase has been correlated as

$$\sigma_e(T) = 100 \exp\left[1268 \left(\frac{1}{303} - \frac{1}{T} \right) \right] (0.005139\xi - 0.00326) \tag{16}$$

where the water content in the membrane, ξ, depends on the water activity, a, according to the following fit of the experimental data

$$\xi = \begin{cases} 0.043 + 17.18a - 39.85a^2 + 36a^3 & 0 \prec a \le 1 \\ 14 + 1.4(a-1) & 1 \le a \le 3 \end{cases} \tag{17}$$

The water activity is in turn calculated by

$$a = \frac{X_{H_2O} p}{p_{sat}} \quad (18)$$

where the saturation pressure of water vapor can be computed from the following empirical equation

$$\ln(p^{sat}) = 70.43464 - \frac{7362.698}{T} + 0.006952T - 9\ln(T) \quad (19)$$

In a fuel cell system, the inlet flow rates are generally expressed as stoichiometric ratios of inlet streams based on a reference current density. The stoichiometric ratios inlet streams are given by the following equations [1].

$$S^{anode} = C_{H_2} v^{anode} \frac{2F}{I_{ref} A} \quad (20)$$

$$S^{cathode} = C_{O_2} v^{cathode} \frac{4F}{I_{ref} A} \quad (21)$$

Water transport through the polymer electrolyte membrane has been investigated by several researchers in different aspects. Most interesting studies in this area includes the determination of water diffusion coefficient and water drag coefficient by Zawodzinski et al. [12, 13] and investigating the diffusion of water in Nafion membranes by Motupally et al. [14].
The electro-osmotic drag coefficient is defined as the number of water molecules transported by each hydrogen proton H^+. The electro-osmotic drag coefficient can be expressed with the following equation:

$$n_d = \frac{2.5\xi}{22} \quad (22)$$

The diffusion coefficient of water in Polymer Membrane is also highly dependent on the water content of the membrane. The relation is given as:

$$D_w^m = \begin{cases} 3.1\times 10^{-7}\xi(\exp(0.28\xi)-1)\exp(-2346/T) & 0 \prec \xi \prec 3 \\ 4.17\times 10^{-8}\xi(1+161\exp(-\xi))\exp(-2346/T) & otherwise \end{cases} \quad (23)$$

Once the electrolyte phase potential is determined in the membrane, the local current density along the axial direction can be calculated as follows

$$I(y) = -\sigma_e^{eff} \left.\frac{\partial \phi_e}{\partial x}\right|_{x = I.F.} \quad (24)$$

where I.F. means the interface between the membrane and cathode catalyst layer. The average current density is then determined by

$$I_{avg} = \frac{1}{L}\int_0^L I(y)dy \quad (25)$$

where L is the cell length.
There are natural boundary conditions of zero-flux prescribed everywhere other than the inlet and outlets of the flow channels. The boundary conditions prescribed at the inlets of the gas channels are:

$u_{in}^{anode} = u_a^0 \qquad u_{in}^{cathode} = u_c^0$

$C_{H_2}^{anode,in} = C_{H_2}^0 \qquad C_{O_2}^{cathode,in} = C_{O_2}^0$

$C_{H_2O}^{anode,in} = C_{H_2O}^{0,a} \qquad C_{H_2O}^{cathode,in} = C_{H_2O}^{0,c}$

A mesh of 325584 nods for rectangular channel, 293056 nods for triangular channel and 274360 nods was found to provide required spatial resolution for different channel geometry. The solution is considered to be converged when the difference between successive iterations is less than 10^{-7} for all variables.

The Electrochemical and Transport Parameters used in these simulations are summarized in Table2, and the operational parameters are presented in Table3.

Table 2. Electrochemical and transport properties

Anode reference exchange current density	A/m^3	1500000000
Cathode reference exchange current density	A/m^3	4000000
Anode transfer coefficient		2
Cathode transfer coefficient		2
Faraday constant	C/mol	96487
H_2 Diffusivity	m^2/s	0.00003
O_2 Diffusivity	m^2/s	0.00003
H_2O Diffusivity at anode	m^2/s	0.00003
H_2O Diffusivity at cathode	m^2/s	0.00003
Anode backing layer porosity		0.5
Cathode backing layer porosity		0.5
Permeability of anode backing layer	m^2	0.000000000001
Permeability of cathode backing layer	m^2	0.000000000001
Equivalent weight of membrane	kg/mol	1.1

Table 3. Operational parameters

Description	Unit	Value
Reference average current density	A/cm^2	1.0
Oxygen flow rate	L/min	0.5
Hydrogen flow rate	L/min	0.3
Anode inlet pressure	Atm	1
Cathode inlet pressure	Atm	1
Cell temperature	C°	60
Anode inlet molar concentration		
Hydrogen	mol/m^3	35.667
Oxygen	mol/m^3	0
Water vapor	mol/m^3	16.121
Cathode inlet molar concentration		
Hydrogen	mol/m^3	0
Oxygen	mol/m^3	7.51
Water vapor	mol/m^3	16.121

3. Results and discussion

In order to show that the program in this study can handle the cell performance of a PEMFC, we apply the present method to solve the whole of a PEMFC as described in previous section and compare it with the experimental results of authors [2].

Figure 3 show the comparison between experimental [2] and numerical results to investigate the influence of the internal flow modification on the overall fuel cell performance for three channel geometries of rectangular, triangular and elliptical shape for $T_{cell} = 60^\circ C$,

$T_{O2} = 55^\circ C, T_{H2} = 55^\circ C, \dot{m}_{O2} = 0.5 L/\min$, $\dot{m}_{I2} = 0.3 L/\min$ and P=2.905 bar. It is clear that there are good agreements between experimental and numerical results. Also it is clear that for whole operating voltage conditions the effect of the internal flow modification on the polarization curves is important that the overall fuel cell performance is at higher value when the geometry of the channel is rectangular but when the geometry of the channels shift to triangular and elliptical, the cell performance decreases that this is due to diffusing the more efficient fuel to gas diffusion layer and increasing the chemical reaction at the catalyst layer surface.

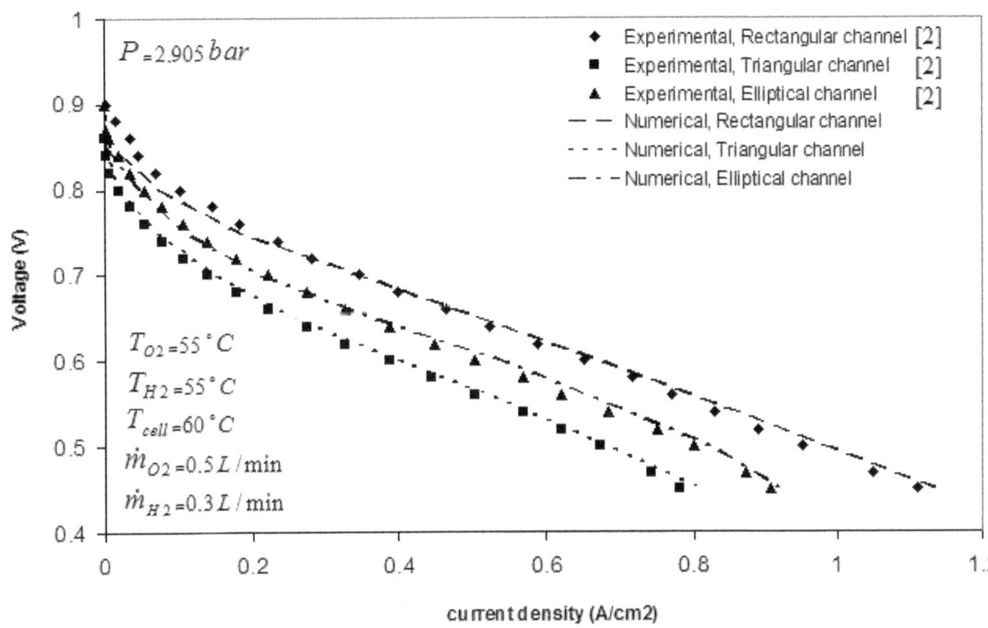

Figure 3. Comparison of numerical and experimental [2] results for three different channels for $T_{cell} = 60^\circ C$, $\dot{m}_{O2} = 0.5 L/\min$, $\dot{m}_{H2} = 0.3 L/\min$, P=2.905 bar, $T_{O2} = 55^\circ C$ and $T_{H2} = 55^\circ C$

Figure 4 and Figure 5 show the comparison between experimental [2] and numerical results to investigate the effect of hydrogen flow rate and oxygen flow rate of the anode and cathode sides at the overall cell performance of the PEM fuel cell for $T_{cell} = 60^\circ C$, $T_{O2} = 55^\circ C, T_{H2} = 55^\circ C$ and P=2.905 bar for a triangular channel geometry. It is clear that there are good agreements between experimental and numerical results. It is also clear that by increasing the hydrogen flow rate from 0.3 L/min to 0.7 L/min and the oxygen flow rate from 0.5 L/min to 0.9 L/min the cell performance enhances but when the flow rate increases from 0.7 L/min to 0.9 L/min for hydrogen and from 0.9 L/min to 1.3 L/min the cell performance decreases. It is due to that by increasing the flow rate of hydrogen and oxygen more fuel and oxidizer transport from GDL to the catalyst layer and the electrochemical reaction enhances but when the flow rate of hydrogen and oxygen increase from 0.7 L/min and 0.9 L/min the transportation of fuel and oxidizer to the GDL decrease and they come out from the channel without an electrochemical reaction.

Figures 6, 7 and 8 show the effect of the number of channels on the performance of a rectangular, triangular and elliptical cross-section channel geometry PEM fuel cell at $T_{cell} = 60^\circ C$, $T_{O2} = 55^\circ C, T_{H2} = 55^\circ C, \dot{m}_{O2} = 0.5 L/\min$, $\dot{m}_{I2} = 0.3 L/\min$ and P=2.905 bar. It is clear that at the conditions of the higher operating voltage (lower over potential), the influence of the internal flow modification on the overall fuel cell performance is negligibly small. At lower operating voltage conditions, on the other hand, the effect of the internal flow modification on the polarization curves becomes important. Also it is clear that when the number of channels at GDL increases from one to four and eight channels, the performance of fuel cell increases about 18%, 15% and 13% for rectangular, elliptical and triangular channel geometry that this is due to better diffusion of gases to GDL and increasing the electrochemical reaction at the catalyst layers. Also it is found from this Figs. that the overall fuel cell performance is at higher value when the geometry of the channel is rectangular but when the geometry of the channels shift to triangular and elliptical, the cell performance decreases that this is due to diffusing the more efficient fuel to gas diffusion layer and increasing the chemical reaction at the catalyst layer surface.

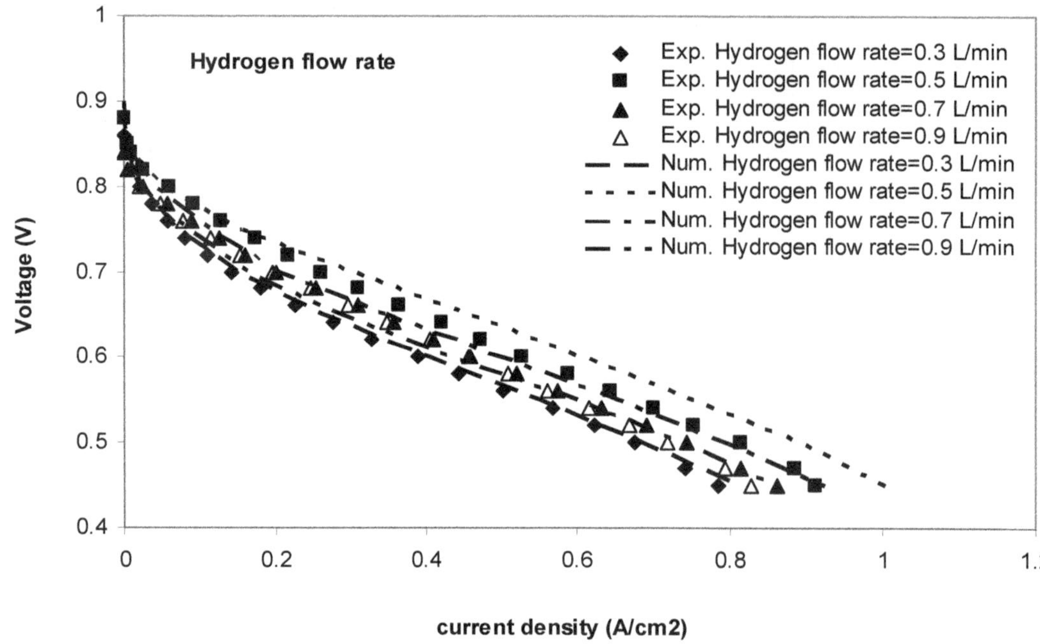

Figure 4. Variation of cell performance at different hydrogen flow rates for $T_{cell} = 60^{\circ}C$, $\dot{m}_{O2} = 0.5L/\min$, P=2.905 bar, $T_{O2} = 55^{\circ}C$ and $T_{H2} = 55^{\circ}C$

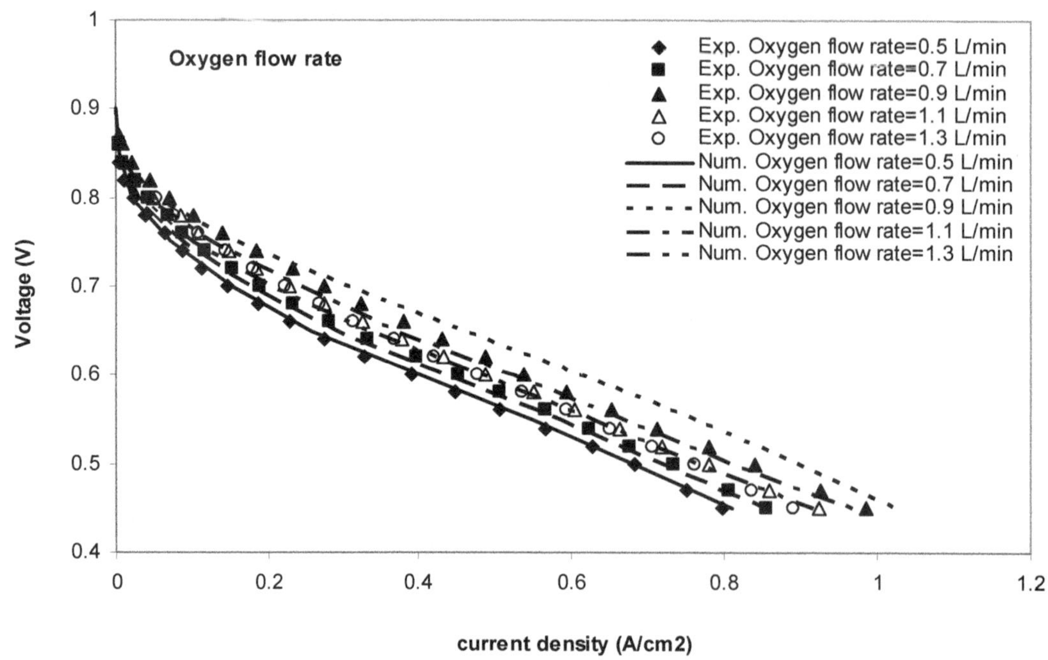

Figure 5. Variation of cell performance at different oxygen flow rates for $T_{cell} = 60^{\circ}C$, $\dot{m}_{H2} = 0.3L/\min$, P=2.905 bar, $T_{O2} = 55^{\circ}C$ and $T_{H2} = 55^{\circ}C$

Figures 9, 10 and 11 show the effect of the area (depth) of channels on the performance of a rectangular, triangular and elliptical cross-section channel geometry PEM fuel cell at $T_{cell} = 60^\circ C$, $T_{O2} = 55^\circ C, T_{H2} = 55^\circ C, \dot{m}_{O2} = 0.5L/\min$, $\dot{m}_{l_2} = 0.3L/\min$ and P=2.905 bar. It is clear that at lower current density the effect of area of channels is small on the performance but at higher current density the effect of it becomes important. Also it is clear that when the area of channels increases from $1mm^2$ to $1.5mm^2$ and $2mm^2$ the performance decreases about 9%, 11% and 13% for triangular, rectangular and elliptical channel geometry that this is due to decreasing the velocity of reactant gases at the channels and decreasing the diffusion of them to the GDL and electrochemical reaction at the catalyst layer.

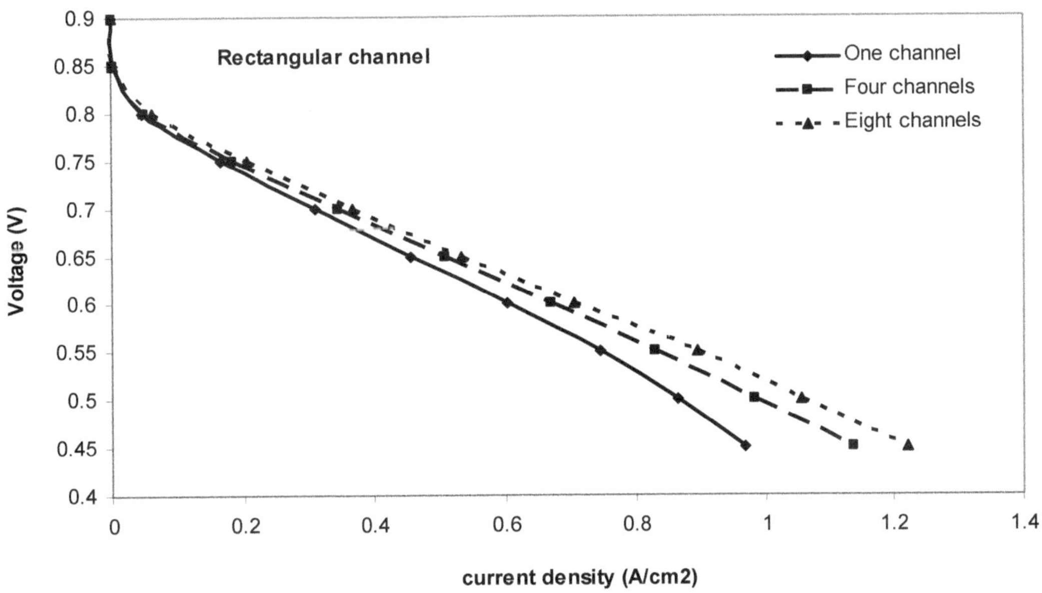

Figure 6. Effect of the number of channels on the performance for rectangular channel

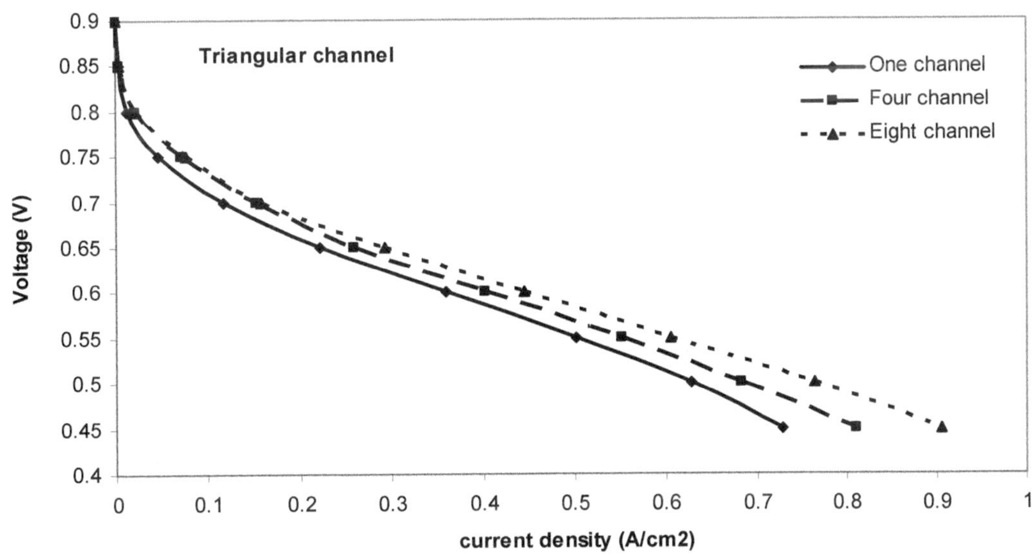

Figure 7. Effect of the number of channels on the performance for triangular channel

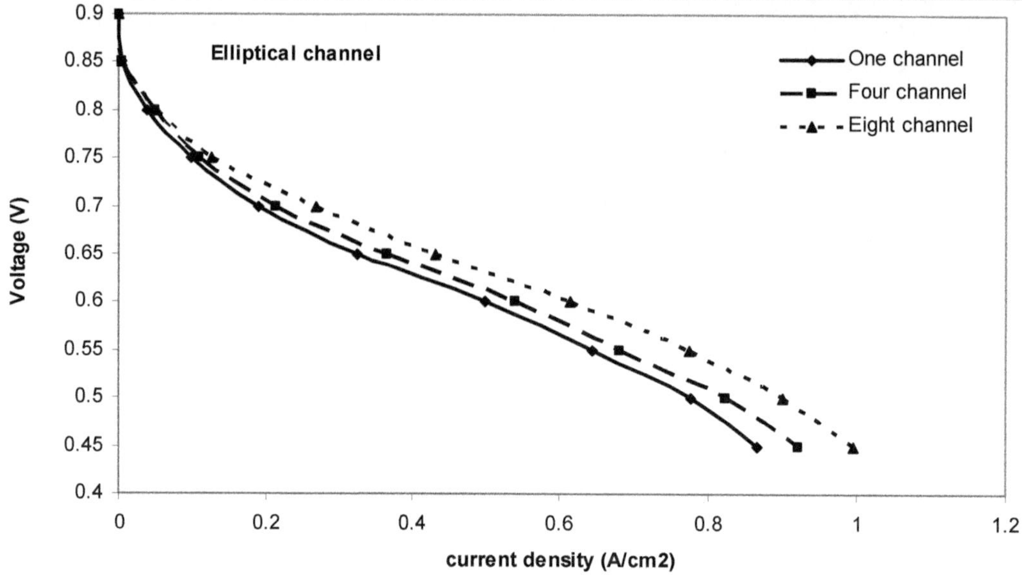

Figure 8. Effect of the number of channels on the performance for elliptical channel

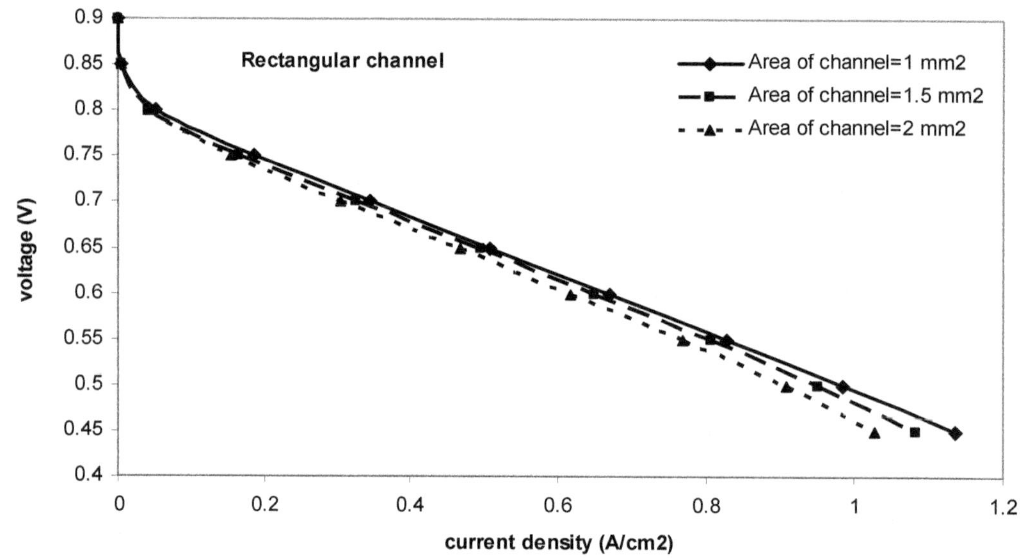

Figure 9. Effect of the area (depth) of channels on the performance for rectangular channel

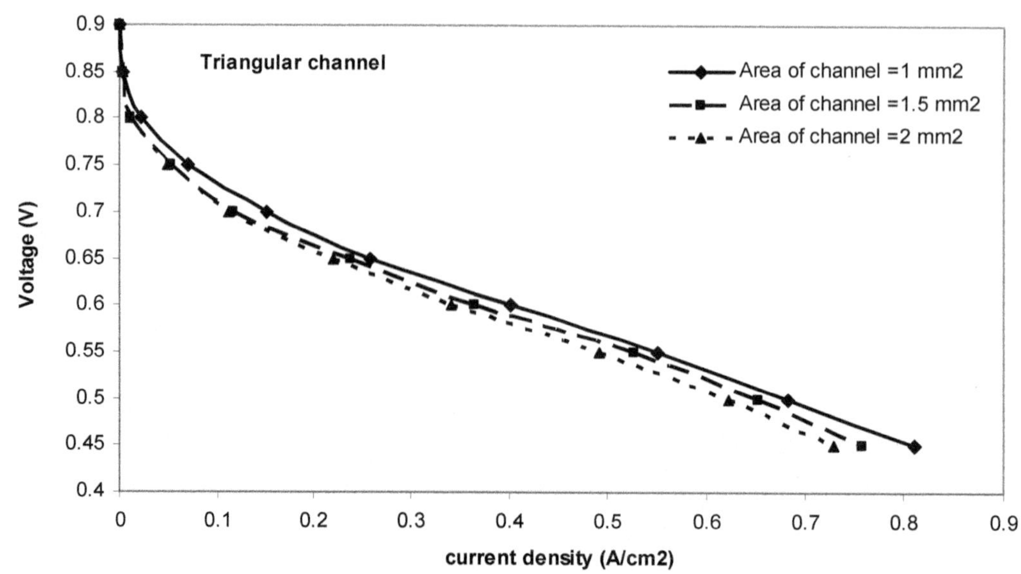

Figure 10. Effect of the area (depth) of channels on the performance for triangular channel

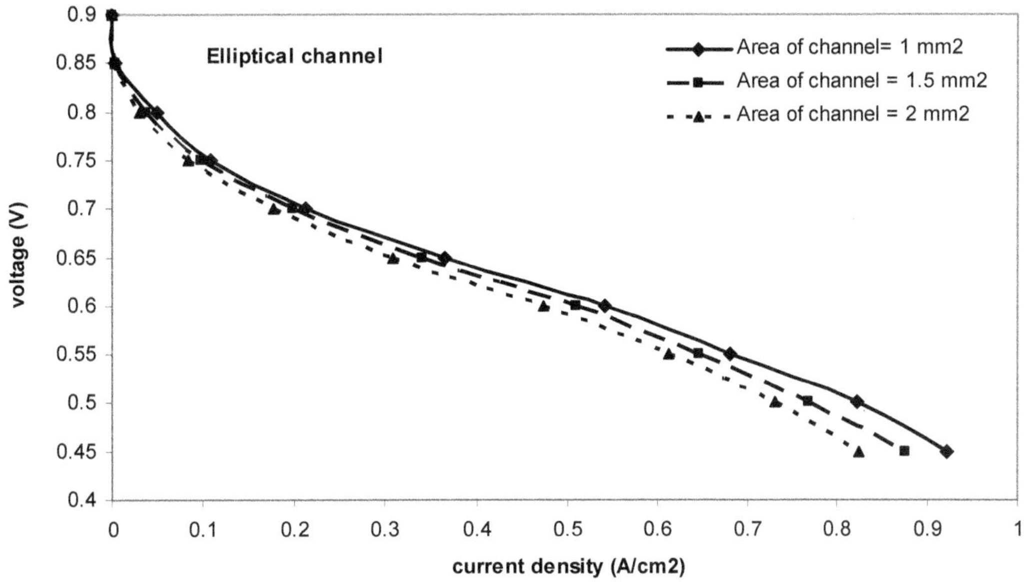

Figure 11. Effect of the area (depth) of channels on the performance for elliptical channel

4. Conclusion

A complete three-dimensional and single phase model for proton exchange membrane (PEM) fuel cells has been used to investigate the effect of increasing the number and cross section area of channels on the performance for three different channel geometry such as rectangular, triangular and elliptical geometry. The complete three-dimensional model for PEM fuel cells based on the two-fluid method was numerically solved with constant-temperature boundary condition at surfaces of anode and cathode current collectors. The results of this paper are in good agreement with experimental results of authors [2]. The results show that when the number of channels at GDL increases from one to four and eight channels, the performance of fuel cell increases about 18%, 15% and 13% for rectangular, elliptical and triangular channel geometry. Also the results show that when the area of channels increases from $1mm^2$ to $1.5mm^2$ and $2mm^2$ the performance decreases about 9%, 11% and 13% for triangular, rectangular and elliptical channel geometry.

Nomenclature

A	Superficial Electrode Area, m^2
C	Molar Concentration, mol/m^3
D	Species Diffusivity, m^2/s
h_k	Enthalpy of species k
I	Current Density, A/m^2
i_0	Reference current density (A/cm^2)
U	Inlet Velocity, m/s
j	Transfer Current Density, A/m^3
$\vec{u}$	Velocity vector, m/s
p	Pressure, Pa
S	Stoichiometric Ratio
T	Temperature, K
V_{oc}	Reference open-circuit potential of an electrode
X_k	Molar fraction of kth species

Greek letters

η	Overpotential, V
ρ	density, kg/m^3
ε	porosity
σ	Ionic Conductivity, S/m
ϕ	Phase Potential, V
v	Volumetric Flow Rate, m^3/s
ξ	Water Content of The Membrane
μ	Viscosity, kg m/s
α	Transfer coefficient for the reaction
τ_{eff}	Effective stress tensor
λ_{eff}	Effective thermal conductivity in a porous material

Acknowledgment

This work was partially supported by Renewable Energy Organization of Iran.

References

[1] Larminie, J. and Dicks, A. "Fuel cell system explained". 2nd ed, Wiley (2003).

[2] Khazaee, I. "Effect of placing different obstacles in flow fields on performance of a PEM fuel cell: Numerical investigation and experimental comparison" Heat and Mass transfer, 49(9), pp. 1287-1298, (2013).

[3] West, A.C. and Fuller, T.F. "The Influence of Rib Spacing in Proton-Exchange Membrane Electrode Assemblies", Journal of applied electrochemistry, 26, pp 557-565 (1996).

[4] Chiang, M.S. and Chu, H.S. "Effect of placing different obstacles in flow fields on performance of a PEM fuel cell: Numerical investigation and experimental comparison Numerical investigation of transport component design effect on a proton exchange membrane fuel cell", Journal of Power Sources, 160, pp 340–352 (2006).

[5] Wang, Z.H., Wang, C.Y. and Chen, K.S. "Two-phase flow and transport in the air cathode of proton exchange membrane fuel cells", Journal of Power Sources, 94, pp 40–50 (2001).

[6] Kuo, J.K., Yen, T.S. and Chen, C.K. "Improvement of performance of gas flow channel in PEM fuel cells" Energy Conversion and Management, 49, pp 2776–2787 (2008).

[7] Khazaee, I., Ghazikhani, M."Performance improvement of proton exchange membrane fuel cell by using annular shaped geometry", Journal of Power Sources, 196, pp.2661-2668 (2011).

[8] Khazaee, I., Ghazikhani, M. and Nasr Esfahani, M. "Effect of gas diffusion layer and membrane properties in an annular proton exchange membrane fuel cell ", Applied Surface Science, 258, pp.2141-2148 (2012).

[9] Khazaee, I., Ghazikhani, M."The effect of material properties on the performance of a new geometry PEM fuel cell", Heat and Mass Transfer, 48, pp.799-807 (2012).

[10] Yan, W.M., Soong, C.Y., Chen, F.L., Chu, H.S. "Effects of flow distributor geometry and diffusion layer porosity on reactant gas transport and performance of PEM fuel cells", Journal of Power Sources, 125, pp.27-39 (2004).

[11] Chu, H.S., Yeh, C., Chen, F., "Effects of porosity change of gas diffuser on performance of proton exchange membrane fuel cell", Journal of Power Sources, 123, pp.1–9 (2003).

[12] Zawodzinski, T.A., Neeman, M., Sillerud, L.O. and Gottesfeld, S. "Determination of water diffusion coefficients in perfluorosulfonate ionomeric membranes", The Journal of Physical Chemistry, 95, pp.6040-6044 (1991).

[13] Zawodzinski, T.A., Davey, J. Valcrio, J. and Gottesfeld, S. "The water content dependence of electro-osmotic drag in proton-conducting polymer electrolytes", Electrochimica Acta, 40, pp.297-302 (1995).

[14] Motupally, S., Becker, A.J. and Weidner, J.W. "Diffusion of Water through Nafion® 115 Membranes", Journal of the Electrochemical Society, 147, pp.3171-3177 (2000).

Control of temperature and energy consumption in buildings - A review

D.W.U. Perera, C. F. Pfeiffer, N.-O Skeie

Faculty of Technology, Telemark University College, Porsgrunn, Norway.

Abstract

Building sector is one of the largest energy consumers in the world and currently it utilizes 40% of the total energy in the European Union. At the beginning of the article, energy crisis related to the buildings is defined with regard to occupant thermal comfort, energy savings and temperature control. Subsequently, a brief presentation of various types of building heating models available for control purposes is given. Afterward, different approaches used for controlling the building thermal comfort and the energy consumption are shown. These strategies are primarily, classical control, advanced control, intelligent control and hybrid control. The proposed survey also provides up-to-date applications of control techniques. The overview hence affords an insight into current control systems used for temperature and energy consumption in buildings. Further, it helps to have a comprehensive understanding about the variety of control techniques in the field of HVAC (Heating, Ventilation and Air conditioning) applications, at the same time delivering information for careful design of suitable controllers.

Keywords: Building heating control; Energy efficiency; Temperature control; Thermal models.

1. Introduction

Buildings are one of the largest energy consumers all over the world. According to the EU Directive on the Energy Performance of Buildings issued in 2010, the building sector in the European Union is continuously expanding and currently uses 40% of total energy consumption in the union [1]. Amongst, residential buildings are the main energy consumer. In 1999, the shares among space heating, water heating and electrical appliances together with lighting were 68%, 14% and 13% respectively [2]. However, during the last 15 years, the space heating energy demand has decreased owing to the national laws, regulations and administrative provisions forced by the governmental bodies [2]. Several policies are mandated by the European Commission because the European Union is dependent on the external oil and gas. In 2000, 50% of the energy supplies were provided by the external countries. Based on the projections it will become increasingly reliant in the future and will reach 70% in 2030 [3]. According to [4], in Norway, household buildings have utilized 34% of the total energy consumption in 2010. Out of that, space heating has consumed two third of the residential building energy and two fifth of commercial building energy [4]. In the early 1990s, Norwegian household energy consumption per square meter was about 210 kWh/m^2, and it has been reduced to 180 kWh/m^2 in 2010, which is a decline in 14% [4]. Recent investigations showed that there is a potential of saving 65 TWh both from residential and industrial buildings in 2020 [5]. Technical potential for savings based on all existing dwelling is 13.4

TWh and possible savings from commercial buildings are 19.5 TWh in 2020 [5]. Norwegian building technical standards, such as TEK10, direct the people for accomplishing these goals.

Even though the buildings constructed before 1970s had enough openings for natural ventilation, later buildings were built with small openings. It helped to overcome the soaring energy cost occurred ought to the oil embargo. This lets low air flow rates into the building. Since people are living inside the buildings for more than half of their life, environmental comfort is important for occupant satisfaction and to obtain a high productivity. There are four main factors that affect the quality of life inside the buildings, and they are: (i) thermal comfort; (ii) visual comfort; (iii) aural comfort; and (iv) indoor air quality [6]. Thermal comfort is measured by Predictive Mean Vote (PMV), which has values from -3 to +3 [7]. Lowest value indicates the coldest condition; the highest value indicates the hottest condition and zero indicates the neutral condition. Comfort conditions are achieved if the PMV belongs to the [-0.5 0.5] range which is desired inside a building [7]. Visual comfort is about having enough luminance level either by solar radiation or lighting. Aural comfort enhances the acoustical environment inside the building which will improve the speech intelligibility and privacy. CO_2 concentration inside a building characterizes the level of indoor air quality (IAQ) [6]. Maintaining IAQ is a major problem with small openings in a building as it reduces the natural ventilation.

If the building is provided with higher amounts of ventilation, more input energy must be supplied to raise the temperature of the incoming air to obtain thermal comfort. When the heating systems are operated continuously throughout the 24 hours of the day, the thermal comfort can be satisfied easily, but it may lead to energy overconsumption as the occupancy is intermittent. Therefore there is a compromise between the air quality and thermal comfort when it comes to energy savings. Further, solar irradiation and lighting helps to increase the indoor air temperature which reduces the external heating requirements. Moreover, the energy consumed to raise the temperature to a different level may be lower compared to preserving the current conditions. Therefore during low outside temperatures, low occupancy levels and at nights temperature can be lowered either by lowering the heater power or by switching it off depending on which is more efficient. Subsequently the temperature can be brought back to the desired level. Hence it is important to manage the three factors ventilation, temperature and lighting simultaneously to obtain a better quality of life while saving energy.

The most often studied building types in the latter discussed research area are residential and office buildings. These buildings range from small rooms to multi-zone buildings with several floors. There are various factors other than the thermal comfort, indoor air quality and lighting that influence the energy consumption in each of these buildings. They are the age of the building, size of the building, envelope construction, weather conditions, efficiency of the equipments and hot water production. Utilization of the best available equipments and components can minimize the impact from these factors, but it will not be an optimal solution. To obtain an optimal energy performance it is necessary to integrate the effects from each and every fact mentioned above. It can be carried out via a precisely selected building control system. A good control system ought to have a lot of benefits such as minimize the energy consumption, reduce the pollution caused by energy usage, improve the comfort, prevent out of hours operation of the equipments, reduce the maintenance cost and limit the excessive wear and tear associated with the building systems. Throughout this article we focus on the control of indoor air temperature for having a better thermal comfort and low energy consumption.

Building heating control systems can be either time controllers (occupancy based controllers) or condition based controllers. Time controlled heating systems response to programmed time frames based on the occupancy of the building. They can be developed either for a 24 hour period with regular daily schedules or for a 7 day period with varying specifications in each day [8]. Further, the time controllers can be either simple on/off type switches or optimal controllers. Optimal time controllers can decide the time needed for reaching the preferred temperature, and they are more efficient compared to the simple on/off time switches [8]. Building heating control by condition means control with respect to temperature, humidity or whatever the control variable. These two control methods may use various strategies such as On/Off control, PID control, predictive control, adaptive control, optimal control and intelligent control or combinations of them [2] to achieve the heating requirements.

1.1 Building heating models

There are three classes of building heating models available for control purposes: (i) mechanistic models (white box models); (ii) black box models; and (iii) grey box models. Mechanistic models are developed based on the physical principles of heat transfer and fluid dynamics. They consist of several equations, which carry numerous coefficients, to represent the building geometry and thermal properties of the building envelope and equipments. A large number of numerical software tools are frequently met for solving such systems of equations [9]. Still, mechanistic models have some deficiencies in the area of control with regard to the calibration of physical parameters. Black box models can be generated for a particular building after measuring the input variables such as temperature, humidity, air flow rates, wind speed and solar irradiation. Examples of black box models are system identification based models, regression models, genetic algorithm, neural network models, fuzzy logic models, neuro-fuzzy models and support vector machine [9-12]. These models do not expect any physical data of the system and also the resulting model coefficients do not have any physical meaning [11]. Black box models may be more reliable compared to physical knowledge based models, but they are valid only for the building where the data is collected. Grey box models are a blend of mechanistic and black box models and hence the information about these systems is partly known [9, 12]. They are most commonly used for parameter estimation and there is only a limited work has been done on them [12].

Simulation programs like Energy Plus, TRNSYS and Fluent provide comprehensive physics based models for buildings. Even though they can be highly accurate, they may have high computational burden specially when applying to on-line control [13]. Black box models can give unrealistic or non-physical results when the inputs are outside the training range. The intended model for building heating control is a simplified, fast enough and in-depth dynamic model which can be integrated with most of the architectural designs. However, to choose the best model for control, model validation is required. Kramer et al. have reviewed contemporary research articles about simplified thermal and hygric building models [10]. They have categorized the simplified models into three collections: (i) neural network models; (ii) linear parametric models; and (iii) lumped capacitance models. Their summing-up concludes that the modeling of sun irradiation and thermal capacitance is not well executed. Besides, it has been hard for them to find good models dealing with both temperature and relative humidity. Another group of researchers has published an article which includes most of the available techniques present in a broad area of building thermal modeling [9]. They have defined physical models, machine learning tools based models and hybrid models as the model classes. In [11], physically based models and data driven models which are predominantly used for control purposes have briefly examined. Zhao and Magoulès [12] follow the same classification produced by [10]. Instead, they use different names: (i) simplified engineering methods; (ii) artificial intelligence methods; and (iii) statistical methods.

1.2 Objective

There are quite a large number of publications dealing with building heating control systems related to thermal comfort and energy saving. In spite of the large amount of research efforts on building heating control strategies, a comprehensive literature review is missing. Still, [2] has briefed the strategies with their advantages and disadvantages. It has classified the indoor building environment control systems into three categories: (i) conventional methods; (ii) computational intelligence systems; and (iii) agent based intelligent control systems. Intelligent control systems are the main focus of their work, and further attention has given to describe the design of agent-based intelligent control systems. Moreover, in [14], available control techniques in HVAC systems are discussed. They have summarized the general details of traditional, advanced and intelligent control techniques.

This article is lined up to give an overview of the existing and most commonly found heating control methods, and appropriate thermal modeling techniques. Section 2 describes theoretical details about control strategies including their advantages, limitations and up-to-date applications. Ultimately, a closure Summary about the building heating control will be presented in section 3.

2. Building control strategies

The primary goal of controlling the building environment (HVAC system) is to maintain the thermal comfort of the occupants and to achieve good energy efficiency. However, in most of the situations, achievement of one of these goals may cause the other goal to be sacrificed to a certain extent. In building control, the usually accepted setpoints for temperature and humidity inside a residential or office building are 22^0C temperature and 45% relative humidity with an operating band of $\pm 2^0$C and $\pm$15% RH [14]. The different strategies used for controlling the building thermal environment within the given ranges are roughly categorized into four classes in this paper. They are: (i) classical control techniques; (ii) advanced control strategies; (iii) intelligent control methods; and (iv) hybrid control methods. However, there exists a certain overlap among the different control technologies.

2.1 Classical control techniques

The two main classical controllers used in building control are two-position control (on/off) and PID control. They have a simple structure and low initial cost, which makes them the most used controllers in HVAC systems in both commercial and residential buildings [14]. Yet, the maintenance cost of the actuator is quite high (because of the intensive oscillations) and the energy efficiency is low for both of these mechanisms that cause high cost [14].

On/off control is one of the oldest techniques that is practiced in buildings for the purpose of energy saving and occupant thermal comfort. It is a simple, fast and inexpensive feedback controller that accepts only binary inputs which is also known as bang-bang control and hysteresis control. This control technique is still being using in domestic and commercial buildings effectively, as the well known thermostat, humidistat and pressure switch [14]. Thermostats are commonly found in home heating systems and domestic refrigerators.

Basic function of a thermostat is to control the heating or cooling equipment according to the target and current conditions. They should not be affected by draughts and heat sources and also advised to fix them in a building halfway up the wall [8]. Thermostat plays a vital role in providing comfort to the people and controlling the most energy intensive systems in dwellings. Furthermore, it can control heating and cooling equipments, humidifiers and dehumidifiers, economizers and ventilation systems [15]. The basic components of a thermostat are: (i) sensor to measure the temperature in the desired environment; (ii) switch/actuator to turn the heating or cooling equipment on and off; (iii) feedback loop to find the offset and decide on/off time; and (iv) user interface to display the current conditions [15]. Thermostat sensor must be accurate, and it should provide a quick response. If the sensor is sluggish, it cannot detect the temperature changes fast enough, which causes the system to consume more energy with less occupant comfort [16]. For example if the thermostat time constant is 1 sec it is fast and if it is 90 sec response is slow [16]. In order to avoid hunting (hunting produces a continuously changing deviation from the normal operating point which is often known as instability, cycling or oscillation), thermostats are introduced with a dead zone of $\pm 1^0$C to $\pm 1.5^0$C. First thermostats were either mechanical or mercury based devices. However they are now replaced with electronic devices, which also have wireless connections instead of wired connections. Nowadays, many new features and functions have emerged to thermostats to facilitate more energy saving [15] such as programmable thermostats [15]. They are simple and trouble free devices [17] and the most advanced thermostats can control even multiple zones and humidity levels [15]. While thermostat is a simple and inexpensive control method, it is often incapable of tracking the setpoint accurately and hence could be inefficient in terms of energy.

Same as on/off control, PID control is also a feedback control mechanism, which does not use knowledge/model of the interested system. It determines the error, which is the difference between the measured process variable and the desired setpoint, and adjusts the control signal according to that value. There are three separate control techniques used in the PID control algorithm: (i) proportional term relates to the present offset; (ii) integral term depends on the accumulation of past errors; and (iii) derivative term predicts the future offsets based on the current rate of change of the process. A control signal is delivered based on a weighted sum of these three actions. The distinct effect of these three terms causes the most important stimulus for the survival of the PID control mechanism, and it also committed to the evolution of modern control approaches [14]. It could be beneficial for certain applications to apply only one or two actions out of the three by setting the other parameters to zero. P control and PI control are two mostly used control algorithms. Thermal process dynamics in a building is usually a slow responding process. Therefore, proportional control can be engaged in building temperature control with a good stability and a reasonable small offset. Also, it is good in building humidity control. Derivative term also contributes to combat the sudden load changes encountered in the system [14]. Still, small amounts of measurement and process noise can cause large variations in the output due to the derivative term present in the PID control.

Even though there are a number of advantages in using PID control such as simplicity of implementation [14], it may not be the most suitable controller for building control due to several reasons [18]. It requires three parameters to be trained for each building zone after the installation. This is quite a time consuming task and re-tuning after the commissioning may be inconvenient. They are unable to handle random disturbances, and therefore large deviations from the setpoint can occur. In buildings, thermal interaction between the zones leads to multi-variable behavior. However, standard PID controller assumes a single-input single-output (SISO) system during the analysis which may cause unacceptable deviations. Since these controllers operate at low energy efficiencies they may not be suitable in the long run [18].

Shein et al. [19] have used a PID controller to control the room temperature using 2 actuators; (i) air conditioner; and (ii) window opening. Kulkarni and Hong have introduced a proportional control system for a residential building heating system, and they have compared it with the two position and PI control schemes with respect to energy consumption and thermal comfort perspectives [20]. They concluded that both two position control and proportional control are pretty similar in the energy consumption perspective, but PI control has benefits over thermal comfort too. Finally, they have stated that proportional control has advantages over on/off control relative to equipment life owing to much smoother control signal.

2.2 Advanced control techniques

Classical control strategies can be applied to building control with certain limitations. They are easy to tune for (SISO) systems but, it is not easy or even impossible to tune them for multiple-input, multiple-output (MIMO) systems. Even though, PID control is one of the best control strategies, it cannot reflect the very important outside temperature effects unless feed forward control is integrated. Moreover, it may not be the best solution for non-linear systems like HVAC systems. Large time delays and higher order dynamics are still hard to control using PID controllers. These requirements can be satisfied with advanced control techniques. These techniques usually use a dynamic model of the system for control purposes, and show non-linear characteristics [2]. As HVAC systems are non-linear, these techniques are extensively used in building control [14]. This paper outlines the advanced control techniques in three different aspects: (i) predictive control; (ii) adaptive control; and (iii) optimal control.

2.2.1Model predictive control (MPC)

Application of predictive control for building automation systems can provide increased energy savings, and they are more cost effective than non-predictive control applications. Several benefits, other than the energy savings, can be attained by using the predictive control framework in building automation. Some of benefits are robustness to disturbances and changes, multivariable control, improvement of the steady state response, future disturbances prediction, future control actions prediction and many more [21, 22]. There are many applications where predictive control is applied to a variety of building HVAC systems. They can be applied to both single zone and multi zone (a zone is generally a completely enclosed or partly open division of a building) buildings including residential, office and public buildings [22]. However, the decision of implementing the predictive control for a particular building depends on the return time of investments [23].

MPC was formulated in 1970s, and at present, it is extensively used in process control applications. It generally provides superior performance in terms of lower energy consumption, better transient response, robustness to disturbances and consistent performance under fluctuating conditions [22]. MPC can also cope with slow moving processes with time delays, and with a large number of manipulated and controlled variables [24]. This multivariable control technique is based on a prediction model which utilizes past information and the future inputs to predict the future output. While controlling the process, using the system model, MPC generates a control vector that minimizes a certain cost function over the prediction horizon in the presence of disturbances and constraints. Model identification is the bottleneck of the whole MPC application procedure, and there are not any stringent requirements on the model structure. It is possible to use any kind of model out of the three types: black box, grey box and white box [23].

Physics based models used in MPC are generally analogous to electrical RC networks. They are dynamic first order models produced using lumped thermal capacitance and resistance of the building. However, finding the first principle model could be time consuming for complex building structures. Data driven models (grey or black box models) fit linear and non-linear mathematical functions to the measured data of the building HVAC system. They can be developed using artificial neural networks, fuzzy logic, support vector machine, first and second order time delay models, statistical models (AR (Autoregressive), ARX (Autoregressive exogenous), ARMA (Autoregressive moving average), ARMAX (Autoregressive moving average exogenous), BJ (Box Jenkins), OE (Output Error)) and combinations of them. Prívara has explained subspace identification methods (4SID), prediction error methods (PEM), MPC relevant identification (MRI), deterministic semi-physical modeling (DSPM) and probabilistic semi physical modeling (PSPM), which are popular statistical methods used in the predictive model formation [23]. MRI, DSPM and PSPM are grey box models while 4SID is a black box model. Accuracy of these data driven models depend on the quality of the measured data. Hence the measurements must have high accuracy, low noise and appropriate sampling frequency depending on the process dynamics [22]. However, in general, precision of the data driven models is high compared to physics based models. Even though, the comprehensive models formulated using programs such as Energy Plus, Fluent and TRNSYS are highly accurate; they are not usually used for controller development [22]. High quality models are very accurate, but quite complex and not good in computational tractability with the loss of physical insight. Use of low quality models may result in less energy saving potential. The best is to use a model with least possible complexity, which is developed on simple physics or simple mathematical formulations, to achieve reasonable accuracy and computational simplicity.

A family house consisting of two rooms and an attic is controlled using MPC in [25]. A linear state space model developed with resistor capacitor network is used as the predictive model to control heating and cooling with a prediction horizon of 24 hours. Kim and Brown have also developed a reduced order state space model with 10 states for controlling a single zone of a building [13]. Reduced order models (ROM) provide less model complexity and less computational requirements while preserving the prediction accuracy. They have successfully used the ROMs in MPC and compared the results with a TRNSYS model.

HVAC systems are non-linear complex systems with delays, and, therefore, it is usually difficult to model them using first principles and apply them in MPC. Back propagation neural networks (NN) with few layers can approximate these non-linear systems with high recognition. A back propagation neural network to establish a prediction model for a HVAC system is used in [26]. They have used four inputs (room temperature and chilled water inlet valve opening at time *t* and *t-1*) to develop a prediction model and used MPC and feedback correction to control the air handling unit. It has been concluded that the used control mechanism can efficiently control the room temperature. Both feed forward neural network prediction model and an ARX model for a single zone building block have been developed by [27], and they have concluded that non-linear NN models with Levenberg Marquardt algorithm gives accurate temperature prediction than linear ARX models. Fuzzy logic techniques can also be applied to building predictive control applications. Fuzzy predictive control scheme has been practiced to control the temperature in an air-conditioning (AC) system in [28]. The controller is based on a Takagi Sugeno fuzzy model, and it performs well in controlling the non-linear dynamic AC unit with less computation. Support vector regression (SVR) has universal approximation ability, and it can be used to model non-linear HVAC systems [29]. It minimizes the prediction error and model complexity simultaneously. Xi et al. used SVR to build a non-linear dynamic model for a HVAC system. Consequently, they developed a non-linear MPC based on SVR model, to control the temperature and relative humidity in a thermal chamber successfully [29]. Lixing et al. also used SVR to predict the cooling load of a building HVAC system [30].

Subspace identification methods with Kalman filter prediction are demonstrated for a two floor residential building by [31]. The building was equipped with a large wireless sensor network to measure the temperatures, humidities and solar radiation. Two statistical approaches, ARMAX and subspace identification methods, to predictively control a ceiling radiant heating system are used in [32]. Collection of data from a large multi-zone office building to build a predictive control model is not straightforward. To overcome this issue, [33] has developed an implicit model of the office building in Energy Plus software. Then they have generated good quality data for predictive model, after exciting the implicit model with specially proposed signals. Later, predictive control of the building is implemented with system identification methods. ARMAX models are also a type of black box linear parametric models preferred in building predictive control. One such a model is presented in [34], which can be used to control the room temperature in office buildings. Furthermore, [35] justified the application of black box linear ARX, ARMAX, BJ and OE models for prediction of room temperature and relative humidity of an office room in a commercial building. For more details about statistical prediction models, interested readers can refer to [36, 37].

2.2.2 Adaptive control

Adaptive control is a specific type of non-linear control system applicable to processes with changing dynamics in normal operating conditions subjected to stochastic disturbances. They control the processes in a closed loop and the information about the system characteristics are obtained online while the system is operating. When the parameters of the plant dynamic model are unknown and/or vary in time the adaptive control system still can obtain or sustain the desired level of control system performance [38]. Conventional control systems use feedback to reject the effect of disturbances upon the controlled variables. Further, they do not determine the control system performance. Instead, adaptive control system measures a particular performance index of the control system using the inputs, the states, the outputs and the known disturbances [38]. After analyzing the measured performance index with the reference PI's, adaptation mechanism modifies the parameters of the adjustable controller in order to maintain the performance index of the control system. Hence the adaptive control system can be interpreted as a feedback system where the controlled variable is the performance index [38]. The primary feedback handles the process signal variations and secondary feedback handles the process parameter changes making the control adaptive. The basic configuration of an adaptive control system is presented in Figure 1.

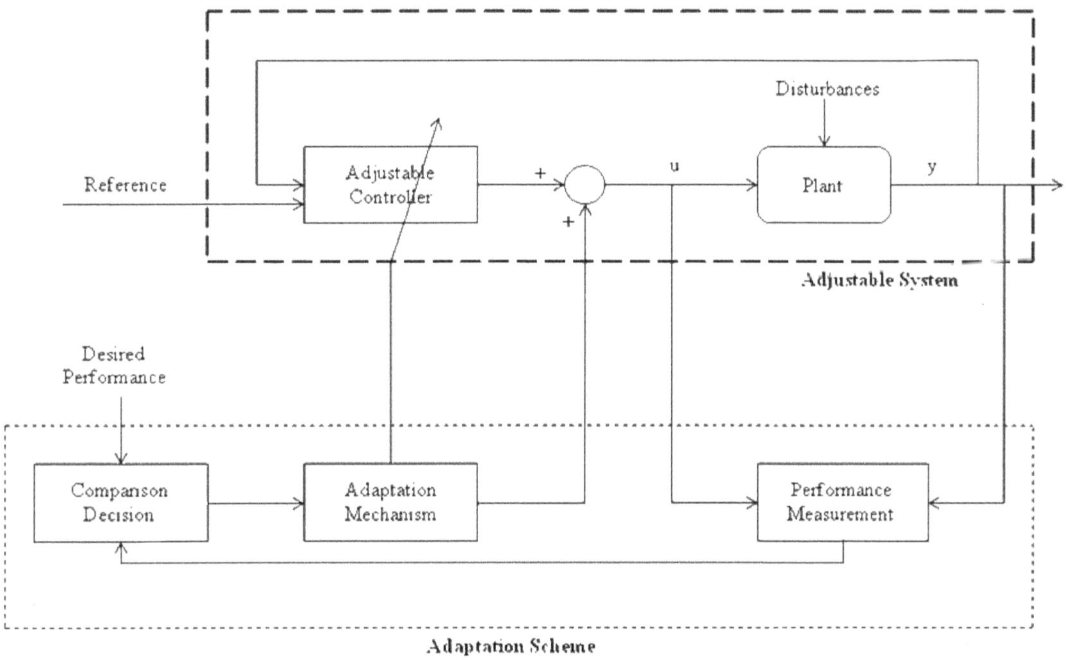

Figure 1. Basic configuration of adaptive control system [38]

Most adaptive control systems can be split into two main groups as feedback adaptive control and feed forward adaptive control. Feedback adaptive controllers are further classified into several categories [39]. Feed forward adaptive control is also said to be open loop adaptive control. This technique considers that there exists a rigid relationship between some measurable variables characterizing the environment, and the parameters of the system model. System performance modifications resulting from the controller parameter changes are not measured and fed back to a comparison-decision block in order to check the efficiency of the parameter adaption [38, 39]. Hence, there is a possibility of system failure if the relationship between the environmental measurements and system model parameters change [38]. However, this type of adaptive control systems react fast to the process changes which is an advantage [39]. There are also some limitations of open loop adaptive control. Negligence of all effects related to unmeasured signals/disturbances, unpredictable changes of the system behavior and the parameter storage requirements to accommodate many operating conditions are some of them [39]. Gain scheduling, a feed forward adaptive control scheme, was used for the first time in 1950s to control aircrafts and they are widely used in variety of situations though they are not fully adaptive [38].

Wen et al. [40] present the use of feed forward adaptive control for a four room building and they have devised a model of the building relating lumped thermal capacitors and resistors. Neural networks can also be used to generate a model for adaptive control that has been illustrated in [41]. Extensive work has been done related to the application of adaptive control in building HVAC systems and they can be observed in [18, 42-46].

2.2.3 Optimal control

In optimal control, control signals are derived to satisfy some physical constraints and at the same time to extremize a chosen performance criterion. Optimal control and its ramifications have found applications in many different disciplines, including aerospace, process control, robotics, bioengineering, economics, finance, and management science, and it continues to be an active research area within control theory

[47]. According to [47], formulation of optimal control problems requires: (i) mathematical model of the system to be controlled; (ii) specification of the performance index; (iii) specification of all boundary conditions on states, and constraints to be satisfied by states and controls; and finally (iv) a statement of what variables are free.

Rather using conventional control methodologies, building control has concentrated on much advanced control techniques. There is an increased interest in applying optimal control for buildings lately to improve the comfort and to minimize the energy consumption. This twofold objective can be accomplished by minimizing a cost function encountered in the optimal control aspect. The Indicated control strategy could rely either on white box, grey box or black box models [48].

Optimal control strategies using mathematical models for building heating control are developed in [49-52]. Grey box models can be used for optimal control that has been illustrated by [53]. They have used measured data from one floor elementary school building to identify a grey box model and later the model has been used to test two optimal control strategies based on setpoint temperature. According to [53], grey box models are well adapted to perform optimization because they run quickly and are liable to constraints.

2.3 Intelligent/Soft control

It is a challenging task to identify a suitable mathematical model for HVAC systems as they are complex, non-linear and MIMO systems. Further, they are exposed to lots of disturbances, contingencies and time delays. Human sensation of thermal comfort is a vague and subjective matter and taking it into account is therefore difficult [54]. In intelligent control, no model is needed for controller configuration and is solely based on the human perception of the thermal comfort. Consequently, it avoids the issues associated with physical model development for distinct building structures. Therefore, intelligent controllers are promising control techniques for HVAC systems [14]. Artificial Intelligence (AI) was started to use in advanced building control in 1930s [55]. However, these techniques were started to use extensively in 1990s [2].

Intelligent control techniques can directly be used in HVAC control and they can also be employed to improve the existing traditional controllers, as well. These intelligent techniques are presented under three fundamental subcategories: (i) fuzzy logic (FL) based controllers; (ii) artificial neural network (ANN) based controllers; and (iii) neuro-fuzzy based controllers. The two last methods are adaptive techniques engaging iterative self tuning process during system operation. Moon et al. have described the three AI based thermal control techniques used for building control [56]. Performance of each control method is tested in a typical two story residential building in USA. They have concluded that adaptive AI based control methods are potentially good in maintaining indoor air temperature more comfortably than conventional methods. However, they have not seen significant energy saving effect over the three techniques.

2.3.1 FL based intelligent control

Fuzzy logic comprised of three basic processes: (i) fuzzification; (ii) fuzzy inference; and (iii) defuzzification. Fuzzification is the process of mapping crisp numbers in the input data matrix into fuzzy sets. Membership functions (such as triangular, trapezoidal, gaussian distribution, bell functions and sigmoidal functions) play the main role throughout this process by mapping each input value to a degree of membership between 0 and 1. Fuzzy inference system maps a fuzzy set into a different fuzzy set using fuzzy rules and logical operations. Fuzzy rules are usually denoted as IF-THEN statements and they can be derived by professionals or be automatically generated from the available numerical data using soft computing techniques and evolutionary algorithms. There are two principal types of fuzzy inference systems: Mamdani fuzzy inference and Sugeno fuzzy inference. Mamdani type is generally accepted and extensively used while Sugeno type is particularly good for dynamic non-linear systems. The aggregate linguistic output value produced from the former processes is the input for the defuzzification process. The target is to have a single crisp value as the output from the final process. Fuzzy logic can explain the thermal comfort linguistically and, therefore, can describe the thermal comfort levels rather than temperature or humidity levels which result in improved thermal comfort. There are several control methods originated based on fuzzy logic control and they are briefly described hereafter.

Fuzzy P controller is a technique where the FL is used in closed loop control. Inputs to the controller are obtained from process measurements and the fuzzy logic system output is used to control the process. In fact, this is a pure FL system which is indicated as fuzzy P controller [2, 55].

PID-like fuzzy controllers are categorized into two main classes. The first type is a typical FLC (Fuzzy Logic Controller), realized as a set of heuristic control rules [2, 55]. This type is usually referred to as PI-like or PD-like fuzzy controllers. Second type consists of a conventional PID controller in conjunction with a set of fuzzy rules and fuzzy reasoning mechanism to tune the PID gains online. This system is, however, model dependent [2].

Application of fuzzy logic control to buildings is effective as this technique is well-suited for non-linear system control [57]. They are capable of uniformly approximating any non-linear function to any degree of accuracy and also provide rapid operation. If the fuzzy rules are designed to be more robust, fuzzy controllers can improve the disturbance response by reducing the overshoot/undershoot present in the controlling variable [57]. Fraisse et al. applied this technique to control discontinuously occupied buildings [57]. Many other researchers have been working on the application of only fuzzy logic to control the thermal comfort of buildings [58-62].

The use of fuzzy-PID, fuzzy-PD and adaptive fuzzy-PD for controlling thermal-visual comfort and indoor air quality is described in [63]. One of their main aims was to reduce the energy consumption and lowest values are obtained with the adaptive fuzzy-PD controller. Moreover, [64] presents the development of a PID-Fuzzy controller for indoor temperature control consent to energy resources management in buildings.

2.3.2 ANN based intelligent control

ANNs are inspired by the human brain, nervous system and its learning process. They consist of a set of interconnected neurons in different layers (input, hidden and output layers) of the network. Output of the network is computed using the inputs, network weights and transfer functions. ANN models have adaptability via a self tuning process and they are increasingly used in advanced thermal control of buildings. Amongst diverse AI methods, ANN method has some advantages over the others: (i) ANNs handle a large number of input variables (fuzzy and neuro-fuzzy methods can use only a limited number of inputs as the increased number of membership functions and fuzzy inference rules make the system more complex to solve); (ii) ANN can handle a large number of input data [65]. With these benefits, application of ANN for predicting the energy consumption of building services claimed to be more reliable. An overview of different neural network architectures in building's energy prediction is presented by [66] with appropriate examples, for a detailed understanding of the concept.

ANN based controllers are very useful in optimizing the energy demand of buildings, especially having high thermal masses [67]. There are vast number of applications covering the neural network control of buildings such as: [65, 67-70]. Moreover, as building HVAC systems are dynamic and non-linear, it is also very usual to use dynamic neural network based controllers. A dynamic neural network based on the idea of a non-linear autoregressive with external input (NARX) to model and control a HVAC system is used in [71].

2.3.3 Neuro-Fuzzy based intelligent control

Fuzzy logic has a challenge in producing optimal fuzzy rules and determining membership functions for better control. In order to overcome this problem and to develop an optimum rule base, a new method coupling both FL and ANN has introduced. Here, the neural network technology is used in fuzzy technology and one such neuro-fuzzy approach is the adaptive neuro fuzzy inference system (ANFIS).

Marvuglia et al. present a demonstration of a combined neuro-fuzzy model for indoor temperature regulation [72]. An autoregressive neural network with external inputs has been used to produce indoor temperature forecasts using the outside temperature, relative humidity, wind speed and past forecasts as inputs. Considerable work has been done in the field of building control using ANFIS [73].

2.4 Hybrid control

Hybrid controllers are formulated by the fusion of soft control techniques (FL and ANN) and classical or advanced control techniques [22]. Some examples of this control strategy are discussed in the previous sections such as fuzzy-PID control, adaptive neuro control and neuro-fuzzy control. Here, the soft control techniques are used at higher levels and the other control technique is used at lower levels of the control structure [22]. Hybrid controllers are beneficial as the combination can solve problems that may not be solved by the individual controller. However, the design of the soft control part obliges user experience and an immense quantity of data for training, whereas the classical or advanced controller is laborious to tune (especially for HVAC systems), which are some constraints of the controller.

Various techniques have been used to improve the performance of PID controllers, and fuzzy-PID control scheme is one of them. A self tuning fuzzy PI controller for the supply air pressure control loop for a HVAC system has been developed in [74]. It performs better under normal operating conditions, and when there exists large parameter variations. Adaptive PI control is also a new technique that can be used to tune PI controllers in HVAC systems [75]. It has faster response, smaller overshoot, higher accuracy, stronger robustness and better stability. Paris et al. have presented the application of three heating control schemes (PID, PID-MPC and PID-FUZZY) for indoor temperature control in multi-energy buildings [76]. They found out that the PID-MPC scheme is versatile over the others as it permits better management of different energy sources. Application of ANN models in thermal control of residential buildings with adaptive and predictive logics is presented in [77]. They take air temperature, humidity and PMV as control variables. These systems are observed to be more comfortable than typical thermostat systems, and have potential of enhancing thermal comfort of residential buildings. Besides, a performance analysis of the ANN based predictive and adaptive models for disturbances in and around residential buildings is given by [78].

3. Summary

Variety of control methodologies as employed in the control of building temperature and energy usage were reviewed in the present study. At the beginning of the paper, energy crisis owing to buildings in all over the world is described. Next, different building control strategies, classical; advanced and intelligent techniques, were presented. Classical control methods are still the first choice in building control today, while advanced and intelligent methods are gaining increased attention. The types of control systems and their advantages and limitations are listed and summarized in Table 1.

Table 1. Advantages and limitations of building heating control strategies

Control strategy	Advantages	Limitations
On/Off control	Low initial cost, Simple structure, Fast response, Feedback type	Accepts only binary inputs, often incapable of tracking the setpoint accurately and hence could be inefficient, Not versatile and effective in the long run
PID control	Feedback type, Derivative term combat with sudden load changes in the system	Little measurement and process noise can cause large variations in the output due to derivative term, Energy inefficient, Tuning is time consuming
MPC	Increased energy savings, Cost effective, Robustness to disturbances, Control of multiple variables, Steady state response improvement, Future disturbance prediction, Prediction of future control actions, Better transient response, Handle slow moving processes with time delays	Need to identify a suitable model of the system, Installation could be expensive
Adaptive control	Increased energy savings, Good stability, Parameters can be changed quickly in response to changes in process dynamics, Easy to apply, React fast	Need to identify a suitable model of the system, The design required for implementation is enormous
Optimal control	Increased energy savings, Rapid response, Multi-variable control	Need to identify a suitable model of the system
Fuzzy control	Non-linear control method and can be applied to HVAC systems effectively, High accuracy, Rapid operation	Can use only a limited number of input variables, Development of optimal number of fuzzy rules and determination of the membership function parameters are not straightforward
ANN control	Handle large number of input variables and data, Reliable predictions	Require large number of data for quality predictions
Neuro-Fuzzy control	Combination can develop optimal fuzzy rules and determine membership function parameters	Can use only a limited number of input variables
Hybrid control	Combination of soft control and classical/advanced control can solve problems that may not be solved by an individual controller	Soft control requires large quantity of data for training, It is difficult to tune the classical/advanced controller

As a whole, the use of classical controllers is because of their low initial cost and simplicity of implementation. However, they admit high maintenance cost and higher energy consumption. Further, they cannot be used in MIMO systems efficiently. Advanced control methodologies could be an alternative approach in building control. These techniques require a good quality dynamic model of the building and they exhibit non-linear characteristics. As HVAC systems are non-linear and time delayed processes, advanced control methods can handle them more smoothly than conventional methods. In intelligent control, no mathematical model is needed and it is solely based on the human perception of the thermal comfort. Hence, it can provide improved thermal comfort to the occupants. Further, these controllers can be used to upgrade the existing traditional controllers.

References

[1] EBPD, On the energy performance of buildings. Official Journal of the European Union, Directive 2010/31/EU of the European Parliament and of the council, 2010: p. 13-34.

[2] Dounis, A.I. and C. Caraiscos, Advanced control systems engineering for energy and comfort management in a building environment-A review. Renewable and Sustainable Energy Reviews, 2009. 13(6-7): p. 1246-1261.

[3] European-Commission, Towards a European strategy for the security of energy supply, green paper., 2000.

[4] Bergesen, B., et al., Energy consumption 2012 - household energy consumption, 2013, Norwegian Water Resources and Energy Directorate.

[5] Valmot, O.R., Enormt potensial for energisparing, in Teknisk Ukeblad2013. p. 30-31.

[6] Krzaczek, M. and J. Tejchman, Indoor Air Quality and Thermal Comfort in Naturally Ventilated Low-Energy Residential Houses, in Air Quality - Monitoring and Modeling, S. Kumar and R. Kumar, Editors. 2012, InTech.

[7] Fanger, P.O., Thermal comfort: Analysis and applications in environmental engineering. 1970: Danish Technical Press.

[8] Carbon-Trust, Building controls - Realising savings through the use of controls, 2007: UK.

[9] Foucquier, A., et al., State of the art in building modelling and energy performances prediction: A review. Renewable and Sustainable Energy Reviews, 2013. 23(0): p. 272-288.

[10] Kramer, R., J. van Schijndel, and H. Schellen, Simplified thermal and hygric building models: A literature review. Frontiers of Architectural Research, 2012. 1(4): p. 318-325.

[11] Spindler, H.C. and L.K. Norford, Naturally ventilated and mixed-mode buildings-Part I: Thermal modeling. Building and Environment, 2009. 44(4): p. 736-749.

[12] Zhao, H.-x. and F. Magoulès, A review on the prediction of building energy consumption. Renewable and Sustainable Energy Reviews, 2012. 16(6): p. 3586-3592.

[13] Kim, D. and J.E. Brown. Reduced-order building modeling for application to model-based predictive control. in 5th national conference of IBPSA. 2012. USA.

[14] Mirinejad, H., et al., Control Techniques in Heating, Ventilating and Air Conditioning (HVAC) Systems 1. Journal of computer science, 2008. 4(9): p. 777-783.

[15] Peffer, T., et al., How people use thermostats in homes: A review. Building and Environment, 2011. 46(12): p. 2529-2541.

[16] Wen, J. and T.F. Smith. Review of thermostat time constant on temperature control and energy consumption. in Sensors for Industry. 2001.

[17] Maheshwari, G.P., et al., Programmable thermostat for energy saving. Energy and Buildings, 2001. 33(7): p. 667-672.

[18] Virk, G.S., J.M. Cheung, and D.L. Loveday. Development of adaptive control techniques for BEMs. in International Conference on CONTROL '91. 1991.

[19] Shein, W.W., Y. Tan, and A.O. Lim. PID controller for temperature control with multiple actuators in cyber-physical home system. in 15th International conference on network based information systems. 2012.

[20] Kulkarni, M.R. and F. Hong, Energy optimal control of a residential space-conditioning system based on sensible heat transfer modeling. Building and Environment, 2004. 39(1): p. 31-38.

[21] Gwerder, M. and J. Todtli. Predictive control for integrated room automation. in 8th REHVA World congress for building technologies. 2005.

[22] Afram, A. and F. Janabi-Sharifi, Theory and applications of HVAC control systems – A review of model predictive control (MPC). Building and Environment, 2014. 72(0): p. 343-355.

[23] Prívara, S., et al., Model predictive control of a building heating system: The first experience. Energy and Buildings, 2011. 43(2–3): p. 564-572.

[24] Bao-Cang, D., Modern predictive control. 2010, Boca Raton, Fla.: CRC Press. xv, 276 s. : ill.

[25] Vasak, M., A. Starcic, and A. Martincevic. Model predictive control of heating and cooling in a family house. in MIPRO. 2011.

[26] Li, S., S. Ren, and X. Wang. HVAC room temperature prediction control based on neural network model. in 5th Conference on Measuring Technology and Mechatronics Automation. 2013.

[27] Thomas, B. and M. Soleimani-Mohseni, Neural network models for predictive climate control in intelligent buildings., 2005, CABA.

[28] Sousa, J.M., R. Babuška, and H.B. Verbruggen, Fuzzy predictive control applied to an air-conditioning system. Control Engineering Practice, 1997. 5(10): p. 1395-1406.

[29] Xi, X.-C., A.-N. Poo, and S.-K. Chou, Support vector regression model predictive control on a HVAC plant. Control Engineering Practice, 2007. 15(8): p. 897-908.

[30] Lixing, D., et al. Support vector regression and ant colony optimization for HVAC cooling load prediction. in International Symposium on Computer, Communication, Control and Automation. 2010.

[31] Toffoli, E., et al. Thermodynamic identification of buildings using wireless sensor networks. in IFAC World Congress. 2008.

[32] Ferkl, L. and J. Siroky, Ceiling radiant cooling: Comparison of armax and subspace identication modelling methods. Building and Environment, 2010. 45: p. 205-212.

[33] Privara, S., et al. Modeling and identification of a large multi-zone office building. in Control Applications (CCA), 2011 IEEE International Conference on. 2011.

[34] Wu, S. and J.-Q. Sun, A physics-based linear parametric model of room temperature in office buildings. Building and Environment, 2012. 50(0): p. 1-9.

[35] Mustafaraj, G., J. Chen, and G. Lowry, Development of room temperature and relative humidity linear parametric models for an open office using BMS data. Energy and Buildings, 2010. 42(3): p. 348-356.

[36] Prívara, S., et al., Building modeling as a crucial part for building predictive control. Energy and Buildings, 2013. 56(0): p. 8-22.

[37] Prívara, S., et al., Building modeling: Selection of the most appropriate model for predictive control. Energy and Buildings, 2012. 55(0): p. 341-350.

[38] Landau, L., et al., Adaptive control. 2011, London: Springer.

[39] Isermann, R., K.H. Lachmann, and D. Matko, Adaptive Control Systems. 1992: Prentice Hall.

[40] Wen, J.T., et al. Building temperature control with adaptive feedforward. in Decision and Control (CDC), 2013 IEEE 52nd Annual Conference on. 2013.

[41] Torres, J.L. and M.L. Martin. Adaptive control of thermal comfort using neural networks. in Argentine Symposium on Computing Technology. 2008.

[42] Chaudhry, S.I. and M. Das. Adaptive control of indoor temperature in a building using a desirable reference temperature profile. in Circuits and Systems (MWSCAS), 2013 IEEE 56th International Midwest Symposium on. 2013.

[43] Chaudhry, S.I. and M. Das. Adaptive control of indoor temperature in a building. in Electro/Information Technology (EIT), 2012 IEEE International Conference on. 2012.

[44] Hongli, L., et al. A novel adaptive energy-effcient controller for the HVAC systems. in Control and Decision Conference (CCDC). 2012.

[45] Leephakpreeda, T., Implementation of adaptive indoor comfort temperature control via embedded system for air-conditioning unit. Journal of Mechanical Science and Technology, 2012. 26(1): p. 259-268.

[46] Zaheer-Uddin, M., Optimal, sub-optimal and adaptive control methods for the design of temperature controllers for intelligent buildings. Building and Environment, 1993. 28(3): p. 311-322.

[47] Becerra, V.M., Optimal control, in Scholarpedia2008.

[48] Wang, S. and Z. Ma, Supervisory and Optimal Control of Building HVAC Systems: A Review. HVAC&R Research, 2008. 14(1): p. 3-32.

[49] Burns, J.A. and E.M. Cliff. On optimal thermal control of an idealized room including hard limits on zone-temperature and a max-control cost term. in Decision and Control (CDC), 2013 IEEE 52nd Annual Conference on. 2013.

[50] Kummert, M. and P. Andre. Building and HVAC optimal control simulation. application to an office building. in 3rd Symposium on Heating, Ventilation and Air Conditioning (ISHVAC 99) conference. 1999.

[51] Yahiaoui, A., et al. Model based optimal control for integrated building systems. in 6th Int. Postgraduate Research Conf. in the Built and Human Environment. 2006.

[52] Rui, Y. and W. Lingfeng. Optimal control strategy for HVAC system in building energy management. in Transmission and Distribution Conference and Exposition (T&D), 2012 IEEE PES. 2012.

[53] Berthou, T., et al. Optimal control for building heating: An elementary school case study. in Conference of International Building Performance Simulation Association. 2013. Chambéry, France.

[54] Mirinejad, H., K.C. Welch, and L. Spicer. A review of intelligent control techniques in HVAC systems. in Energytech, 2012 IEEE. 2012.

[55] Song, Y., S. Wu, and Y.Y. Yan, Control strategies for indoor environment quality and energy efficiency—a review. Int. J. Low-Carbon Tech., 2013.

[56] Moon, J.W., et al., Comparative study of artificial intelligence-based building thermal control methods – Application of fuzzy, adaptive neuro-fuzzy inference system, and artificial neural network. Applied Thermal Engineering, 2011. 31(14–15): p. 2422-2429.

[57] Fraisse, G., J. Virgone, and J.J. Roux, Thermal control of a discontinuously occupied building using a classical and a fuzzy logic approach. Energy and Buildings, 1997. 26(3): p. 303-316.

[58] Shahnawaz Ahmed, S., et al., Fuzzy logic based energy saving technique for a central air conditioning system. Energy, 2007. 32(7): p. 1222-1234.

[59] Eftekhari, M.M. and L.D. Marjanovic, Application of fuzzy control in naturally ventilated buildings for summer conditions. Energy and Buildings, 2003. 35(7): p. 645-655.

[60] Gouda, M.M., S. Danaher, and C.P. Underwood, Thermal comfort based fuzzy logic controller. Building services engineering research and technology, 2001. 22(4): p. 237-253.

[61] Homod, R.Z., et al., Gradient auto-tuned Takagi–Sugeno Fuzzy Forward control of a HVAC system using predicted mean vote index. Energy and Buildings, 2012. 49(0): p. 254-267.

[62] Kristl, Ž., et al., Fuzzy control system for thermal and visual comfort in building. Renewable Energy, 2008. 33(4): p. 694-702.

[63] Kolokotsa, D., et al., Advanced fuzzy logic controllers design and evaluation for buildings' occupants thermal–visual comfort and indoor air quality satisfaction. Energy and Buildings, 2001. 33(6): p. 531-543.

[64] Paris, B., et al., Hybrid PID-fuzzy control scheme for managing energy resources in buildings. Applied Soft Computing, 2011. 11(8): p. 5068-5080.

[65] Moon, J.W., S.-H. Yoon, and S. Kim, Development of an artificial neural network model based thermal control logic for double skin envelopes in winter. Building and Environment, 2013. 61(0): p. 149-159.

[66] Kumar, R., R.K. Aggarwal, and J.D. Sharma, Energy analysis of a building using artificial neural network: A review. Energy and Buildings, 2013. 65(0): p. 352-358.

[67] Argiriou, A.A., I. Bellas-Velidis, and C.A. Balaras, Development of a neural network heating controller for solar buildings. Neural Networks, 2000. 13(7): p. 811-820.

[68] Liang, J. and R. Du, Design of intelligent comfort control system with human learning and minimum power control strategies. Energy Conversion and Management, 2008. 49(4): p. 517-528.

[69] Moon, J.W., et al., Determining optimum control of double skin envelope for indoor thermal environment based on artificial neural network. Energy and Buildings, 2014. 69(0): p. 175-183.

[70] Castilla, M., et al., Neural network and polynomial approximated thermal comfort models for HVAC systems. Building and Environment, 2013. 59(0): p. 107-115.

[71] Kusiak, A. and G. Xu, Modeling and optimization of HVAC systems using a dynamic neural network. Energy, 2012. 42(1): p. 241-250.

[72] Marvuglia, A., A. Messineo, and G. Nicolosi, Coupling a neural network temperature predictor and a fuzzy logic controller to perform thermal comfort regulation in an office building. Building and Environment, 2014. 72(0): p. 287-299.

[73] Soyguder, S. and H. Alli, An expert system for the humidity and temperature control in HVAC systems using ANFIS and optimization with Fuzzy Modeling Approach. Energy and Buildings, 2009. 41(8): p. 814-822.

[74] Pal, A.K. and R. Mudi, Self-tuning fuzzy PI controller and its application to HVAC systems. International journal of computational cognition, 2008. 6.

[75] Bai, J. and X. Zhang, A new adaptive PI controller and its application in HVAC systems. Energy Conversion and Management, 2007. 48(4): p. 1043-1054.

[76] Paris, B., et al., Heating control schemes for energy management in buildings. Energy and Buildings, 2010. 42(10): p. 1908-1917.

[77] Moon, J.W. and J.-J. Kim, ANN-based thermal control models for residential buildings. Building and Environment, 2010. 45(7): p. 1612-1625.

[78] Moon, J.W., Performance of ANN-based predictive and adaptive thermal-control methods for disturbances in and around residential buildings. Building and Environment, 2012. 48(0): p. 15-26.

Permissions

All chapters in this book were first published in IJEE, by International Energy and Environment Foundation (IEEF); hereby published with permission under the Creative Commons Attribution License or equivalent. Every chapter published in this book has been scrutinized by our experts. Their significance has been extensively debated. The topics covered herein carry significant findings which will fuel the growth of the discipline. They may even be implemented as practical applications or may be referred to as a beginning point for another development.

The contributors of this book come from diverse backgrounds, making this book a truly international effort. This book will bring forth new frontiers with its revolutionizing research information and detailed analysis of the nascent developments around the world.

We would like to thank all the contributing authors for lending their expertise to make the book truly unique. They have played a crucial role in the development of this book. Without their invaluable contributions this book wouldn't have been possible. They have made vital efforts to compile up to date information on the varied aspects of this subject to make this book a valuable addition to the collection of many professionals and students.

This book was conceptualized with the vision of imparting up-to-date information and advanced data in this field. To ensure the same, a matchless editorial board was set up. Every individual on the board went through rigorous rounds of assessment to prove their worth. After which they invested a large part of their time researching and compiling the most relevant data for our readers.

The editorial board has been involved in producing this book since its inception. They have spent rigorous hours researching and exploring the diverse topics which have resulted in the successful publishing of this book. They have passed on their knowledge of decades through this book. To expedite this challenging task, the publisher supported the team at every step. A small team of assistant editors was also appointed to further simplify the editing procedure and attain best results for the readers.

Apart from the editorial board, the designing team has also invested a significant amount of their time in understanding the subject and creating the most relevant covers. They scrutinized every image to scout for the most suitable representation of the subject and create an appropriate cover for the book.

The publishing team has been an ardent support to the editorial, designing and production team. Their endless efforts to recruit the best for this project, has resulted in the accomplishment of this book. They are a veteran in the field of academics and their pool of knowledge is as vast as their experience in printing. Their expertise and guidance has proved useful at every step. Their uncompromising quality standards have made this book an exceptional effort. Their encouragement from time to time has been an inspiration for everyone.

The publisher and the editorial board hope that this book will prove to be a valuable piece of knowledge for researchers, students, practitioners and scholars across the globe.

List of Contributors

Hao Yu
Department of Industrial Engineering, Narvik University College, Postboks 385 Lodve gate 2, NO-8505 Narvik, Norway

Wei Deng Solvang
Department of Industrial Engineering, Narvik University College, Postboks 385 Lodve gate 2, NO-8505 Narvik, Norway

Chen Chen
Department of Industrial Engineering, Narvik University College, Postboks 385 Lodve gate 2, NO-8505 Narvik, Norway
Northern Research Institute Narvik AS.Postboks 250, NO-8504 Narvik, Norway

I. Khazaee
Faculty of Mechanical and Energy Engineering, Shahis Beheshti University, A.C., Tehran, Iran

Hui Zhu
School of Energy Science and Engineering, Central South University,NO.932, South Lushan Road, Changsha, Hunan Province, 410083, China

Hanqing Wang
School of Civil Engineering, Hunan University of Technology, West Taishan Road, Zhuzhou, Hunan Province, 412007, China

Guangxiao Kou
School of Civil Engineering, Hunan University of Technology, West Taishan Road, Zhuzhou, Hunan Province, 412007, China

I. Khazaee
Department of Mechanical Engineering, Torbat-e-jam branch, Islamic Azad University, Torbat-e-jam, Iran

Zelong Zhang
Institute of Thermal Science and Power Engineering, Naval University of Engineering, Wuhan 430033, China
Military Key Laboratory for Naval Ship Power Engineering, Naval University of Engineering, Wuhan 430033, China
College of Power Engineering, Naval University of Engineering, Wuhan 430033, China

Lingen Chen
Institute of Thermal Science and Power Engineering, Naval University of Engineering, Wuhan 430033, China
Military Key Laboratory for Naval Ship Power Engineering, Naval University of Engineering, Wuhan 430033, China
College of Power Engineering, Naval University of Engineering, Wuhan 430033, China

Yanlin Ge
Institute of Thermal Science and Power Engineering, Naval University of Engineering, Wuhan 430033, China
Military Key Laboratory for Naval Ship Power Engineering, Naval University of Engineering, Wuhan 430033, China
College of Power Engineering, Naval University of Engineering, Wuhan 430033, China

Fengrui Sun
Institute of Thermal Science and Power Engineering, Naval University of Engineering, Wuhan 430033, China
Military Key Laboratory for Naval Ship Power Engineering, Naval University of Engineering, Wuhan 430033, China
College of Power Engineering, Naval University of Engineering, Wuhan 430033, China

Maher A.R. Sadiq Al-Baghdadi
Fuel Cell Research Center, International Energy and Environment Foundation, Najaf, P.O.Box 39, Iraq

Devander Kumar
The Technological Institute of Textile & Sciences, Bhiwani, Haryana-127021, India

Sudhir Kumar
National Institute of Technology, Kurukshetra, Haryana-136118, India

Yi Zhang
College of Mechanical Engineering, Zhejiang University of Technology, Hangzhou, 310014, China

Lingeng Chen
Institute of Thermal Science and Power Engineering, Naval University of Engineering, Wuhan 430033, China
Military Key Laboratory for Naval Ship Power Engineering, Naval University of Engineering, Wuhan 430033, China
College of Power Engineering, Naval University of Engineering, Wuhan 430033, China

Guozhong Chai
College of Mechanical Engineering, Zhejiang University of Technology, Hangzhou, 310014, China

Xiao Zhang
Department of Mechanical Engineering & Materials Science, Washington University in St. Louis, 1 Brookings Drive, St. Louis, MO 63130, USA

Subhodeep Banerjee
Department of Mechanical Engineering & Materials Science, Washington University in St. Louis, 1 Brookings Drive, St. Louis, MO 63130, USA

Ling Zhou
Department of Mechanical Engineering & Materials Science, Washington University in St. Louis, 1 Brookings Drive, St. Louis, MO 63130, USA

Ramesh Agarwal
Department of Mechanical Engineering & Materials Science, Washington University in St. Louis, 1 Brookings Drive, St. Louis, MO 63130, USA

Suraj K. Puvvada
Department of Mechanical Engineering & Materials Science, Washington University in St. Louis, 1 Brookings Drive, St. Louis, MO 63130, USA

Ramesh K. Agarwal
Department of Mechanical Engineering & Materials Science, Washington University in St. Louis, 1 Brookings Drive, St. Louis, MO 63130, USA

Henrique Marin Campos
Post-graduation Program on Civil Engineering, Technological Federal University of Paraná, Curitiba, Brazil

Ana Katherine Rodríguez Manrique
Post-graduation Program on Civil Engineering, Technological Federal University of Paraná, Curitiba, Brazil

Bruno Victor Kobiski
Post-graduation Program on Civil Engineering, Technological Federal University of Paraná, Curitiba, Brazil

Eloy Fassi Casagrande Júnior
Post-graduation Program on Civil Engineering, Technological Federal University of Paraná, Curitiba, Brazil

Jair Urbanetz Junior
Post-graduation Program on Civil Engineering, Technological Federal University of Paraná, Curitiba, Brazil

N Salami
Energy institute, Faculty of Mechanical Engineering, Brno University of Technology, Technická 2896/2, 616 69 Brno, Czech Republic

Z Skala
Energy institute, Faculty of Mechanical Engineering, Brno University of Technology, Technická 2896/2, 616 69 Brno, Czech Republic

Tijo Joseph
School of Engineering, University of Guelph, Guelph, Ontario, Canada

Animesh Dutta
School of Engineering, University of Guelph, Guelph, Ontario, Canada

Maher A.R. Sadiq Al-Baghdadi
Fuel Cell Research Center, International Energy and Environment Foundation, Najaf, P.O.Box 39, Iraq

Abdul Jabbar N. Khalifa
Al-Nahrain University, College of Engineering, Mechanical Engineering Department, Jadiriya, P.O. Box 64040, Baghdad, Iraq

Marwa AW. Ali
Al-Nahrain University, College of Engineering, Mechanical Engineering Department, Jadiriya, P.O. Box 64040, Baghdad, Iraq

M. Nassereddine
School of Computing, Engineering & Mathematics, University of Western Sydney, Australia

J. Rizk
School of Computing, Engineering & Mathematics, University of Western Sydney, Australia

M. Nagrial
School of Computing, Engineering & Mathematics, University of Western Sydney, Australia

A. Hellany
School of Computing, Engineering & Mathematics, University of Western Sydney, Australia

Bo Yang
Institute of Thermal Science and Power Engineering, Naval University of Engineering, Wuhan 430033, P. R. China
Military Key Laboratory for Naval Ship Power Engineering, Naval University of Engineering, Wuhan 430033, P. R. China
College of Power Engineering, Naval University of Engineering, Wuhan 430033, P. R. China

Lingen Chen
Institute of Thermal Science and Power Engineering, Naval University of Engineering, Wuhan 430033, P. R. China
Military Key Laboratory for Naval Ship Power Engineering, Naval University of Engineering, Wuhan 430033, P. R. China
College of Power Engineering, Naval University of Engineering, Wuhan 430033, P. R. China

Yanlin Ge
Institute of Thermal Science and Power Engineering, Naval University of Engineering, Wuhan 430033, P. R. China
Military Key Laboratory for Naval Ship Power Engineering, Naval University of Engineering, Wuhan 430033, P. R. China
College of Power Engineering, Naval University of Engineering, Wuhan 430033, P. R. China

Fengrui Sun
Institute of Thermal Science and Power Engineering, Naval University of Engineering, Wuhan 430033, P. R. China
Military Key Laboratory for Naval Ship Power Engineering, Naval University of Engineering, Wuhan 430033, P. R. China
College of Power Engineering, Naval University of Engineering, Wuhan 430033, P. R. China

Elena Eftimie
Department of Product Design, Mechatronics and Environment, Transilvania University of Braşov, Eroilor 29, 500036, Romania

Yi Zhang
College of Mechanical Engineering, Zhejiang University of Technology, Hangzhou, 310014, China

Lingeng Chen
Institute of Thermal Science and Power Engineering, Naval University of Engineering, Wuhan 430033, China
Military Key Laboratory for Naval Ship Power Engineering, Naval University of Engineering, Wuhan 430033, China
College of Power Engineering, Naval University of Engineering, Wuhan 430033, China

Guozhong Chai
College of Mechanical Engineering, Zhejiang University of Technology, Hangzhou, 310014, China

I. Khazaee
Department of Mechanical Engineering, Torbat-e-jam branch, Islamic Azad University, Torbat-e-jam, Iran

D.W.U. Perera
Faculty of Technology, Telemark University College, Porsgrunn, Norway

C. F. Pfeiffer
Faculty of Technology, Telemark University College, Porsgrunn, Norway

N.-O Skeie
Faculty of Technology, Telemark University College, Porsgrunn, Norway